Springer Series in Geomechanics and Geoengineering

Editors: Wei Wu · Ronaldo I. Borja

Magued Iskander

Modelling with Transparent Soils

Visualizing Soil Structure Interaction and Multi Phase Flow, Non-Intrusively

Springer

Professor Wei Wu, Institut für Geotechnik, Universität für Bodenkultur, Feistmantelstraße 4, 1180 Vienna, Austria, E-mail: wei.wu@boku.ac.at

Professor Ronaldo I. Borja, Department of Civil and Environmental Engineering, Stanford University, Stanford, CA 94305-4020, USA, E-mail: borja@stanford.edu

Author

Magued Iskander, PhD, PE
Polytechnic Institute of NYU
6 Metrotech Ctr.
Brooklyn, NY 11201
USA
E-Mail: iskander@poly.edu

ISBN 978-3-642-02500-6 e-ISBN 978-3-642-02501-3

DOI 10.1007/978-3-642-02501-3

Springer Series in Geomechanics and Geoengineering ISSN 1866-8755

Library of Congress Control Number: 2010927008

Typeset & Cover Design: Scientific Publishing Services Pvt. Ltd., Chennai, India.

Printed on acid-free paper

5 4 3 2 1 0

springer.com

To Lolli & Kito

Contents

Acknowledgements

This manuscript is the first of its kind. It represents a summary of over a decade of research. The research, and the book, would not have taken place without the support of the United States National Science Foundation. I am grateful to NSF for funding my early work on transparent soils under NSF grant No. CMS 9733064 CAREER: Modeling 3D Flow and Soil Structure Interaction using Optical Tomography. Continued NSF funding under grants No. DGE 0337668 and DGE 0741714 is also gratefully acknowledged. I am also thankful to Southwest research Institute for its support of preliminary work on this topic.

Research is by nature a collaborative endeavor. I am fortunate to have enjoyed the fruits of collaboration with so many talented individuals, who are acknowledged at the beginning of each chapter. Samer Sadek and Jinyuan Liu, my first two doctoral students, helped shape my views on transparent soil modeling. Jay Lo, Kazunori Tabe, Louis Ge, and Honghua Zhao made important contributions to the research and manuscript. Previous collaborations with Richard Mannheimer, Chuck Oswald, and Jiunnren Lai have been seminal to the work that appears in this manuscript.

A large number of professional colleagues contributed formally and informally to the success of the research effort in transparent soil modeling. Professors David Frost and Carlos Santamarina of the Georgia Institute of Technology and Professors Jay Meegoda and Priscilla Nelson of the New Jersey Institute of Technology served as external committee members to the dissertations of students working on transparent soils at Polytechnic Institute. The following Polytechnic Institute faculty contributed valuable insights throughout this research: Walid Aboumoussa, Bruce Garetz, Nikhil Gupta, Vikram Kapila, Konstantinos Kostarelos (now at University of Cyprus), Masoud Ghandehari, Iwao Teraoka, Xiaodong (Sheldon) Wang (now at NJIT), and Sungho Yoon. The following graduate students proof read the entire manuscript: Andrew Dimmond, Saumil Parikh, Mahmoud Ahmed, and Sridhar Chandran.

I am grateful to Dr. Thomas Ditzinger for adopting this manuscript into the Springer Series on Geomechanichs and Geoengineering. I am also thankful to the series editors: Prof. Wei Wu from Universität für Bodenkultur in Austria and Prof. Ronaldo Borja from Stanford University for recommending publication of this manuscript. Credit is also due to Mrs. Heather King from Springer for her assistance during the production of this work.

Last but not least, I would like to express my gratitude to my wife. This work could not have been finished without her support, encouragement, and love.

About the Author and Contributors

Magued Iskander, PhD, PE is an Associate Professor at Polytechnic Institute of New York University (formerly Polytechnic University or Brooklyn Poly). He specializes in geotechnical and foundation engineering. He is widely recognized as the leader of two areas of geotechnical research: (1) modeling of soil structure interaction and multiphase flow using transparent soil surrogates and (2) long-term behavior of piling made of fiber reinforced polymers. He has 85 publications, including 7 edited books, and 25 journal articles. He conducted over $6 million in grants and contracts and graduated 31 doctoral and masters' students. He can be reached at Iskander@poly.edu.

Jinyuan Liu, PhD, PE, PEng is an Assistant Professor of geotechnical engineering at Reyerson University in Toronto, Canada. He received a PhD from Polytechnic University in 2003 where he wrote a dissertation titled *Visualization of 3D Deformations using Transparent Soils.* His research interests include underground excavation and tunnelling, soil-structural interaction, and physical modeling. He can be reached at jinyuan.liu@ryerson.ca.

Samer Sadek, PhD, PE is an Associate Vice President and Principal Tunnel Engineer at HNTB. He received a PhD from Polytechnic University in 2001, where he wrote a dissertation titled *Soil Structure Interaction in Transparent Synthetic Soils using Digital Image Correlation.* He can be reached at ssadek@hntb.com.

Kazunori Tabe, PhD recently received a PhD from Polytechnic Institute of NYU where he wrote a dissertation titled *Aquabeads to Model the Geotechnical Behavior of Natural Soils.* He can be reached at kazunoritabe@gmail.com.

Louis Ge, PhD, PE is an Assistant Professor at Missouri S&T. He received a Ph.D. in Civil Engineering at the University of Colorado at Boulder in 2003. His research interests focus on constitutive modeling and innovative geotechnical testing. He can be reached at geyun@mst.edu.

Hung-Chieh (Jay) Lo is a geotechnical engineer at Sinotech Engineering Consultants in Taiwan. He received an M.Sc degree from Polytechnic University where he wrote a thesis titled *Modeling 2D Flow and Surfactant Flushing using Transparent Aquabeads.* He can be reached at jaylo1976@gmail.com.

Honghua Zhao, PhD is a Lecturer in the Department of Engineering Mechanics at Dalian University of Technology in China. She received a Ph.D. from Missouri S&T in 2007 where she wrote a dissertation titled *Development and Assessment of Transparent Soil and Particle Image Velocimetry in Dynamic Soil-Structure Interaction.* She can be reached at honghuazhao@yahoo.com.cn.

Chapter 1
Introduction to Transparent Soils

Magued Iskander

Abstract. Experimental models to measure spatial deformation patterns within a soil mass are typically limited by the fact that soil sensors do not provide a continuous image of the measured continuum. Additionally, soil sensors exhibit static and dynamic characteristics that are different from those of the surrounding soils, and therefore can change the response of the measured continuum. The fundamental premise of this research is that transparent synthetic soil surrogates can be used to represent the behavior of natural soils in model tests in order to overcome these difficulties. This book presents experimental methods that are fundamentally different from previous experiment studies, where spatial flow patterns and deformations are obtained non-intrusively and continuously without interruption from sensors. In the short term, transparent synthetic soils and the proposed optical set-up and image processing technique are expected to be easily adjusted and applied in geotechnical engineering research. In the long term, transparent synthetic soils should prove to be a powerful tool in solving many geotechnical and geoenvironmental engineering problems and become helpful in the design of many new structures.

1.1 Background

Visualization of flow, deformation, and strain in soils can provide insight for solving many geotechnical problems. Several methods have been attempted to measure continuous spatial deformations in soils. In the late sixties and early seventies, X-ray of embedded lead particles was used for measuring spatial deformations (Roscoe et al 1963, Lorenz and Heinz 1969, Bransby and Milligan 1975). The results, however, were influenced by the fact that lead exhibit different properties from the surrounding soil. It is also a discrete measurement that cannot provide a continuous image of the measured continuum. More recently, new techniques such as Computerized Axial Tomography (CAT scan) and Magnetic Resonance Imaging (MRI) were also utilized for experimental modeling of geotechnical problems (Mandava et al (1990), Desrues et al (1991), Orsi et al (1992), and Posadas et al

M. Iskander: Modeling with Transparent Soils, SSGG, pp. 1–4.
springerlink.com

(1996), Ng et al (1996), Wong (1999). However, the broad application of these techniques is limited by the high cost of the apparatus, and difficulties in experimental set up and in interpreting the results. At the same time, recent developments in computer imaging, and digital photography resulted in wide spread use of optical techniques in industrial and academic settings.

Optical methods have been used to measure the response of transparent materials, which model specific soil properties. For example, translucent media made of crushed glass and a matched refractive index fluid have been used to study stress distribution under simple shear (Allersma 1998), as well as the response of saturated embankments under seismic loading (Konagai et al 1992, 1994). Transparent surrogates made of glass beads or quartz powder and matched refractive index pore fluids have also been used to help visualizing flow inside models (Chen and Wada 1986, Brock and Orr 1991). However, glass and quartz surrogates are limited by their inability to represent the geotechnical properties of a wide range of natural soils and also by their poor transparency.

Transparent slurry, made of amorphous silica powder and pore fluids with a matching refractive index, was originally developed for modeling Non-Newtonian slurry flow problems by Mannheimer (1990). This material was consolidated and tested and found to exhibit macroscopic geotechnical properties of natural clays (Iskander et al 1994, Mannheimer and Oswald 1993). Amorphous silica powder has already been used in model tests with simple optical techniques to study flow of contaminants into perforated wick drains (Welker et al 1999, 2000) pile penetration in clays (Gill and Lehane 2001 and Lehane and Gil 2004), and deformations fields near penetrometers and vanes (Toiya et al 2007, and Stanier 2006). The developed transparent soils were also used in centrifuge tests to model the behavior of suction embedded plate anchors (Song and Hu 2009).

1.2 Available Transparent Soils

So far, three families of transparent materials have been developed for modeling sand and clay. The first family is made of amorphous silica powder to model the geotechnical properties of natural clays (Iskander 1997 and 1998, Iskander et al 2002a and Liu et al 2003). The second family that models sand is made of transparent silica gels (Iskander et al 2002b). Both families have the same refractive index and use the same pore fluids, thus permitting their use in the same model (Sadek et al 2002). The third family is made of Aquabeads and is suitable for representing flow in many soils and strength of very weak sediments. Aquabeads have a different refractive index from silica gels and powders, and thus cannot be combined with silica in the same transparent model.

The transparent soil families described in this book are truly transparent, not merely translucent, thus permitting complete light intrusion and use of simple optical visualization methods to probe the interior and measure the spatial distribution of stress and flow.

1.3 Objectives

The objectives of this work are to:

1. Present the properties of several families of transparent materials that are suitable for representing the behavior of soils in model tests.
2. Present feasible methods to obtain 3D deformation and flow inside transparent soil models non-intrusively.
3. Illustrate the possibility of using transparent synthetic soils in studying geotechnical problems.

1.4 Organization of This Book

This book consists of mainly five parts. First, the subject is introduced along with previous research and fundamentals of imaging in chapters one to four. Second, investigations of geotechnical properties of transparent synthetic soils are presented in chapter five to seven. Third, the image processing algorithm, digital image cross-correlation, and its advanced form, adaptive cross-correlation, are presented and validated in chapters eight to ten. Fourth, three-dimensional deformation under a model footing is presented in chapter eleven. Fifth, two dimensional flow tests are presented in chapter twelve. Finally, a postscript containing the author's reflections is presented at the end. An appendix dealing with camera calibration is also included.

References

Allersma, H.: Using imaging technologies in experimental geotechnics. In: Proc. 2nd Int. Conf. on Imaging Technologies: Techniques and Applications in Civil Engr., ASCE, pp. 1–9 (1998)

Brock, D., Orr, F.: Flow visualization of viscous fingering in heterogeneous porous media. In: SPE Annual Technical Conference and Exhibition, pp. 211–222 (1991)

Bransby, P., Milligan, G.: Soil deformation near cantilever sheet pile. Geotechnique 25(2), 175–195 (1975)

Chen, J., Wada, N.: Visualization of immiscible displacement in a three-dimensional transparent porous medium. Experiments in Fluids 4(6), 336–338 (1986)

Desrues, J., Mokni, M., Mazerolle, F.: Tomodensitometrie et la localisation sur les sables. In: Proc. X Int. Conf. Soil Mech. & Found. Eng., vol. 1, pp. 61–64 (1991) (in French)

Gill, D., Lehane, B.: An optical technique for investigating soil displacement patterns. ASTM Geotechnical Testing Journal, GTJODJ 24(3), 324–329 (2001)

Iskander, M., Lai, J., Oswald, C., Mannheimer, R.: Development of a transparent material to model the geotechnical properties of soils. ASTM Geotechnical Testing Journal, GTJODJ 17(4), 425–433 (1994)

Iskander, M.: A transparent material to model the geotechnical properties of soils. In: Proceedings 14th ICSMFE, vol. 1, pp. 315–319 (1997)

Iskander, M.: Transparent soils to image 3D flow & deformation. In: Proc., 2nd Int. Conf. Imaging Technologies: Techniques and Applications in Civil Engr., ASCE, pp. 255–264 (1998)

Iskander, M., Liu, J., Sadek, S.: Transparent amorphous silica to model clay. ASCE J. of Geotechnical & Geoenvironmental Engr. 128(3), 262–273 (2002a)
Iskander, M., Sadek, S., Liu, J.: Optical measurement of deformation using transparent silica gel to model sand. Int. J. of Physical Modeling in Geotechnics 2(4), 13–26 (2002b)
Konagai, K., Tamura, C., Rangelow, P., Matsushima, T.: Laser-aided tomography: a tool for visualization of changes in the fabric of granular assemblage. Proc. JSCE No: 455 I-21, Structural Engr. / Earthquake Engr. 9(3), 193–201 (1992)
Konagai, K., Rangelow, P., Sato, T.: Real-time observation of dynamic changes in the fabric of granular material structures through laser-aided tomography. In: Proc. 10th European Conf. on Earthquake Engineering, pp. 459–466 (1994)
Lehane, B., Gil, D.: Displacement fields induced by penetrometer installation in an artificial soil. Int. J. Physical Modeling in Geotechnics 4(1), 25–36 (2004)
Liu, J., Iskander, M., Sadek, S.: Consolidation and permeability of transparent amorphous silica. ASTM Geotechnical Testing Journal 26(4), 390–401 (2003)
Lorenz, H., Heinz, W.: Change of density in sands due to loading. In: Proc. 7th Int. Conf. Soil Mech. & Found. Engr., vol. 1, pp. 267–273 (1969)
Mandava, S., Watson, A., Edwards, C.: NMR imaging of saturation during immiscible displacements. AIChE Journal 36(11), 1680–1686 (1990)
Mannheimer, R.: Slurries you can see through. Technology Today, p. 2 (1990)
Mannheimer, R., Oswald, C.: Development of transparent porous media with permeabilities and porosities comparable to soils, aquifers, and petroleum reservoirs. Ground Water 31(5), 781–788 (1993)
Ng, T., Kelley, M., Sampson, J.: MRI studies of direct shear tests on round particles. In: Proc. Engr. Mechanics, ASCE, pp. 572–575 (1996)
Orsi, T., Anderson, A., Leonard, J., Bryant, W., Edwards, C.: Use of X-ray computed tomography in the study of marine sediments. In: Proc. CEO V, ASCE, College Station, Texas, pp. 968–982 (1992)
Posadas, D., Tannus, A., Panepucci, H., Crestana, S.: Magnetic resonance imaging as a non-invasive technique for investigating 3-D preferential flow occurring within stratified soil samples. Computers and Electronics in Agriculture 14(4), 255–267 (1996)
Roscoe, K., Arthur, J., James, R.: The determination of strains in soils by an X-ray method. Civil Engineering and Public Works Review 58, 873–876, 1009–1012 (1963)
Sadek, S., Iskander, M., Liu, J.: Geotechnical properties of transparent silica. Canadian Geotechnical Journal 39(1), 111–124 (2002)
Stainer, S.: Geotechnical Modeling using a transparent synthetic soil. M.Sc. Report, University of Sheffield (Supervisor Dr. C. Hird) (2006)
Song, Z., Hu, Y.: Plate anchor keying under inclined pullout in clay: observation and estimation. In: Proc. XVII Int. Conf. Soil Mechanics & Geotechnical Engr., pp. 708–711 (2009)
Toiya, M., Hettinga, J., Losert, W.: 3D imaging of particle motion during penetrometer testing, from microscopic to macroscopic soil mechanics. Granular Matter 9, 323–329 (2007)
Welker, A., Bowders, J., Gilbert, R.: Applied research using transparent material with hydraulic properties similar to soil. ASTM Geotechnical Testing J. 22(3), 266–270 (1999)
Welker, A., Bowders, J., Gilbert, R.: Using a reduced equivalent diameter for a prefabricated vertical drain to account for smear. Geosynthetics Int. 7(1), 47–57 (2000)
Wong, R.: Mobilized shear strength components of Athabasca sand in Triaxial compression. Canadian Geotechnical J. 23(4), 413–422 (1999)

Chapter 2
Optical Techniques in Geotechnical Engineering

Magued Iskander and Samer Sadek

Abstract. Imaging and its applications in geotechnical engineering are growing enormously. This has been fueled by advances in digital imaging along with the increasing power of computing. In this chapter, a literature review of the advances of imaging in some civil engineering applications is presented. The applications are categorized according to their disciplines, with focus on geotechnical engineering applications.

2.1 Introduction

Imaging techniques have been used in civil engineering research for a long time. However, the application of imaging techniques in civil engineering research has recently increased dramatically. In the past, pictures were used in most research work to qualitatively demonstrate or to explain a phenomenon. For example, in the early 1940's we could see pictures of bearing capacity failure in plane strain models, clay particles in scanning electron microscope images, failure in Triaxial soil samples, crack propagation in concrete samples, and many other applications in civil engineering research. Beginning the late eighties and early nineties, imaging techniques have been fueled by significant advances in computation and digital photography, which permitted the use of imaging quantitatively not only in laboratory studies but also in field measurement. Today imaging techniques are becoming an essential tool in most civil engineering disciplines. In this chapter, a literature review of the advances of imaging in some civil engineering applications is presented. The applications are categorized according to their disciplines, with focus on geotechnical engineering applications. This chapter is not intended as a bibliography of all imaging research in civil engineering, it simply presents a variety of representative applications that are now used in civil engineering.

2.2 Imaging Applications in Geotechnical Engineering

Imaging in geotechnical engineering has been used for a wide array of problems. We can classify imaging uses in this area into: (1) soil stress measurement, (2) soil

M. Iskander: Modeling with Transparent Soils, SSGG, pp. 5–18.
springerlink.com

deformation measurement, (3) soil fabric and void characterization, (4) soil classification and grain size analysis, and (5) geo-environmental applications, as follows:

2.2.1 Soil Stress Measurements

Application of image analysis in stress measurement has been limited to photoelasticity. Many researchers attempted to utilize photoelasticity in granular media. In one of the early attempts Drescher and De Jong (1972) tried to verify a mathematical model by analyzing internal stresses of a plane strain granular model using sensitive photoelastic discs. Later, Allersma (1982) also used photoelastic particles for measurement of stresses in plane strain problems. The technique is based on using transparent solid particles that have birefringence property. The voids in between the particles were filled with a liquid that has the same refractive index as the particles. Spatial information about the principle stress, stress orientation and principal stress difference were obtained using images captured at the boundaries. This technique is limited to plane strain problems. In addition to this limitation, the boundary values of the state of stress must be determined.

Paikowsky et al (1993) studied an array of photoelastic discs that had a photoelastic property to resemble two-dimensional stresses in a particulate media. Particles or discs were confined between two glass plates; loads were applied at the top of the array. The discs used had a diameter of 30 mm (1.2 inches) and thickness of 10 mm (3/8 of an inch). Images of the setup were captured for photoelasticity analysis. The focus of the study was on particles contact stresses and internal stresses within the particles.

2.2.2 Soil Deformation Measurements

Many imaging techniques have been implemented for soil deformation measurement. Perhaps the oldest imaging technique used, was using photographs to track specific targets on the boundaries of a soil model. An example of these early applications is Butterfield and Andrawes (1971). Advanced optical techniques have also been utilized in soil mechanics research for deformation measurements. Woods (1974) used holographic interferometry for studying the effectiveness of slurry filled trenches for dynamic isolation of footings. Recently the use of imaging techniques in soil deformation measurement expanded enormously; from tracking pre-embedded grids to the use of highly complicated imaging algorithms for deformation measurements. In addition, some imaging techniques such as target tracking have been expanded to field applications and centrifuge experiments. This growth is driven by the advent of computerized tracking, which provides faster and more accurate analysis.

2.2.2.1 Computerized Aggregate-Based Target Tracking

Micro deformation of granular soils was studied by Raschke et al (1996), Gustafsson and Gustafsson (1996), and Guler et al (1999). These researchers analyzed very small areas of sand, approximately 5 cm^2, for deformation analysis on a particle by particle basis. Raschke et al developed two image processing algorithms that depend mainly on edge detection of the particle and tracking the particles using image matching function. This method is limited to small areas. Gustafsson and Gustafsson (1996) used different colored particles for different particle sizes, and used a threshold separation technique. The movement of each group of particles was monitored separately using a tracking algorithm similar to Raschke et al. Guler et al customized commercial software and hardware for measurement of granular soil particle. The imaging algorithm used by the authors employs a "block-based" match to analyze the displacement of sand particles. This technique eliminates the use of the user judgment in analysis, which is required in most target tracking techniques.

Horii et al (1997) used image-matching technique to measure the displacement field in rock samples under plane strain loading. An artificial contrast was applied to the rock surface. To increase the resolution of the digitized images, an image of the sample was captured using a regular camera and a sensitive film and then the image was digitized using a CCD camera and step motor. However, the matching algorithm was not published by the authors. The technique was reported to allow for measurements of strains as small as 5μm.

2.2.2.2 Computerized Grid-Based Target Tracking

Alshibli and Sture (1999) measured shear band thickness of sandy samples under plane strain conditions. The researchers studied shear bands at failure using two different techniques. The first technique was based on a pre-drawn grid on the latex membrane of the samples at failure. In the second technique, images of sliced epoxy-impregnated samples were used for verifying the results obtained from the first method. Saada et al (1999) studied shear band propagation in sand by tracking an embedded grid in sand sample. The authors studied both torsional and axial stresses using a grid that was imprinted on a sand sample after freezing the sample.

Xing et al (1997) proposed a target-tracking system for measuring two-dimensional deformation in a large soil model (0.5m x 1.22m x 9.1m). The technique is based on tracking white polypropylene markers, which were pre-placed in the model. Gill (1999) also used target tracking for studying pile penetration in transparent soils using an imaging system and embedded markers. Gill used a transparent synthetic soil, which is used also in this research and was published by Iskander et al (1994). An extensive discussion of the target tracking technique is presented in section 4.2 of this book.

Digital imaging was used by Macari-Pasqualino et al (1993) for volume change measurement in triaxial samples. The triaxial tests were performed in zero-gravity to study the mechanical behavior of granular materials at low confining pressure. The tests were monitored using three cameras and images were digitized using frame grabbers. Images were used for measuring the volume change of the samples and their failure modes.

2.2.2.3 Deformation Analysis Using Advanced Techniques

New techniques based on X-ray radiography such as X-ray Computerized Tomography (CT) and Magnetic Resonance Imaging (MRI) were used for deformation measurement. Bourdeau (1993) used radiography to monitor markers placed in soil model, for study of soil interaction with geo-grids under cyclic loading. Shi et al (1999) used CT for monitoring evolution of failure in soil samples during direct shear testing. Also, Wong (1999) used computer tomography scanning, X-ray imaging and scan electron microscope to examine sheared Triaxial samples of dense sand. In his study, images of post-testing cemented samples were used for examining the interlocking structure, shear-band pattern and porosity distributions inside and outside shear bands. Ng et al (1996) used MRI for monitoring samples formed of round particles during direct shear tests; the focus of the study was volumetric change, initial particle arrangement, and the fabric evolution during shear.

Laser tomography was used by Konagai et al (1992) and (1994) for monitoring soil embankment deformation. In order to use laser light, Konagai et al used crushed glass with matching refractive index pore fluid. A simple target tracking technique was used for monitoring particles; thus very coarse particles were used for the analysis. The change of the refractive index of the pore fluid was a serious problem in this experiment, because of the evaporation of one the components of the pore fluid mix.

2.2.2.4 Field Measurements

Allersma (1997) presented a study of landslide movement due to water infiltration. Measurements were done primarily by simple target tracking of marked labels on the slope. Images were captured from a fixed camera. Photos were analyzed to produce time verses displacement diagrams of the labeled targets. This technique enabled the researchers to conclude that the failure started at the ground surface. This technique is limited to recording surface measurements only. Additionally the camera has to be located outside the deforming area, and is subjected to unforeseen factors such as wind and temperature, which could limit the accuracy of the measurements.

This technique can be extended to large areas using aerial photography. For example, Gokceoglu & Aksoy (1996) captured the landslide mechanism for a an area of 120 km^2.

Ground thermal variations have also been used to capture aerial infrared images. For example, Shikada et al (1993) proposed the use of infrared images to predict the occurrence of landslides. The researchers showed that the surface temperature in the landslide area is higher than other areas.

2.2.2.5 Centrifuge Experiments

Charreir et al (1997) developed a target tracking software for image analysis of centrifuge camera images to study a number of problems such as displacement of shallow tunnels, and bearing capacity of footings near slopes. All problems presented were plane strain problems. Displacement was measured by target tracking of embedded markers in the soil model at the boundaries. A transparent wall for the model enabled the centrifuge camera to capture continuous images of an enhanced grid in the soil sample. The complete 2-dimensional field of deformation for the modeled problem was obtained by calculating the grid displacements at the model boundaries. The same technique was used by Chambon et al (1997). The researchers used specially developed software "pctrack" for displacement measurement using target tracking. The results were compared to a stereo-photogrammetry setup, in which two cameras were used for analysis. A tunnel excavation problem was studied. The authors presented and anticipated sources of errors for such measurements, such as; (1) misalignment of the camera optical axis, (2) index bias resulting from the difference in refractive index between air and the transparent wall of the model, and (3) optical bias resulting from the camera lens.

2.2.3 Soil Fabric and Void Characterization

Internal porosity and void ratio distribution through a soil quantum were studied using image analysis by many researchers. In most cases, not only the voids and their distribution were studied, but also natural soil fabric and its anisotropy.

2.2.3.1 Study of Soil Fabric Using Digital Photography

Early researchers studied soil fabric and void distribution using image analysis qualitatively (Bhatia and Soliman 1990, Leng et al 1993). Muhunthan and Chameau (1992) proposed a unified framework for describing the fabric of granular media based on classical stereology principles. This framework led to the development of fabric tensors, which describe the discrete granular media in a continuum sense. The research outlined a procedure based on fixing the specimens initial fabric using resin impregnation. The specimen is then sliced and analysis is performed on captured images of the slices. At the microscopic level, the analysis measures the variation of porosity within the image; and at the macroscopic level, it measures the variation of the porosity for different sections.

Kuo and Frost (1993) proposed a modified technique for quantifying the soil porosity tensor, in order to minimize the bias in the original procedure. The technique was applied to study the uniformity of fabric of Ottawa sand. Muhunthan and Masad (1997) proposed a different modified analysis to minimize the bias. The researchers presented a comparison between the porosity tensors calculated using all three techniques. Frost and Kuo (1996) proposed a fully automated imaging system for evaluation and study of frequency distribution of local void ratio for granular materials. Later, Kuo et al (1998) extended the technique by studying surface area, mean free path and porosity tensors to characterize the distribution of directional data from microscopic granular materials arrays.

Moore and Donaldson (1995) used the theory of fractals to quantify soil porosity and void distribution from captured images. Fractals were also used by

Gimenez et al (1997) for estimating hydraulic conductivity based on porosity analysis of soil samples.

Fourier analysis of particle images enabled the study of morphological descriptors, which were used originally by Luerkens (1991) for fine particle morphology. Wang et al (1997) proposed descriptors that were back calculated by applying Fourier-Legendre series on binary images of coarse particles obtained through an optical microscope. The first step in this method is acquiring images of aggregate particles. Next, the images are transformed to binary images. Particle outlines are digitized and transformed to polar coordinates (the center of gravity of the particle is the origin). Fourier-Legendre analysis is then applied onto the particle layout for quantifying a series of parameters that are considered morphology descriptors.

2.2.3.2 Soil Fabric Analysis Using SEM

Scan Electron Microscope (SEM) images have been used for studying the morphology of clay and sand particles. For example, Kuo et al (1996) used SEM images of Kaolinite particles to investigate particle texture. Penumadu and Hazan (1997) studied voids ratio and void fabric of clay samples at both macro and micro scales. Analysis of SEM binary images (Black and White) of the Kaolinite samples, showing the particles in black and voids in white was used. The authors were able to quantify the area of voids and its distribution in the captured sample image by counting the number of white pixels. This technique can be applied in more than one portion of the histogram of the original image to study both, micro voids and macro voids, because they can be differentiated at different gray levels. Yudhbir and Rahim (1993) used SEM images for analyzing the effect of sand morphology and micro-fabric on the mechanical behavior sand. Bai and Smart (1997) used SEM to study the change in microstructure of Kaolin during consolidation and undrained shear. Shi et al (1998) used SEM images accompanied by a special software "CIPS" to study the orientation of the clay aggregate and its effect on strength and permeability. Lange et al (1997) used SEM images for estimating the pore distribution at different stress levels of cement base materials by metal intrusion technique.

2.2.3.3 Recent Trends for Studying Soil Fabric

Cousset (1998) suggested a new formulation for pore size measurement of engineering materials such as sand, clays, rocks, concrete and wood using images produced by Nuclear Magnetic Resonance (NMR). Ketchman and Carlson (2001) proposed the use of Computed Tomography (CT) for estimating the porosity of natural soils and rocks. Soil behavior is influenced by the interactions between soil particles. The influence of particle eccentricity on soil shear properties and wave propagation was measured by Santamarina and Fam (1995), Aloufi and Santamarina (1995), and Santamarina et al (2002). These studies at the particle level helped in understanding the macro-geotechnical properties of natural soils and rock mass.

2.2.3.4 Geotextile Fabric

Not only soil fabric and morphology were studied using image analysis, but also geotextiles have been under investigation. Bhatia et al (1993) used images for morphological study of nonwoven geotextiles. Images were analyzed for pore size, fiber diameter, porosity, and reduction of porosity due to soil clogging.

2.2.4 Soil Classification and Grain Size Distribution Analysis

Raschke and Hryciw (1997) have shown that the grain size distribution of non-uniform soil can be determined by collecting and analyzing images of a soil specimen spread out over a back-lit glass plate at a series of magnification levels. Hryciw et al (1997) proposed three different methods for soil characterization of collected in-situ images by Vision Cone Penetrometer (VisCPT). The first method is based on converting the images to the frequency domain using Fourier Transform (FT), thus the dominant frequencies will correlate with the dominant particle sizes. Additionally, the first moment of the normalized power spectrum of uniform soils with respect to the ordinate can be used to characterize the soil texture. The second method is based on edge detection techniques, which is based on the Hough Transformation. The grain size distribution calculated by this method showed an excellent agreement with the actual grain size distribution. The third method is edge density method, in which the edge pixels for small window (kernel) of soil image are counted. The edge pixel density (EPD), which is the ratio between the edge pixels and total number of pixels in the window, is calculated. By studying the variation of EPD throughout the image and its convergence as the window size increases, the average size of soil particles can be estimated. Ghalib and Hryciw (1999) proposed another method for particle size distribution analysis; they used mosaic imaging and watershed algorithms for analysis. The advantage of this recent technique is that it reduces the required number of magnification of the images and thus it allows the analysis of bigger sample areas.

Three-dimensional analysis for characterization of granular material samples was studied by Kuo et al (1996). The researchers based their analysis on capturing images in two perpendicular positions. After initial images are captured, the particles attached to a tray are rotated 90 for another image. The analysis was performed on coarse aggregates of size range 0.38-0.5 inch (10-13 mm) impregnated with a resin and thin sliced.

Image analysis was also used by Burns et al (1997) for evaluating the size distribution of air micro-bubbles. Floatation of air micro-bubbles in clayey suspensions is used for separation of clay minerals or tiny particles that have lower density than that of liquid used. Images of generated bubbles were captured and analyzed for measurement of the area of the air bubbles. A reference scale was added in the captured images for calibrating the measured particles sizes.

2.2.5 Imaging Techniques in Geoenvironmental Studies

Ground flow problems have been studied using imaging techniques for a long time. Many two-dimensional problems were modeled in labs using transparent containers and dyed water indicating flow pattern. Recently, advances in imaging inspired researchers for more imaging applications. For example, porosity identification using images of soil were analyzed by Blair et al (1993) and, Berge et al (1997) to estimate the hydraulic conductivity.

Weil et al (1993) used ground penetrating radar images for field detection of contaminant transport. Although the proposed method is considered cost effective, it can only provide qualitative results. This technique was also applied by Dodt et al (1993) to detect contamination using GPR aerial photos. Heiler et al (1993) used ground penetrating radar images for locating buried drums containing hazardous material.

Transparent synthetic soils used in this research have been used in for modeling flow problems. Welker et al (1999) used color saturation to study flow of contaminants into perforated vertical (wick) drains in a transparent synthetic soil model.

New techniques such as Particle Image Velocimetry (PIV), Magentic Resonance Imaging (MRI) Neutron Radiography (NR), X-ray Computerized Tomography (CT), and Single Photon Emission Computed Tomography (SPECT) were also used in the field of geo-environmental engineering. For example, Cheng et al (2000) used PIV to count polystyrene spherical particles and measure their size and shape uniformity, which facilitate aggregation modeling. Pervizpour et al (1999) used MRI for determination of the spatial distribution and mass fraction of hydrocarbon liquids in contaminated sites, by comparing site images with images of well-characterized laboratory specimens of porous material containing hydrocarbon liquids. Wong and Wibowo (2000) used CT to study the uniformity of air spatial saturation along with the porosity variation throughout in soil column. Perret et al (2000) used SPECT to measure flow in an undisturbed soil column by quantifying the spatial distribution of the tracer inside soil column samples.

2.3 Example Applications of Imaging Techniques in Civil Engineering

A detailed description of all imaging application in civil engineering is beyond the scope of this chapter, only a brief description of example applications are presented in the following sections.

2.3.1 Pavement Crack Measurement

In the last decade, image analysis has been used extensively for assessment of pavement and road condition due to its simplicity. A video camera attached to an inspection van captures a continuous image of the road. Many researchers came up with automated systems for assessment of road condition. For example, Haas et al (1997) developed an automated algorithm for detecting cracks and joints condition. Smith and Lin (1997) developed a fuzzy logic classification scheme for pavement distress condition. Oh et al (1997) developed iterative algorithm for overcoming noisy images of roads due to shadows and low light conditions. Koustsopoulos and Mishalani (1997) presented a model for distress assessment in a local (microscopic) and global (macroscopic) level using captured images of pavement. Lee (1993) presented a comparison between 15 different imaging algorithms used in crack detection.

Ground Penetration Radar (GPR) has also been used for pavement assessment. Special computer algorithms were developed for quick analysis of GPR data (Adeli & Hung 1993 and Maser 1996). Heiler and McNeil (1997) proposed a modified system for analyzing the GPR data using an artificial neural network (ANN).

2.3.2 *Traffic Analysis and Control*

Currently imaging systems provide essential data for transportation and traffic engineering planning (Anon 1999). Machine vision techniques were introduced to intersection traffic signal control in the late 1970's (Chou and Sethi 1993). Nowadays, many systems have been developed all over the world for traffic analysis and control applications, in addition to image based systems for traffic violations. Nallamathu and Wang (1997) developed one of the first automated systems for license plate recognition using character recognition algorithm for the use in monitoring violators at toll stations and many other traffic applications.

2.3.3 *Concrete Morphology and Micro-cracks*

Imaging techniques are widely used in studying concrete, grain size analysis of its components, its morphological characteristics, hydration process, and micro-cracks development and propagation (Chermant 2001). Researchers use digital photographs (Ammouche et al 2000 and Diamond 1993), SEM (Yang and Buenfeld 2001), tomography (Hall et al 2000) and, interferometry (Jia and Shah 1994). Moreover, the interaction between concrete and steel rebar was studied using advanced optical techniques and image processing such as speckle interferometry (Ghandehari et al 1999).

2.4 Summary

In summary, imaging and its applications in geotechnical engineering are growing enormously. This has been fueled by advances in digital imaging along with the increasing power of computers. It is expected that these factors will result in more use of imaging in geotechnical engineering research in the near future.

References

Adeli, H., Hung, S.: A fuzzy neural network learning model for image recognition. Integrated Computer-Aided Engineering 1(1), 43–55 (1993)

Allersma, H.: Photo-elastic Stress Analysis and Strains in Simple Shear. In: Vermeer, P.A., Luger, H.J. (eds.) Proc. IUTAM Symposium on Deformation and Failure of Granular Materials, Delft, pp. 345–353 (1982)

Allersma, H.: Using imaging technologies in experimental geotechnics. In: Proc. 2nd Int. Conf. on Imaging Technologies, ASCE, pp. 1–9 (1997)

Aloufi, M., Santamarina, C.: Low and high strain behavior of grain masses - the effect of particle eccentricity. Transactions of ASAE 38, 877–887 (1995)

Alshibi, K., Sture, S.: Sand shear band thickness measurement by digital imaging techniques. J. of Computing in Civil Engineering 13(2), 103–109 (1999)

Ammouche, A., Breysse, D., Hornain, H., Didry, O., Marchand, J.: A new image analysis technique for the quantitative assessment of microcracks in cement-based materials. Cement and Concrete Research 30(1), 25–35 (2000)

Anon: Video technology in traffic engineering and transportation planning. J. of Transportation Engineering, ASCE 125(3), 169–175 (1999)

Bai, X., Smart, P.: Change in microstructure of kaolin in consolidation and undrained shear. Geotechnique 47(5), 1009–1017 (1997)

Berge, P., Berryman, J., Blair, S., Pena, C.: Scaler properties of transversely isotropic tuff from images of orthogonal cross sections. In: Proc. 2nd Int. Conf. on Imaging Technologies: Techniques and Applications in Civil Engineering, ASCE, pp. 185–196 (1997)

Bhatia, S., Soliman, A.: Frequency distribution of void ratio of granular materials determined by an image analyzer. Soil and Foundation 30(1), 1–16 (1990)

Bhatia, S., Huang, Q., Smith, J.: Application of digital image processing in morphological analysis of geotextiles. In: Proc. of the Conference on Digital Image Processing: Techniques and Applications in Civil Engineering, Kona, Hawaii, USA, ASCE, pp. 95–108 (1993)

Blair, S., Berge, P., Berryman, J.: Estimates of permeability of porous materials using image analysis of cross sections. In: Proc. of the Conference on Digital Image Processing: Techniques and Applications in Civil Engineering, Kona, Hawaii, ASCE, pp. 109–116 (1993)

Bourdeau, P.: Radiographic visualization in experimental soil mechanics. In: Proc. of the Conference on Digital Image Processing, ASCE, pp. 125–134 (1993)

Burns, S., Yiacoumi, S., Frost, D., Tsouris, C.: Application of digital image analysis for distribution measurement of microbubbles. In: Proc. 2nd Int. Conf. on Imaging Technologies: Techniques and Applications in Civil Engineering, ASCE, pp. 100–107 (1997)

Butterfield, R., Andrawes, K.: The Visualization of planer displacement field. In: Parry, R. (ed.) Proc. Roscoe Memorial Symposium: Stress Strain Behavior of Soils, Cambridge University, pp. 467–475 (1971)

Chambon, P., Molaiard, L., Garnier, J.: An attempt to compare image processing and stereo photogrammetry in geotechnical centrifuge models. In: Proc. 2nd Int. Conf. on Imaging Technologies: Techniques and Applications in Civil Engineering, ASCE, pp. 30–39 (1997)

Charrier, J., Moliard, J., Garnier, J.: Numerical image processing in centrifuge testing. In: Proc. 2nd Int. Conf. on Imaging Technologies, ASCE, pp. 20–29 (1997)

Cheng, C., Atkinson, J., Van Benschoten, J., Bursik, M., De Pinto, J.: Image-based system for particle counting and sizing. J. of Environmental Engineering V 126(3), 258–266 (2000)

Chermant, J.: Why automatic image analysis? An introduction to this issue. Cement and Concrete Composites 23(2-3), 127–131 (2001)

Chou, E., Sethi, V.: Machine vision based traffic-adjusted intersection signal control. In: Proc. of the Conference on Digital Image Processing, Hawaii, ASCE, pp. 150–156 (1993)

Cousset, P.: Pore size NMR imaging. Magnetic Resonance Imaging 16(5/6), 621–623 (1998)

Diamond, S.: Interactive digital image feature analysis as applied to cement and concrete microstructural investigation. In: Proc. of the Conference on Digital Image Processing Techniques and Applications in Civil Engineering, ASCE, pp. 158–164 (1993)

Dodt, J., Genske, D., Kappernagel, T., Noll, P.: Digital evaluation of contaminated sites. In: Proc. of the Conference on Digital Image Processing, ASCE, pp. 165–172 (1993)

Drescher, A., De Jong, D.: Photoelastic verification of a mechanical model for the flow of granular material. J. of Mechanics and Physics of Solids 20, 337–351 (1972)

Frost, J., Kuo, C.: Automated determination of the distribution of local void ratio from digital images. ASTM Geotechnical Testing J., GTJODJ 19(2), 107–117 (1996)

Ghalib, A., Hryciw, R.: Soil particle size distribution by mosaic imaging and watershed analysis. J. of Computing in Civil Engineering 13(2), 80–87 (1999)

Ghandehari, M., Krishnaswamy, S., Shah, S.: Technique for evaluating kinematics between rebar and concrete. J. of Engineering Mechanics, ASCE 125(2), 234–241 (1999)

Gill, D.: Experimental and Theoretical Investigations of Pile and Penetrometer Installation in Clay, Ph.D. Thesis, Trinity College, Dublin, UK (1999)

Gimenez, D., Allmaras, R., Huggins, D., Nater, E.: Prediction of the saturated hydraulic conductivity –porosity dependence using fractals. Soil Science Society of America J. 61(5), 1285–1292 (1997)

Gokceoglu, C., Aksoy, H.: Landslide susceptibility mapping of the slopes in the residual soils of the Mengen region (Turkey) by deterministic stability analyses and image processing techniques. Engineering Geology 44(4), 147–161 (1996)

Guler, M., Edil, T., Bosscher, P.: Measurement of particle movement in granular soils using image analysis. J. of Computing in Civil Engineering 13(2), 116–122 (1999)

Gustafsson, L., Gustafsson, P.: Studying mixed granular flows by image analysis. In: Proc. 11th Conference ASCE Engineering Mechanics Division, pp. 614–617 (1996)

Haas, C., Kim, Y., Greer, R.: A model for imaging assisted automation of infrastructure maintenance. In: Proc. 2nd Int. Conf. on Imaging Technologies, ASCE, pp. 108–117 (1997)

Hall, C., Colston, S., Jupe, A., Jacques, S., Livingston, R., Ramadan, A., Amde, A., Barnes, P.: Non-destructive tomographic energy-dispersive diffraction imaging of the interior of bulk concrete. Cement and Concrete Research 30(3), 491–495 (2000)

Heiler, M., McNeil, S.: Pavement condition assessment using ground penetration radar. In: Proc. 2nd Int. Conf. on Imaging Technologies, ASCE, pp. 128–137 (1997)

Heiler, M., Motazed, B., McNeil, S.: Ground penetration radar: images of subsurface characterization. In: Proc. of the Conference on Digital Image Processing, ASCE, pp. 173–178 (1993)

Horri, H., Takamatsu, K., Inoue, J., Sasaki, N.: Measurement of displacement field by matching method and observation of strain localization in soft rock. In: Proc. 2nd Int. Conf. on Imaging Technologies: Techniques and Applications in Civil Engr., ASCE, pp. 10–19 (1997)

Hryciw, R., Ghalib, A., Rasckle, S.: Methods for soil characterization for images of grain assemblies. In: Proc. 2nd Int. Conf. on Imaging Technologies, ASCE, pp. 88–99 (1997)

Iskander, M., Lai, J., Oswald, C., Mannheimer, R.: Development of a transparent material to model the geotechnical properties of soils. ASTM Geotechnical Testing J., V. 17(4), 425–433 (1994)

Jia, Z., Shah, S.: Two-dimensional electronic-speckle-pattern interferometry and concrete-fracture processes. Experimental Mechanics, SEM 34(3), 262–270 (1994)

Ketchman, R., Carlson, W.: Acquisition, optimization and interpretation of X-ray computed tomographic imagery: application to geosciences. Computer and Geoscience 27(4), 381–400 (2001)

Konagai, K., Tamura, C., Rangelow, P., Matsushima, T.: Laser-aided Tomography: a tool for visualization of changes in the fabric of granular assemblage. In: Proc. JSCE No: 455 I-21, Structural Engr. / Earthquake Engr., vol. 9(3), pp. 193s–201s (1992)

Konagai, K., Rangelow, P., Sato, T.: Real-time observation of dynamic changes in the fabric of granular material structures through laser-aided tomography. In: Proc. Tenth European Conference on Earthquake Engineering, pp. 459–466 (1994)

Koustsopoulos, H., Mishalani, R.: Interpretation and aggregation of spatially detailed sensor data on pavement condition. In: Proc. 2nd Int. Conf. on Imaging Technologies, ASCE, pp. 148–159 (1997)

Kuo, C., Frost, J., Chameau, J.: Image analysis determination of stereology based fabric tensors. Geotechnique, Thomas Telford, London 48(4) (1998)

Kuo, C., Frost, J., Lai, J., Wang, L.: Three-dimensional image analysis of aggregate particles from orthogonal projections. Transportation Research Record No. 1526, National Research Council, Washington, DC, USA, pp. 98–103 (1996)

Kuo, C., Frost, J.: Revised methodology to minimize bias in determining the porosity and void tensor of particulate media. In: Proc. of the Conference on Digital Image Processing: Techniques and Applications in Civil Engineering, ASCE, pp. 186–194 (1993)

Lange, A., Abbell, A., Willis, K., Powell: Characterization of cement pore structure using wood's metal and mercury porosimetry. In: Proc. 2'nd Int. Conf. on Imaging Technologies: Techniques and Applications in Civil Engineering, ASCE, pp. 197–206 (1997)

Lee, H.: Survey: fundamental pavement crack imaging algorithms. In: Proc. Digital Image Processing: Techniques and Applications in Civil Engineering, Hawaii, ASCE, pp. 195–202 (1993)

Leng, X., Hounslow, M., Bai, X., Luo, D., Costa, D., Xue, X., Tovey, N., Smart, P.: Image analysis of clay microstructure. In: Proc. Digital Image Processing: Techniques and Applications in Civil Engineering, Hawaii, ASCE, pp. 77–86 (1993)

Luerkens, D.: Theory and Application of Morphological Analysis: Fine Particles and Surfaces. CRC Press, Boca Raton (1991)

Macari-Pasqualino, E., Costes, N., Parker, J.: Digital image techniques for vol. change measurements in triaxial tests. In: Proc. Digital Image Processing: Techniques and Applications in Civil Engineering, Hawaii, ASCE, pp. 211–219 (1993)

Maser, K.: Condition assessment of transportation infrastructure using ground penetration radar. J. of Infrastructure Systems, ASCE 2(2) (1996)

Moore, C., Donaldson, C.: Quantifying soil microstructure using fractals. Geotechnique 45(1), 105–116 (1995)

Muhunthan, B., Chameau, J.: Mathematical characterization of fabric and its use in mechanics of geomaterials. In: Proc. of Engineering Mechanics, College Station, TX, ASCE, pp. 725–728 (1992)

Muhunthan, B., Masad, E.: Determination of void fabric tensor of soils without radial sampling bias. In: Proc. 2nd Int. Conf. on Imaging Technologies: Techniques and Applications in Civil Engineering, ASCE, pp. 40–55 (1997)

Nallamathu, S., Wang, K.: Automated license plate recognition with a neural net based accelerator. In: Proc. 2nd Int. Conf. on Imaging Technologies: Techniques and Applications in Civil Engineering, ASCE, pp. 166–174 (1997)

Ng, T., Kelley, M., Sampson, J.: MRI studies of direct shear tests on round particles. In: Proc. of Engineering Mechanics, Boca Ractor, Florida, ASCE, vol. 1, pp. 572–575 (1996)

Oh, H., Garrick, N., Achenie, L.: Segmentation algorithm using iterative clipping for processing noisy pavement images. In: Proc. 2nd Int. Conf. on Imaging Technologies: Techniques and Applications in Civil Engineering, ASCE, pp. 138–147 (1997)

Paikowsky, S., Di Rocci, K., Xi, F.: Inter-particulate contact force analysis and measurement using photoelastic techniques. In: Proc. Int. Conf. Discrete Element Methods, Williams and Mustoeed, pp. 449–462 (1993)

Penumadu, D., Hazan, I.: Geo material characterization using digital micrograph analysis. In: Proc. 2nd Int. Conf. on Imaging Technologies: Techniques and Applications in Civil Engineering, ASCE, pp. 66–75 (1997)

Perret, J., Prasher, S., Kontazas, A., Hamilton, K., Longford, C.: Preferential solute flow in intact soil columns measured by SPECT scanning. Soil Science Society of America J. 64(2), 469–477 (2000)

Pervizpour, M., Pamukcu, S., Moo-Young, H.: Magnetic resonance imaging of hydrocarbon-contaminated porous media. J. of Computing in Civil Engineering 13(2), 96–102, ASCE

Raschke, S., Hryciw, R.: Vision cone penetrometer (V-CPT) for direct subsurface soil observation. J. of Geotechnical and Geoenvironmental Engineering, ASCE 123(11), 1074–1076 (1997)

Raschke, S., Hryciw, R., Donohoe, G.: Microdeformations in sands by digital image processing and analysis. Transportation Research Record 1548, 31–37 (1996)

Saada, A., Liang, L., Figueroa, J., Cope, T.: Bifurcation shear band propagation in sands. Geotechnique 49(3), 367–385 (1999)

Santamarina, J., Fam, M.: Changes in dielectric permittivity and shear wave velocity during concentration diffusion. Canadian Geotechnical J. 32(4), 647–659 (1995)

Santamarina, J., Klein, K., Wang, Y., Prencke, E.: Specific surface: determination and relevance. Canadian Geotechnical J. 39(1), 233–241 (2002)

Shi, B., Murakami, Y., Wu, Z.: Orientation of aggregates of fine-grained soil: quantification and application. Engineering Geology 50(1-2), 59–70 (1998)

Shi, B., Murakami, Y., Wu, Z., Chen, J., Inyang, H.: Monitoring of internal failure evolution in soils using computerization X-ray tomography. Engineering Geology 54(3-4), 321–328 (1999)

Shikada, M., Kusaka, T., Kawata, Y., Miyakita, K.: Extraction of characteristics properties in landslide areas using surface temperature. In: International Geoscience and Remote Sensing Symposium (IGARSS), vol. 1, pp. 103–105. IEEE (1993)

Smith, G., Lin, C.: Fuzzy logic distress classifier for pavement imaging. In: Proc. 2nd Int. Conf. on Imaging Technologies: Techniques & Applications in Civil Engr., ASCE, pp. 118–127 (1997)

Wang, L., Lai, J., Frost, D.: Fourier morphological descriptors of aggregate profiles. In: Proc. 2nd Int. Conf. on Imaging Technologies: Techniques and Applications in Civil Engineering, ASCE, pp. 76–87 (1997)

Weil, G., Graf, R., Forister, L.: Non-destructive remote sensing of hazardous waste sites. In: Proc. of the Conference on Digital Image Processing: Techniques and Applications in Civil Engineering, Hawaii, ASCE, pp. 279–284 (1993)

Welker, A., Bowders, J., Gilbert, R.: Applied research using transparent material with hydraulic properties similar to soil. Geotechnical Testing J., GTJODJ 22(3), 266–270 (1999)

Wong, R.: Mobilized strength components of Athabasca oil sand in triaxial compression. Canadian Geotechnical J. 36(4), 718–735 (1999)

Wong, R., Wibowo, R.: Tomographic evaluation of air and water flow patterns in soil column. Geotechnical Testing J., GTJODJ 23(4), 413–422 (2000)

Woods, R., Barnett, N., Sagesser, R.: Holography – a new tool for soil dynamics. J. of Geotechnical Engineering Division, ASCE 100(GT11), 1231–1247 (1974)

Xing, Z., Wells, L., Yaping, J., Shearer, S.: Using image analysis to measure two-dimensional soil deformation. J. of Terramechanics 34(2), 73–82 (1997)

Yang, R., Buenfeld, N.: Binary segmentation of aggregate in SEM image analysis of concrete. Cement and Concrete Research 31(3), 437–441 (2001)

Yudhbir, Rahim, A.: Image Analysis Applied to Geotechnical Engineering. In: Proc. of the Conference on Digital Image Processing: Techniques and Applications in Civil Engineering, Hawaii, ASCE, pp. 291–298 (1993)

Chapter 3
Introduction to Light and Optics

Magued Iskander and Samer Sadek

Abstract. An introduction to light and optics is presented in this chapter, with emphasis on light properties that are related to transparent soil modeling. Simple theories which are required for understanding the light behavior in transparent synthetic soils, including propagation of light, refraction, reflection, and speckle effects are presented.

3.1 Introduction

When a light ray strikes an object, it is reflected or refracted either partially or totally. In fact, that is what makes an object transparent or opaque. To explain these phenomena, there are many theories that justify light behavior. These theories have multiple levels of explanation. There are theories that predict the behavior of light with little or no explanation of the phenomena (*e.g.* Snell's Law) and theories that try to follow the phenomena at the micro level and then expand it or integrate it to justify well-known observations. In this chapter, a brief explanation of light properties that are related to this research will be addressed using simple theories. However, more details are also presented in some parts for clarification of important concepts. The discussion presented in this chapter is considered an introduction to advanced optics, which is required for understanding the light behavior in transparent synthetic soils, including refraction and reflection of light.

3.2 Nature of Light

The nature of light has been the subject of debate that started as early as the sixteenth century with an attempt to set the laws of refraction and reflection. This debate was mainly between two major opinions, which viewed light as either a stream of particles or as a propagating wave. Newton assumed that light is a stream of particles (corpuscular). Later in the nineteenth century, Maxwell proposed that light is an electromagnetic wave propagating in an imaginary space that he referred to as "Aether." Later in the twentieth century with the development of Max Plank's quantum mechanics, Einstein proposed that light consists of globs or particles of energy, each particle or *"photon"* has energy proportional to its

M. Iskander: Modeling with Transparent Soils, SSGG, pp. 19–25.
springerlink.com

frequency. Accordingly, photons have both particle and wave manifestations. By the end of 1920s, Einstein's theory was accepted by most scientists. Nevertheless Albert Einstein once said "*Every physicist thinks that he knows what a photon is, I spent my life to find out what a photon is and still don't know it*" (Hecht 1998). In this chapter, the description of light as a wave is sufficient to explain most of the light properties discussed. A light wave is an electromagnetic wave, which has two components, electric wave and magnetic wave. Both waves are propagating in the same direction of the original wave, but oscillating in two perpendicular planes. These planes are perpendicular to the wave front plane of the moving electromagnetic wave.

Light waves propagate in vacuum with a speed "*c*" approximately equal to $3\text{x}10^8$ m/sec. But the speed of light in any matter is a variable, which depend on the wavelength of the light wave and the properties of the medium where the light wave is propagating.

3.3 Propagation of Light in a Matter

Most transparent materials are *dielectric* which means that they are poor conductors of electricity. The electrons located in dielectric atoms are bounded to the atom, *i.e.* they aren't free as for the case of metals. When a light beam strikes a transparent medium it goes through with a different speed than its speed in vacuum. This is mainly due to the interference of the light wave with the electron oscillators (oscillating electrons in the atom) inside the transparent material. Each electron inside a dielectric transparent material oscillates with a certain frequency in the atoms and it was found by "Ewald-Oseen" that the electrons-oscillators generate an electromagnetic wave having essentially two components (Hecht 1998). One component cancels the primary light wave within the medium and the other moves through the dielectric material at speed "*v*" less than the speed of light in vacuum. The ratio between the speed of light in a matter and in vacuum is expressed by the refractive index of the material "*n*", which can be calculated from the following equation:

$$n = v/c \tag{3.1}$$

The refractive index of light in air is 1.0003, *i.e.* the velocity of light in air is approximately equal to its velocity in vacuum. The refractive index of light depends on the natural frequency of the material "ω," so in many cases n is expressed as

$$n = f(\omega) \tag{3.2}$$

The dependency of the refractive index on the natural frequency of the transparent material is reflected also from the variation of refractive index with the different frequencies of light; *i.e.* same material has different values of n for different light colors.

On the other hand, the main characteristic of metals is the presence of free electrons. Unlike the situation for transparent dielectrics, the free electrons in the metals oscillate with a frequency out of phase of natural light, canceling the light wave. That is why most metals have a silvery gray appearance since most of the light rays are reflected. In fact further studies showed that metal have a complex refractive index in the following form:

$$n^2(\omega) = 1 - (\omega_p/\omega)^2 \tag{3.3}$$

Where, ω_p is the natural frequency of the metal, (also known as the *plasma frequency)* and ω is the frequency of the electromagnetic wave.

From Eq. 3.3 it can be shown, that if ω_p is greater than ω, the refractive index is a complex number, indicating that most of the light rays will be reflected. Note that the frequency of some electromagnetic waves, such as X-rays, is higher than the natural frequency of some metals, which permits the electromagnetic wave to penetrate the metal. Thus, for such electromagnetic wave some metals are transparent.

3.4 Refraction of Light

The change in the velocity of light through transparent materials, results in a change in the direction of the transmitted wave as shown Fig. (3.1). This can be explained with more than one theory. A simple approach is Fermat's principle, which states, *"the actual path between two points taken by a beam of light is the one that is traversed in the least time."* Using this theory Snell's law (Eq. 3.4) of refraction can be derived as follows[1].

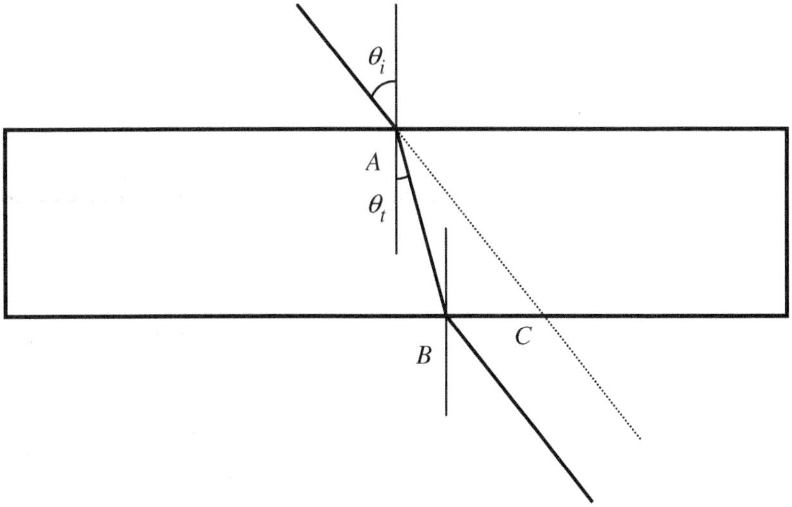

Fig. 3.1 Refraction of Light

[1] Historically, Snell's law was reported a long time before Fermat's principle.

$$\frac{\sin\theta_i}{\sin\theta_t} = \frac{v_i}{v_t} \tag{3.4}$$

where, v_i is the velocity of Light in the first medium (from which the incident ray is coming),.and v_t is the velocity of light in the second medium (in which the light is transmitted). In case of the first medium is air ($v_i \approx c$) Snell's law can be re-written as:

$$\frac{\sin\theta_i}{\sin\theta_t} = n \tag{3.5}$$

where, "n" is the refractive index of the transparent medium.

During the transmission of light, some of the light energy is absorbed by the transmitting media. For example, water molecules absorb red light, which causes the blue appearance for deep water. In fact at a depth of 30 meter in sea water the red component of light is almost completely removed (Hecht 1998).

3.5 Reflection of Light

Applying the electromagnetic waves theory on incident beams of light reveals that there must be reflection of some of the transmitted beam at the boundaries. In fact, it can be proved that, if the incident angle "θ_i" equals to 0°, the *Reflectance* (the ratio between the incident power and the reflected power), which is expressed as R, can be calculated as follows:

$$R = \frac{n_t - n_i}{n_t + n_i} \tag{3.6}$$

where n_i and n_t are the refractive indices of the incident medium and the transmitting medium, respectively.

The amount of reflection of a glass/air boundary can be calculated from Eq. 3.6 to be 4% assuming the refractive index of glass and air to be 1.5 and 1 respectively. Further analysis shows that reflectance increases with the increase of the incident angle.

3.6 Light in a Granular Medium

Although water is transparent, water vapor has a white appearance. Water consists of small grains or particles, which have diameter larger than the wavelength of light. When light passes through water vapor it reflects and refracts at the boundaries of the particles, which results in the white appearance of the whole medium. This phenomenon can be observed in many examples such as, sugar, salt, and snow. All these particles are transparent but as soon as particles are put together a white appearance is observed. But if the voids between the particles are filled with a material that has the same refractive index, the whole medium appears transparent, because no reflection or refraction occurs at the particle boundaries. It can also be shown that if these voids are filled with a material that has a different refractive index, the whole medium will be white in color. For example, white paint is simply a suspension of transparent particles in transparent oil with a different refractive index.

The basic idea behind transparent synthetic soils is using a matched refractive index liquid to fill up the voids in between transparent solid particles. The main source of transparency degradation in the transparent synthetic soil mix is air particles trapped in the solid-liquid mix. Good de-airing can produce highly transparent samples. The other source of transparency degradation is the exact matching of the refractive indices of solids and pore fluid. Temperature variation may affect the refractive index match because the refractive index of fluids is more sensitive to temperature than solids. Furthermore, changes in the properties of the pore fluid with time can affect transparency. For example, Konagai et al (1992) reported that the change in the refractive index due to evaporation of one component of the pore fluid mix changed the sample to be completely translucent.

3.7 Basic Definitions

In this section, a brief discussion of some basic terms is presented. These terms will be used later throughout the book.

3.7.1 Speckle Effect

Speckle effect is a natural phenomenon that can be described as a granular appearance of a rigid body if illuminated by a coherent light such as laser. It also appears if a coherent light is transmitted through scattering objects such as ground glass or particles in a liquid (case of transparent synthetic soils). More discussion is presented in section 4.3 in chapter 4.

3.7.2 Coherent Light

Light waves that maintain a fixed and predictable phase relationship with each other over a period of time are called coherent light waves. The coherent light can be produced by a laser source and it differs from ordinary light because it is made up of waves of the same wavelength and all in phase where ordinary light contains many different wavelengths and phase relations.

3.7.3 Polarized Light

As discussed in section 3.2, light can be considered as energy moving in the form of electromagnetic waves (Cloud 1998). The electromagnetic wave has two components, electric and magnetic waves. Both waves are propagating in the same direction of the original light wave, but oscillating in two perpendicular planes. These planes are also perpendicular to the wave front plane of the moving electromagnetic wave. In case of a light wave, the direction of the oscillation of the electric field is considered to be the polarization direction. In case of ordinary light source, such as light bulb, the direction of polarization varies randomly. When a light wave has one plane of polarization, i.e. the electric field has one plane of oscillation, this light wave is said to be a "*plane-polarized light wave*", or "*polarized light*". Any laser beam is usually polarized in the vertical direction when the beams travel in horizontal direction.

3.7.4 Polarizers

A polarizer is a plate or sheet that allows the electrical component of the light wave to pass in one direction only. Thus, natural light can be polarized if it passes through a polarizer. The inclination of the plane of propagation of the passing wave's electric field can be changed by rotating the polarizer plate. If for instance, two polarizers are aligned on one axis and the directions of polarization are perpendicular to each other, light will not pass through this axis. In this case these polarizers are called "cross polarizers".

Sunlight is scattered in all directions. When it meets a horizontal surface, like a road or water, a large portion of the light is reflected horizontally. This horizontally polarized light is seen as white glare and masks visible light thus reducing visibility. Polarized eye glasses or polarizing filters, used in photography, alleviate glare by removing light polarized at 90° to the filter's polarization axis. These filters consist of many microscopic crystals embedded in a transparent polymer. The crystals are aligned during manufacture such that the filter absorbs light which is polarized parallel to the direction of the crystal alignment, but transmits light which is polarized perpendicular to it.

Polarizers are widely used in photoelastic measurements,[2] where experimental setup usually uses 2 polarizers. The sample, which has the photoelastic property, is placed in between the two polarizers. The first plate (the plate that light crosses before the sample) is called polarizer and the second plate, (the plate that light crosses after the crossing the sample) is called the analyzer plate.

3.7.5 Birefringence

Some materials are optically anisotropic, which means that they have different optical properties in different directions. The birefringence property of a material refers to having different refractive indices in different directions of polarization of light passing through. Most crystalline substances are birefringent. Though amorphous solids, such as glass and plastics, are usually optically isotropic, they can display birefringence under certain circumstances. For example, if stresses are

[2] Photoelasticity is presented in chapter 4.

applied to glass particles, the particles show some changes in their refractive indices. This is the idea behind photoelastic measurements, which is discussed in more details in the following chapter.

References

Cloud, G.: Optical Methods of Engineering Analysis. Cambridge University Press, Cambridge (1998)

Hecht, E.: Optics, 3rd edn. Addison-Wesley Longman, Inc., New York (1998)

Konagai, K., Tamura, C., Rangelow, P., Matsushima, T.: Laser-aided tomography: A tool for visualization of changes in the fabric of granular assemblage. Proc. JSCE No: 455 I-21, Structural Engr. / Earthquake Engr. 9(3), 193–201 (1992)

Chapter 4
Optical Measurement of Strain and Stress

Samer Sadek and Magued Iskander

Abstract. A review of optical techniques that can be used for measurement of deformations and stresses in transparent synthetic soils is presented in this chapter. The presented techniques include methods based on (1) target tracking and image cross correlation, (2) speckle and holographic interferometry, (3) photoelasticity, and (4) cross tomography. All techniques are presented with a brief background for basic understanding of the major experimental issues, as well as a discussion of how they can be applied for measuring stress and strain in transparent soils. The advantages and disadvantages of using each technique are also discussed.

4.1 Introduction

Several methods have been used to measure continuous spatial deformations and stress fields in experimental models. X-rays, lasers, ultrasounds, and electromagnetic waves have all been found to be useful in performing these measurements. Increasingly, non-intrusive techniques have become widely used in many engineering applications. For example, in geotechnical engineering X-rays were used for measuring deformation and flow at discrete points in soils in the late 60s and early 70s (Bransby and Milligan 1975; Kirpatrick and Belshaw 1968). New techniques, such as Computerized Axial Tomography (CAT scan) and Magnetic Resonance Imaging (MRI) were also utilized by Orsi et al (1992) and Dersues et al (1991) for experimental modeling of geotechnical problems. However, the broad application of X-ray based techniques is limited by the high cost and difficulties in experimental set up. At the same time, during the past two decades the use of simple optical techniques has been growing in civil engineering laboratories. This is due to recent developments in the computing, imaging, and digital photography industries. A review of optical based techniques that can be used for deformations and stress measurement in a transparent quantum such as the transparent synthetic soils is presented in this book. Every technique is presented with a brief background necessary for basic understanding of its advantages and disadvantages, as well as a discussion of how it can be applied to our problem.

M. Iskander: Modeling with Transparent Soils, SSGG, pp. 27–44.
springerlink.com

The techniques explained in this chapter can be classified into four major categories:

1. Target tracking based techniques, which is the base of image cross correlation and will be explained in more details in chapter eight of this book.
2. Interferometry-based techniques, which are increasingly growing in all engineering applications.
3. The photoelastic properties of silica gels made the discussion on the photoelasticity-based techniques essential as a third category.
4. Cross tomography-based techniques, which are considered as an extension to x-ray techniques that were used earlier in civil engineering laboratories.

4.2 Target Tracking

Target tracking is an old technique which has been used in many disciplines. It has been implemented in laboratories as well as field measurements. It mainly identifies an object in an image and tracks the object movement in consecutive images. Object tracking can be done manually or using specialized software. Most traditional applications of target tracking usually track a limited number of objects. Most of the available software depends on matching one object in many images and tracking its movement. Target tracking is the base for many advanced techniques, which are used extensively in engineering applications, such Stereo Photogrametry, Geographic Information System (GIS), Particle Image Velocimetry (PIV) ...etc. The target tracking system is refined to suit the special need of the application (GIS, PIV, etc.). For example in PIV, the software is based on tracking a large number of particles to satisfy the need for studying velocity distribution in fluid mechanics problems. Though the idea of target tracking is simple, the refinement that is used in each application is fairly complicated.

Simple target techniques can be applied in transparent soils by inserting colored markers in the transparent quantum and tracking these markers, before and after deformation. This technique was used by Gill (1999) to study pile penetration problems. Targets can be tracked using two-cameras for 3D measurements. The advantages of this method are: (1) Its setup is relatively simple, and (2) This technique does not need special experimental skills. This simple setup has two disadvantages, (a) It provides discrete measurements for the spatial deformation of the transparent quantum, and (b) The technique is limited by the total number of markers that can be inserted in the model without impairing its transparency.

4.2.1 Digital Image Correlation

One of the most recent techniques used for target tracking is Digital Image Correlation (DIC). DIC is explained in details in Chapter 8. DIC techniques are widely used for experimental measurements in mechanical engineering, mainly for:

(1) surface displacement analysis (SDA) in solid mechanics, and (2) particle image velocimetry (PIV) in fluid mechanics. DIC is based on using a correlation function in locating the best matching position of two images and thus predicting the body movements. DIC is a relatively simple and flexible technique that can be used in different applications.

Conceptually DIC is suitable for application in transparent synthetic soil models because transparent synthetic soil application combines both applications the SDA and the PIV. In SDA, surface deformation of solid bodies is measured by correlating images of the speckles. In PIV, the flow and velocity profiles of turbulent flow are measured using highlighted longitudinal laser sheet in the flow field of the fluid. Preliminary investigations of using DIC in transparent synthetic soil models showed a great potential for further use. Thus, DIC was chosen for detailed investigation and application in transparent synthetic soil models. A detailed explanation of the technique and an example application are presented in Chapter 8 and 9 of this book.

4.3 Interferometry

Interferometry is a method that measures the difference in distance traveled between two beams of light through the interference of these beams. If two light waves travel different distances, they are no longer in phase. The analysis of the phase shift between these two light beams results in quantifying such distance. This distance is commonly named as the Path Length Difference (PLD). The absolute phase relationship in interferometry experiments is lost, which means that a single experiment cannot indicate whether the two waves are out of phase by one wavelength or by many wavelengths. This is an important limitation of many interferometry-based techniques for large displacement measurements.

If two coherent, out of phase wave fronts interfere (Fig. 4.1), bands of interference will appear on any projection plane. These bands are called “interference fringes” and they appear as adjacent dark and light bands. Interference fringes are utilized by many interferometric techniques for deformation measurements.

A detailed discussion of interferometry is available in Jones & Wykes (1989) Shchepinov et al (1996) and Cloud (1998). A summary of interfrometric applications as they relate to transparent soil modeling is presented next.

Many interferometry techniques are now used in various engineering applications. Most of these applications utilize one of two major interferometry techniques: holographic interferometry and speckle interferometry. Historically, holographic interferometry started earlier; and still, holographic interferometry has very wide applications. However, nowadays speckle interferometry applications are becoming increasingly more important. In the following sections, the basics of both techniques are presented.

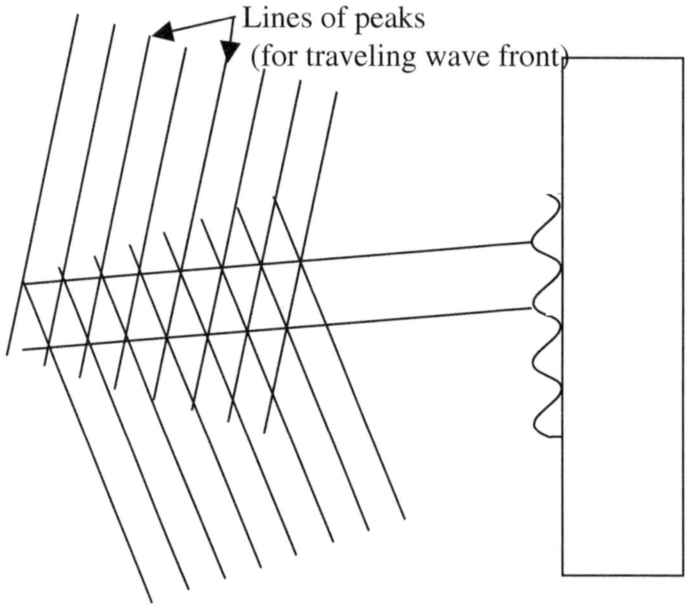

Fig 4.1 Two Interfering Coherent Wave Fronts

4.3.1 Holographic Interferometry (HI)

Holography can be defined as a method by which a wave front can be recorded on a hologram and subsequently can be reconstructed in the absence of the original wave front. A hologram is a special photographic plate that can record the reflected light intensity and its phase shift. Observation of this reconstructed wave front will give exactly the same physical effect as the observation of the original wave front, i.e. a three dimensional image is observed just as if the object was still present and being illuminated in the same way as when the holographic recording was made.

Holographic interferometry (HI) is the oldest technique of the interferometry-based techniques. The oldest study involving holography in geotechnical engineering was published by Woods et al (1974). HI utilizes a hologram in order to produce interference fringes by which the body movement or strains can be calculated. This can be done by superposition of the light from the reconstructed image (original un-deformed shape) and the real object (while it is being deformed) through the hologram. The interference of the two wave-fronts results in a uniform field of view of interference fringes. This condition requires that (1) the processed hologram must be precisely placed in its original position with an accuracy better than ¼ of the wavelength, and (2) the hologram suffer minimum emulsion shrinkage during development.

Holographic fringes are formed because of the interference between the scattered light waves from two identical surfaces placed at slightly different positions in space. This technique is considered one of most accurate techniques for out of plane motion (Cloud 1998).

Holographic interferometry (HI) can be preformed as real-time HI or frozen fringe HI. In real time HI (live HI), the original body shape is recorded on a hologram and the object live image is viewed through the original shape hologram. When the real object undergoes a small static displacement, a small variation in the relative phase of the 2 wave fronts, a fringe pattern can be observed. In frozen fringe HI (Double exposure HI), two hologram sheets are used to record 2 different position and both holograms are used to compare the light fields corresponding to 2 different positions to obtain fringe pattern.

HI might be used in transparent synthetic soils by illuminating the whole soil mass using a laser source and recording it in a hologram. The next step is to record the deformed mass of the transparent synthetic soil. The two recorded hologram will show the 3D spatial deformation of the soil mass. This requires a strong laser beam to cross the sample uniformly and have clear speckles throughout the sample. Analyzing the recorded fringes might the face problem of separating the in plane and out of plane movement. Currently all published HI results for 3D deformation are limited to monolithic models or samples.

4.3.1.1 Limitations of Holographic Interferometry

Jones and Wykes (1989) briefly discussed limitations of HI to be:

1. The laser power must be steady and sufficient to ensure that the interference fringes formed on the holographic recording medium between the object and the reference beams are stationary during the recording of the hologram.
2. The path difference between the object and reference beams must be less than the coherence length. In other words, the light beams reflected from the body must be able to interfere.
3. Rigid body movement should be avoided as much as possible, this might cause the holographic fringes to be localized away from the object surface and it is then more difficult to relate the fringe positions to the object coordinates.

In addition to these limitations, HI setup may be practically too complicated for modeling geotechnical problems and the scale of measurement may be too small for actual soil deformations in experimental models.

4.3.2 Speckle Interferometry

Interference fringes can be produced using a speckle effect. The speckle effect is a natural phenomenon that can be described as a granular appearance of a rigid body if illuminated by a coherent light such as laser. It also appears if a coherent light is transmitted through scattering objects such as ground glass or particles in a liquid (case of transparent synthetic soils). There are two types of speckles, objective speckles and subjective speckles. In objective speckle the speckle are formed due to the interference of light waves directly reflected from the object (Fig. 4.2)

The speckle size formed in objective speckle is dependent on (1) the length between the observation plane and the object "L", and (2) the illuminated area of the object, assumed to have a diameter "D". The average size of the speckle patches can be estimated statistically to be (Goodman 1975):

$$S_{obj} = 1.22 \frac{L}{D} \lambda \tag{4.1}$$

where, λ, is the wavelength of light used for illumination.

Subjective speckle is creating an image of the speckle using an optical system. This can be done by a lens, as shown in Fig. 4.3. The size of individual subjective speckle can be calculated from the following equation:

$$S_{subj} \approx 1.22(1+M)\lambda \frac{f}{a} \tag{4.2}$$

where, "*f*" is the focal length, "*a*" is aperture of the lens, "*M*" is the lens magnification, and λ is the wavelength of light.

In any case, the speckle is a unique signature derived from the local characteristics of a small area on the object surface. So if this small area moves, the speckle will also move, indicating the same displacement. Speckle phenomenon is the base idea for all speckle interferometry techniques, which can be categorized into speckle photography, speckle correlation and electronic speckle pattern correlation.

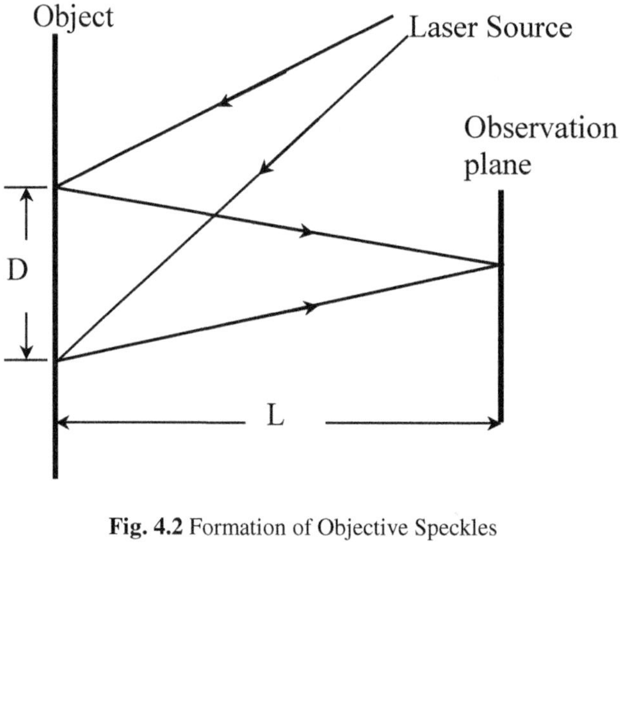

Fig. 4.2 Formation of Objective Speckles

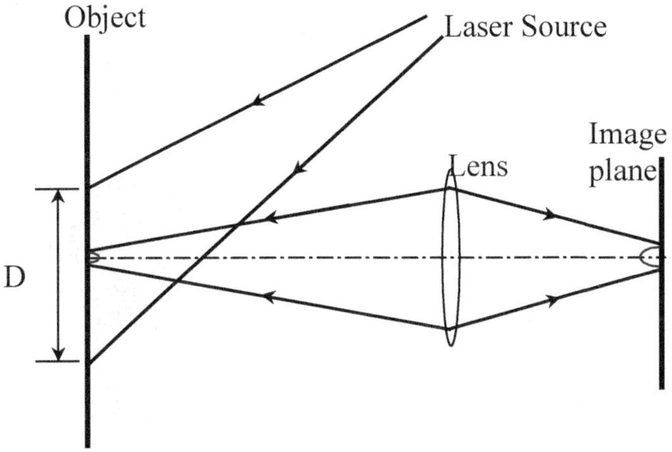

Fig. 4.3 Formation of Subjective Speckles

4.3.2.1 Speckle Photography

The method utilizes a high-resolution photograph of a specimen that is illuminated with a coherent light. The following steps explain one of the methods (single-aperture speckle recording) that utilize this technique:

1. The object is illuminated with an expanded laser beam.
2. An image is captured of the illuminated object.
3. After displacement is applied, the same photo plate (film) is exposed to the displaced object.
4. The processed plate is illuminated by a pinhole light source; the transmitted image will show complete spatial fringes of the displaced object.

The result of this technique is a permanently stored whole-field record of interference fringes, which yield to a map of displacement field of the object. The fringe pattern is localized in the image of the object being studied and two components of the displacement vector can be obtained individually.

Fringes are typically produced by back illuminating the double exposure negative, and then the produced fringes are analyzed for displacement calculation on a point-by-point basis (Fig. 4.4) or a whole field of displacement is analyzed by more sophisticated procedure.

This technique is relatively easier than other interferometry techniques but it has its own difficulties. These difficulties can be summarized as follows:

- The sensitivity of such technique is limited by the fixed speckle size due to aperture ratio of the recording optical system.
- Method is considered time consuming for developing and analyzing pictures.
- Whole field analysis is difficult, due to limitation of (1) the speckle size, and (2) the area that can be illuminated and captured for speckle analysis.

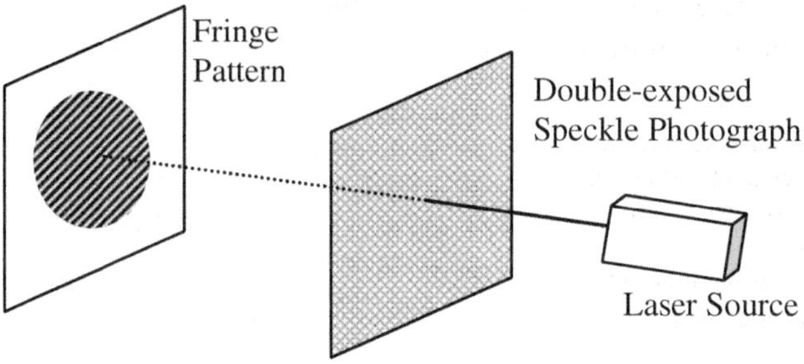

Fig. 4.4 Point-by-point Double-Exposure Speckle Photograph

4.3.2.2 Speckle Correlation Interferometry

In this method, interference fringes are formed using two incident beams on the object. The interference of these two beams will results in a certain speckle distribution. Any displacement in the object surface will result in movement of the object speckle thus, changing the interference fringe patterns. There are two main setups for this technique; the first is sensitive to out of plane displacement and, the second is more sensitive for in plane movements. The out-of-plane setup (Fig. 4.5) utilizes a reference beam recorded directly on the image plane to track the phase change of the incident beam on the object. In the second setup, (Fig. 4.6) interference fringes are formed directly on the object to reduce the sensitivity to out-of-plane movements.

Speckle pattern correlation is the most difficult experimental technique of all the interferometry techniques because it is more sensitive than any other technique

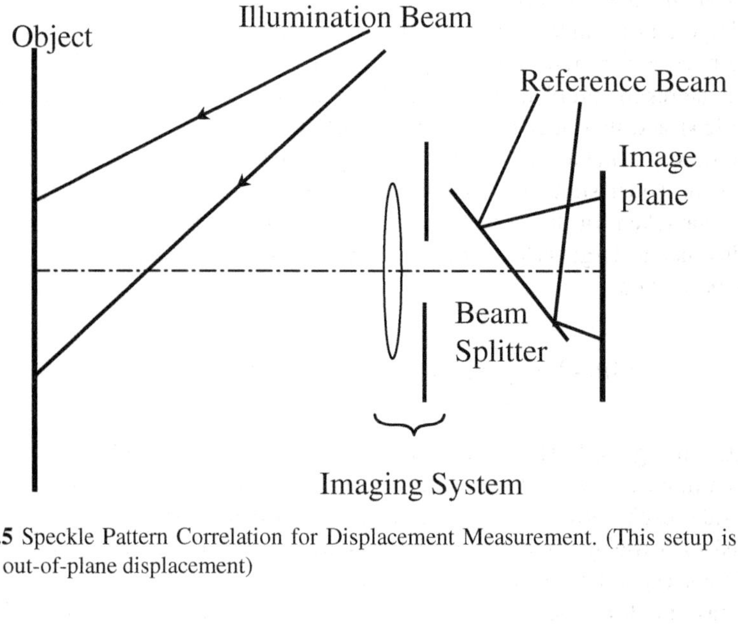

Fig. 4.5 Speckle Pattern Correlation for Displacement Measurement. (This setup is sensitive to out-of-plane displacement)

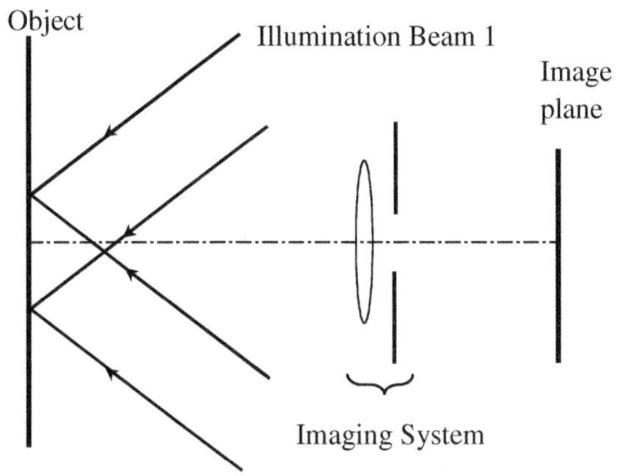

Fig. 4.6 Speckle Pattern Correlation for Displacement Measurement (Setup is less sensitive to out of plane measurement, thus preferred for in-plane measurements)

to vibration and rigid body movements. Accordingly, researchers often use optical isolation tables to perform speckle pattern correlation measurements.

4.3.2.3 Electronic Speckle Pattern Interferometry (ESPI)

This technique produces the fringe interference pattern electronically; images are superimposed digitally for producing the interference patterns. First, a picture of the laser-illuminated object is acquired and stored in a computer system. A second image is taken after displacement. Fringes are produced by superposition of the images taken. Fringes are obtained either way by spatial filtering and rectifying the resultant image. Fringes for two cases, before and after movement are compared for displacement calculation.

This technique is considered as advancement in the speckle photography techniques. ESPI shortens the time needed for speckle photography analysis, but is still subject to the other limitation of speckle photography.

4.3.2.4 Limitation of Speckle Interferometry

Given the fact that all speckle interferometry methods are based on speckle configuration on an image surface, a key factor is the speckle size, which is discussed earlier in this chapter. Displacements measured by speckle interferometry methods can be roughly estimated to be in the same order of magnitude of speckles formed on the object surface.

If displacements are more than the speckle size most likely, a de-correlation will come into view and displacement cannot be measured, or in some cases the measured displacements will not be the actual ones.

Speckle interferometry methods have an object size limitation, because fringe patterns can be at a high frequency such that large objects cannot be analyzed (too many fringes, with an image resolution limitation). This can be partially adjusted by selecting the proper camera resolution (case of ESPI) or magnification level of the optical system (case of speckle photography).

All speckle interferometry methods except speckle photography are very sensitive to vibration. For reasonable results complete vibration isolation is required, which is typically done using an optical isolation table.

Shchopinove et al (1996) remarks that many researchers encountered that speckle interferometry techniques are very sensitive to rotational movement, which may cause de-correlation for the experiment, i.e. interference fringes cannot indicate the movement any more.

4.3.2.5 Speckle Interferometry in Transparent Synthetic Soils

The high accuracy of interferometry measurement made many researchers use interferometry setups in civil engineering labs, (Jacquot and Facchini 1997a &b and Ghandehari et al 1999). Most setups involve speckle interferometry for surface deformation measurements. However, using speckle interferometry techniques in a granular media such as transparent synthetic soils is quite challenging. A light sheet of laser may be used to illuminate a slice of the transparent synthetic soil model (Fig. 4.7). The resulting speckles of the granular media may be used for speckle interferometry analysis of transparent synthetic soils.

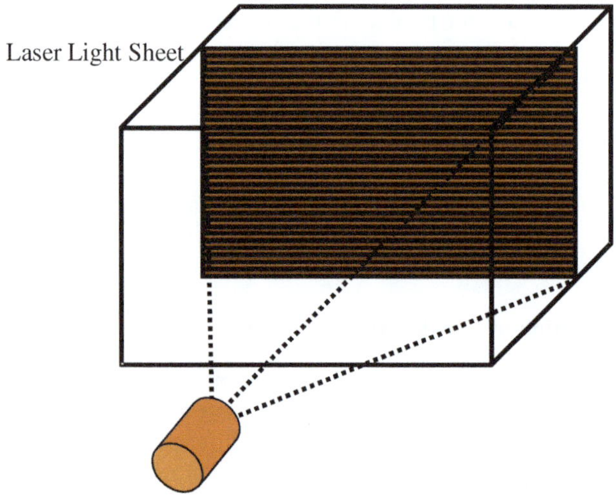

Fig. 4.7 Interference Fringes in an Imaginary Plane of a Transparent Synthetic Soil Model

Analysis of speckle interferometry in transparent synthetic soil is limited by the following factors:

1. Large deformation may cause de-correlation of the analysis.
2. Many setups of the speckle interferometry are very sensitive to out of plane movements, a $\lambda/2$ out-of-plane movement of the object may result in decorrelation of the analysis.
3. Granular particles in a transparent media are free not only to move out of plane but also to rotate. All these movements combined in one image may cause de-correlation and will make measurements difficult.
4. In many geotechnical applications, model testing must be performed in a centrifuge to satisfy similitude requirements. Preventing vibration in centrifuge testing may be impossible, thus limiting the use of any interferometry setups to 1-g models.
5. For large deformation the use of white light is recommended in interferometry analysis due to the laser power limitation (Shchopinove et al 1996). This way, another limitation comes up; scattering of white light in granular media such as transparent synthetic soil will limit the penetration depth of the light sheet.

4.4 Photoelasticity

4.4.1 Theory

Photoelasticity is a stress measurement technique, which utilizes the birefringence property of most transparent materials under stresses. The birefringence property of a material can be explained as the anisotropy of the material's refractive index. Most transparent materials develop birefringence if their state of stress is anisotropic. Thus, photoelasticity is induced birefringence by the anisotropic state of stresses.

Stress-induced birefringence of the material under investigation cannot be detected by the human eye. Polarized light passing through the material will go out in two separate plane-polarized waves. These waves are out of phase, which indicates that they experienced different modes of retardation in the material due to the anisotropy of the refractive index. The human eye will not be able to detect this phase difference, as the human eye is neither sensitive to phase shift nor change in polarization direction of the light wave. The human eye can only detect such phenomena if a mixing device is used to convert the phase shift into an intensity difference. This can be done by passing the light beam through a polarizer sheet (usually named "analyzer sheet") downstream of the transparent material (Fig. 4.8), and thus, interference fringes of light beams become visible. These interference fringes are stress dependent, which are utilized for calculating the stress throughout the material.

Photoelasticity and its application are thoroughly explained in many textbooks such as Cloud (1998) and Dally and Riley (1991) but will be briefly explained next.

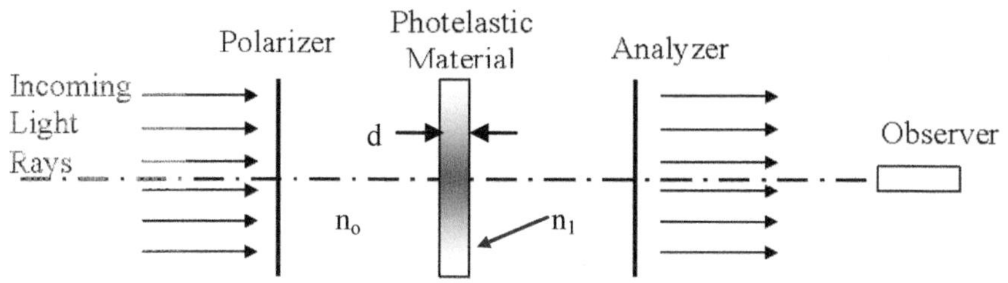

Fig. 4.8 Simple Linear Setup For Photoelastic Investigation

Assume a material of thickness "d" with a refractive index n_1 in a media of a refractive index n_o (Fig 4.8). Light passing through the material will experience retardation "R_1" which can be calculated as:

$$R_1 = \left(\frac{n_1 - n_0}{n_0} \right) d \tag{4.3}$$

Thus for an isotropic material with double refractive indices n_1 and n_2 (birefringence) two light beam will split out of the material each with a different retardation "R_1" and "R_2". The relative retardation can be calculated as:

$$R = R_1 - R_2 = \left(\frac{n_1 - n_2}{n_0} \right) d \tag{4.4}$$

R can be expressed as "m" multiple of light wave length "λ"

$$m = \frac{R}{\lambda} = \left(\frac{n_1 - n_2}{n_0} \right) \frac{d}{\lambda} \tag{4.5}$$

If σ_1 and σ_3 are the principle stresses in the transparent material, the retardation R can be related to the principal stresses as follows:

$$R = C_\sigma (\sigma_1 - \sigma_3) d \tag{4.6}$$

where, "C_σ" is an empirical coefficient known as the "stress-optic coefficient".

The state of stress in the transparent material can be predicted if the relative retardation is calculated experimentally. Analysis of the interference fringes in the captured image of a transparent body under stress provides the information required for stress prediction. There are two types of interference bands derived from the analysis of the fringes. The first is "isoclinic fringes", which are loci of points of constant inclination of the principle axis of refraction. The second is "isochromatic fringes", which are the loci of points where relative retardation is an integral multiple of the wavelength of retardation. Thus the isoclinic fringes indicate the direction of the principle stresses and the isochromatic fringes are indication to the magnitude of the stress. The isochromatic fringes are more important for the stress analysis than the isoclinic fringes. The problem is that both fringes appear superimposed in one image of the material under investigation. Separating these fringes is necessary for stress analysis and can be done by adding two special polarizers that have relative retardation of λ/4 (Fig, 4.9). These polarizers are placed at 45° to axis of the original polarizers. This technique eliminates the isoclininc fringes leaving only the isochromatic fringes (Fig. 4.10).

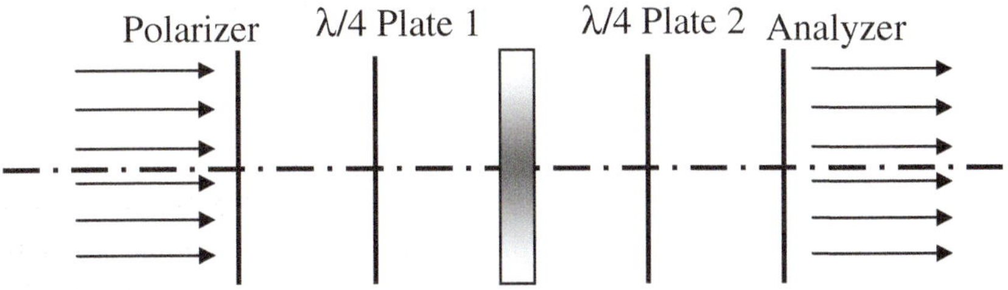

Fig. 4.9 Adding Retarding Plates to Exclude The Isoclinic Fringes

Fig. 4.10 Fringe Patterns on the Surface of a Photoelastic Notched Beam Loaded in 4 Point Bending

4.4.2 Photoelasticity of Transparent Synthetic Soils

Many researchers attempted to utilize photoelasticity in geotechnical labs. For example, Paikowsky et al (1993) studied contact stresses of two-dimensional array of particles; photoelastic cylindrical particles were placed between two glass plates and contact stresses were quantified. Also, Allersma (1982) studied plane strain problems using photoelasticity.

In an attempt to investigate the possibility of using transparent synthetic soils as a photoelastic media for stress measurement, the photoelastic properties of silica gel were investigated using the following simple procedure: The sample was placed between 2 cross polarizers, preventing light from passing through the sample, i.e. sample appeared black. Theoretically, if the sample color changes under any change in the state of stress means that silica gel possess a photoelastic property.

Stresses were applied on one silica gel particle (diameter ≈ 2.5 mm) manually using a vice; fringe pattern appeared as shown in Fig. 4.11. Silica gel particles show birefringence property under stress. The same procedure was repeated for fused silica particles, and it also showed that it too exhibit a photoelastic property. It was observed that the fringes were very clear in silica gel particle at very high stress levels such that particles broke just after the development of clear fringes. This observation may be because silica gel[1] particles are not solid, thus it is difficult to monitor fringes at low stress levels.

Fig. 4.11 Investigation of a Silica Gel Particle For Photoelastic Property

This finding can be utilized in measuring stresses in a transparent synthetic soil model but there are special considerations that must be accounted for:

1. In most photoelasticity applications, experiment setup is either, a plane strain or plane stress problem. Additionally, the analyzed problems are always 2D. For 3D analysis, the common procedure is stress freezing and then slicing the model and analyzing the stresses, sheet by sheet. This method cannot be applied in transparent synthetic soils because the soil samples are not monolithic. A new technique, which was discussed by Dupre and Lagarde (1997), in which a photoelasticity analysis is performed by optically slicing the model using laser sheets, can potentially be used. This technique is promising and can be used in

[1] Properties of silica gels and fused silica are discussed in Chapter 6.

transparent synthetic soils but the analysis is much more complicated than the monolithic model used by Dupre and Lagarde, because soil particles will develop three-dimensional fringe envelops at the particle level.
2. In estimating the stress-optic coefficient "C_σ", it is assumed that the material is linearly elastic and the correlation between the change in stress and birefringence property is linear. Thus, the fact that stress fringes in silica gel appears at high stress level, violates this assumption.
3. Photoelastic analysis of plastic, viscoelastic, and viscous behavior is very complex. Most of the published formulations are for simple problems.
4. Photoelasticity in transparent synthetic soil has a great potential for studying contact stresses of particles at the micro scale level.

4.5 Cross Tomography

4.5.1 Theory

Cross Tomography is widely used in the medical field in the form of the CAT scan (Computerized Axial Tomography). In CAT scan, X-rays crossing the human tissues are analyzed to determine the location and consistency of different tissues. Computerized tomography was utilized by some researchers using laser light such as Pei (1999). In Pei's research, laser beams transmitted through circular sample were detected using light sensors. Analysis was performed to locate the internal configuration (geometry and location) of an intruding body of different optical properties inside the sample.

The analysis in cross tomography is mathematically intensive. It usually involves forward and inverse analysis. The forward problem is concerned with predicting how energy will propagate in a medium according to a physical model appropriate to the types of interactions occurring at the intruding wavelength. The input in forward analysis includes the properties of the investigated media including geometry, size, physical properties, and boundary conditions. The forward analysis predicts the outcome of the experiment. The inverse analysis can be explained as follows: Based on the detectors reading, what should the internal composition of the model under investigation be? Iterations are performed mathematically for analyzing the forward and inverse analysis.

Tomographic analysis imposes a practical difficulty, in the amount of calculation required for simple problems. Thus for practical purpose, simplified assumptions are used to analyze such problems. For example, in x-ray computed tomography analysis, the scattering and diffraction of the x-rays are ignored. However, for laser light tomography these assumptions may not be valid, as rays can easily be scattered. Thus, the number of iteration may be large and in some cases the analysis may be ill-posed, meaning there is no unique solution for the given input and output.

4.5.2 Cross Tomography in Transparent Synthetic Soils

The simplicity of the experimental setup of cross tomography is very attractive for application in transparent synthetic soil models. Transparent synthetic soil surrounded by multiple laser sources and detectors can be easily setup as shown in Fig. 4.12. The mismatch between the refractive indices of the solid particles and pore fluid can provide a suitable contrast for prediction of localized densification throughout the sample. In addition, the refractive index of the pore fluid can be adjusted (by either dilution or concentration) to provide the required contrast.

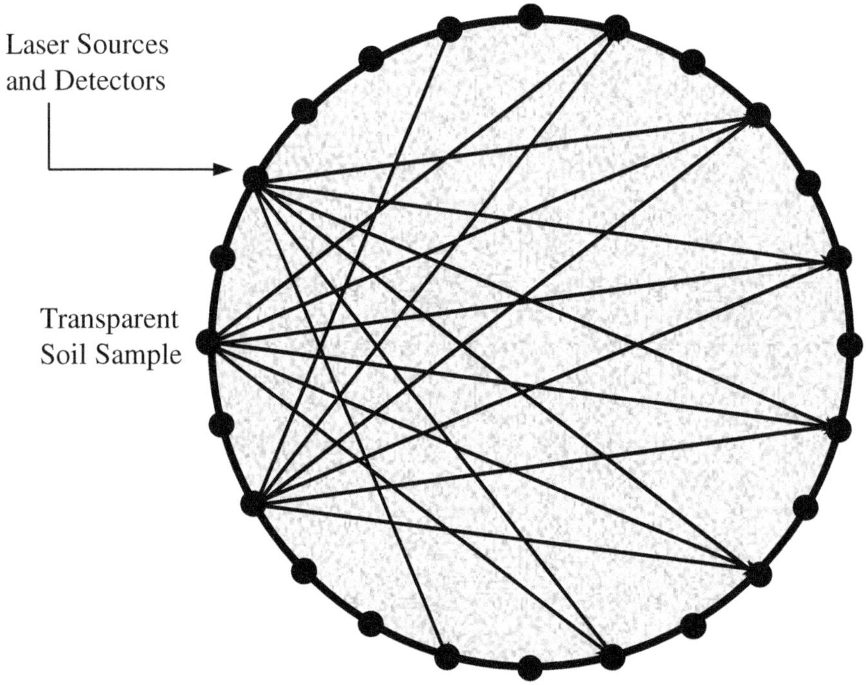

Fig. 4.12 Proposed Cross Tomography Setup For a Cylindrical Sample of Transparent Synthetic Soil

Further investigation for laser cross tomography in transparent synthetic soils showed a number of difficulties which can be summarized as follows:

1. Most cross tomography analyses were done to locate a few intruding particles in a uniform mass. In many cases, the technique is utilized for detecting the geometry and location of one particle such as tumor in a human tissue (Pei 1999). So, in the case of the transparent synthetic soils, the analysis may be mathematically intensive or at least can result in a very coarse resolution due to large number of soil particles, and small number of laser paths.

2. Analysis may be based on localized densification, i.e., not based on a detection of particles location, which may lead to coarse resolution and correlation error.
3. The size of sample may be limited by the scattering of the laser beam. In our trials, red laser beams[2] penetrated 10 cm (4 inches) into the sample, with acceptable scattering (entering beam thickness = 0.7mm, at the other end the thickness reached 3 mm). On the other hand, when the sample size increased to 30.5 cm (12 inches), the scattering of the beam was high, and the thickness of the emerging beam was 3 cm.
4. In some study cases, such pile penetration problems, the incident laser will not be able to penetrate the pile material if the pile material is opaque. If an opaque pile is used, the reflected laser beams will mislead the indicators readings. If a transparent pile is used, the difference in refractive index of the pile material must be taken into consideration.

In conclusion, cross tomography in transparent synthetic soils using laser beams is a tough challenge that has its advantage in simple setup and low cost.

4.6 Summary

A review of optical methods currently in use for stress and deformation measurements indicates that:

1. The speckle effect caused by the interaction of a laser and transparent soils can be used to interrogate the cross section of interest within a transparent soil model.
2. Digital image correlation of the speckle image is the simplest and most straight forward approach for measuring deformations within transparent soil models.
3. Photoelasticity holds some promise for measuring contact stresses between soil particles.
4. Tomography and classic interferometry methods are too complex for measuring spatial deformations and stresses with transparent soil models. Further, they will most likely yield unsatisfactory results.

References

Allersma, H.: Photo-elastic stress analysis and strains in simple shear. In: Vermeer, P.A., Luger, H.J. (eds.) Proc. IUTAM Symposium on Deformation and Failure of Granular Materials, Delft, pp. 345–353 (1982)

Bransby, P., Milligan, G.: Soil deformation near cantilever sheet pile. Geotechnique 25(2), 175–195 (1975)

Cloud, G.: Optical Methods of Engineering Analysis. Cambridge University Press, Cambridge (1998)

---

[2] Wavelength of 633 nm, laser power is 35 mW.

Dally, J., Riley, W.: Experimental Stress Analysis, 3rd edn. McGraw Hill, New York (1991)

Desrues, J., Mokni, M., Mazerolle, F.: Tomodensitometry and localization in sands. In: Proc. X ECSMFE: Deformation of Soils and Displacements of Structures, Florence, pp. 61–64 (1991) (in French)

Dupre, L., Lagarde, A.: Photoelastic analysis of a three-dimensional specimen by optical slicing and digital image processing. Experimental Mechanics 37(4), 393–397 (1997)

Fomin, N.A.: Speckle Photography for Fluid Mechanics Measurements. Springer, Heidelberg (1998)

Ghandehari, M., Krishnaswamy, S., Shah, S.: Technique for Evaluation Kinematics between Rebar and Concrete. Journal of Engineering Mechanics 125(2), 234–241, ASCE (1999)

Gill, D.: Experimental and Theoretical Investigations of Pile and Penetrometer Installation in Clay, Ph.D. Thesis, Trinity College, Dublin, UK (1999)

Goodman, J.: Statistical properties of laser speckle patterns. In: Dainty, J. (ed.) Laser Speckle and Related Phenomena. Topics in Applied Physics, ch. 2, vol. 9. Springer, Heidelberg (1975)

Jacquot, P., Facchini, M.: Interferometric imaging using holographic and speckle techniques, I) fundamentals and basic characteristics. In: Proceedings of 2nd Int. Conf. on Imaging Technologies: Techniques and Applications in Civil Engineering, ASCE, pp. 217–234 (1997a)

Jacquot, P., Facchini, M.: Interferometric imaging using holographic and speckle techniques, II) impact in civil engineering. In: Proceedings of 2nd Int. Conference on imaging technologies: techniques and applications in Civil Engineering, ASCE, pp. 235–254 (1997b)

Jones, R., Wykes, C.: Holographic and Speckle Interferometry, 2nd edn. Cambridge University Press, Cambridge (1989)

Kirpatrick, W., Belshaw, D.: On interpretation of triaxial test. Geotechnique 18(3), 336–350 (1968)

Orsi, T., Anderson, A., Leonard, J., Bryant, W., Edwards, C.: Use of x-ray computed tomography in the study of marine sediments. In: Proc. CEO V, ASCE, College Station, pp. 968–982 (1992)

Paikowsky, S., Di Rocci, K., Xi, F.: Inter-particulate contact force analysis and measurement using photoelastic techniques. In: Proc. Int. Conf. Discrete Element Methods, pp. 449–462 (1993)

Pei, Y.: Optical Tomographic Imaging Using Finite Element Method, Ph.D. Dissertation, Polytechnic University, Brooklyn, New York (1999)

Shchepinov, V., Pisarev, V., Novkov, N., Balalov, V., Odintsev, I., Bondarenko, M.: Strain and Stress Analysis by Holographic and Speckle Interferometry. Wiley, New York (1996)

Woods, R., Barnett, N., Sagesser, R.: Holography – a new tool for soil dynamics. J. of Geotechnical Engineering Division 100(11), 1231–1247, ASCE (1974)

Chapter 5
Geotechnical Properties of Transparent Silica Powders

Magued Iskander and Jinyuan Liu

Abstract. This chapter presents a comprehensive study of the geotechnical properties of transparent amorphous silica powders including shear strength, permeability, and consolidation. The stress-strain behavior under both normally consolidated and over consolidated conditions are similar to those of natural clays. The shear strength parameters and permeability of this material are within the reported values for natural clays. The consolidation behavior of transparent amorphous silica is similar to that of natural peat and organic clays, where secondary consolidation dominates the behavior. The material can be consolidated at various pressures to meet specific model test requirements in terms of strength, deformation, and permeability. It provides an excellent media to simulate macroscopic deformation and flow in clays using an optical set-up.

5.1 Introduction

5.1.1 What Is Amorphous Silica?

Amorphous solids are any non-crystalline materials, in which the atoms and molecules are not organized in a definite lattice pattern. Amorphous solids are usually known for not having a definite melting temperature, instead they have a melting range (Levine 1995). Such solids include glass, plastics, and gels. Usually amorphous solids are prepared by cooling from either their liquid or gas state. That is probably the reason the commercial name "precipitated silica" is used for amorphous silica powders.

Amorphous silica, like most other silica, is inert and insoluble in water. It is available in an infinite variety ranging from dense, hard, impervious mineral opal to extremely light, porous, synthetic products (Iler 1955). Amorphous silica powders and gel are both known to be hygroscopic. Some of the powders and gels have this high power for adsorbing water, commonly known as hydrophilic amorphous silica and some others are chemically treated to have their high affinity of adsorption towards oils and they are commonly known as hydrophobic amorphous silica.

M. Iskander: Modeling with Transparent Soils, SSGG, pp. 45–84.
springerlink.com

Amorphous silica powders are available in a wide array of shapes and particles are in both opaque and transparent forms. The amorphous silica powders used in this research are manufactured by PPG Industries, Inc. It has various industrial applications in painting, coating, rubber industry, oral care industry (toothpastes), tire industry, and in printing.

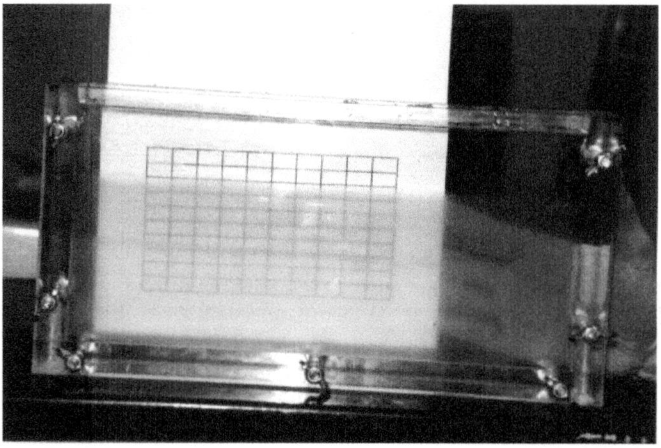

Fig. 5.1 Target Grid Viewed through 5 cm (2 inch) Thick Layered Transparent Soil Model

5.1.2 Use of Amorphous Silica in Experimental Modeling

The development of transparent materials that can be customized to meet model test requirements is the necessary foundation for utilizing optical techniques to study spatial deformation patterns and flow characteristics in 3D transparent soil models. Three families of transparent materials have been developed for modeling sand and clay. The first family, discussed in this chapter, is made by consolidating suspension of amorphous silica powder and liquids with matched refractive indices and was found to model the geotechnical properties of natural clays (Iskander et al 1994, and Iskander 1997, 1998). The second family, which models sand, is made of transparent silica gels and matched refractive index pore-fluid is discussed in Chapter 6. Both families have the same refractive index and use the same pore fluids, thus permitting their use in the same model (Fig. 5.1), where the bottom three rows are viewed through amorphous silica powder and the middle 4.5 rows are viewed through silica gel. The top 2.5 rows are viewed through pore fluid. The third family which is discussed in Chapter 7 (Aquabeads) has a different refractive index and cannot be combined with the first two.

The material described in this chapter has been used, with simple optical techniques, to measure the response of transparent soils in model tests. Welker et al (1999) used color saturation to study flow of contaminants into perforated vertical drains. Gill (1999), Song et al (2009), and Ni et al (2010) used target tracking to study foundations in clays.

This chapter is a comprehensive study of the geotechnical properties of transparent amorphous silica powders, including shear strength, permeability, and consolidation. Additionally, the strength and deformation properties of amorphous silicas are compared to a wide range of natural clays.

5.2 Material Description

When a beam of light strikes a boundary surface separating two different materials, such as a glass–air interface, part of the light energy is reflected and part enters the second medium. Transparent materials are those materials where little energy is reflected and most of the light energy enters through the boundary. The change in the direction of the transmitted ray (*refraction*) depends on the ratio of the refractive indices of both materials. The transparent clays used in this study are made by matching the refractive indices of transparent amorphous silica particles and pore fluid, thus preventing refraction. The clarity of transparent soils depends on the perfect matching of the refractive indices of silica and pore fluid, and the absence of entrapped air and impurities, which tend to refract light resulting in a translucent or partially transparent medium.

5.2.1 Physical Properties of Amorphous Silica

Amorphous silica powder used in this study is a commercial product of PPG, Inc., used without further processing. It consists of ultra fine particles with individual diameters on the order of 0.02μm, as shown in Fig. 5.2. These particles combine to form larger porous aggregates. Amorphous silica powder is well suited for producing transparent clay-like material because 1) it is hygroscopic, thereby adsorbing pore fluid and displacing air, and 2) it has a large surface area resembling clays. In principle, a wide range of amorphous silicas can be obtained commercially and matched with suitable liquids to produce transparent soils.

Four different silicas ranging in aggregate size from 1.4 μm to 175μm were tested. The tested sizes encompass the range of available amorphous silica powder sizes. The physical properties of the tested silicas are shown in Table 5.1. The bulk density is lighter than that of most natural soils due to the internal porosity of the aggregates. Most physical properties were controlled by the properties of the fine particles. For example, surface area and oil absorption varied in a small range relative to the variation in aggregate sizes.

Amorphous silica powders used in the research have white appearance. In reality, the particles are colorless. The reason for the white appearance is the diffraction and refraction of light as discussed extensively in Chapter 3. The reported refractive index of the colorless silica powders ranges from 1.41 – 1.46 (Weast 1986). Amorphous silica obtained from PPG is reported to have refractive index of 1.442.

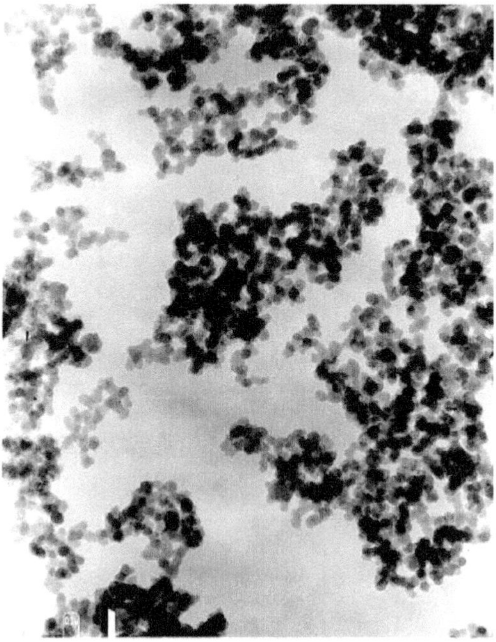

Fig. 5.2 Electronic Microscopic Image of FGSP Structure

Table 5.1 Physical Properties of Amorphous Silica*

Material	Median Aggregate Size μm	Surface Area m^2/g	Oil Absorption ml/100g	Specific gravity	Bulk Density kg/m^3
Hi-Sil T600 (HST600)	1.4	150	150	2.1	56
Flo-Gard FF (FGFF)	10	180	210	2.0	128
Flo-Gard SP (FGSP)	25	220	260	2.0	144
Hi-Sil SC-72 (SC72)	175	150	200-350	2.0	230

* Data from PPG Industries (1996).

5.2.2 Matched Refractive Index Pore Fluids

Two pore fluids were used. The first was a 1:1 blend of Drakeol® 35 mineral oil and Norpar® 12 paraffinic solvent, by weight. The refractive index, viscosity, and density of the oil blend at room temperature (24°C) were 1.447, 5.0 cP, and 800 kg/m^3, respectively. The second was a mixture of calcium bromide and water to have a refractive index of 1.448 at 25°C. The viscosity and density of the brine mixture at room temperature were 3.6 cP and 1572 kg/m^3, respectively. The values of the refractive indices were those that resulted in the clearest specimens.

5.3 Sample Preparation

First, amorphous silica powder was dispersed in the pore fluid at concentrations of 9–20% by weight. Smaller concentrations were needed for finer sizes to maintain workability. Second, vacuum was used to de-air the mix until the suspension became transparent. Four to eight hours were required; the larger the aggregates size, the longer the required time. The suspension was then moved into a one-dimensional consolidometer, shown in Fig. 5.3, which consisted of two porous discs inside a Teflon cylinder. Two cylinders were used to produce specimens 3.8 cm in diameter for triaxial testing and 6.4 cm in diameter for consolidation testing. The system was suitable for consolidation pressure up to 140 kPa. Additional consolidation was completed in a triaxial cell under hydrostatic pressure. Large transparent clay models were made by consolidating the de-aired suspension to the desired size in a transparent Plexiglas mould (Fig. 5.1).

The prepared specimens exhibit a high apparent total void ratio, *e*, due to the internal porosity of the transparent aggregates. The inter-aggregate void ratio, e_i, is more representative than the total void ratio, *e*, for geotechnical purposes, since it only considers the volume in-between the aggregates. The inter-aggregate void ratio can be derived as follows:

$$e_i = \frac{V_v - V_{vi}}{V_s + V_{vi}} = \frac{e - a\gamma_s}{1 + a\gamma_s} \tag{5.1}$$

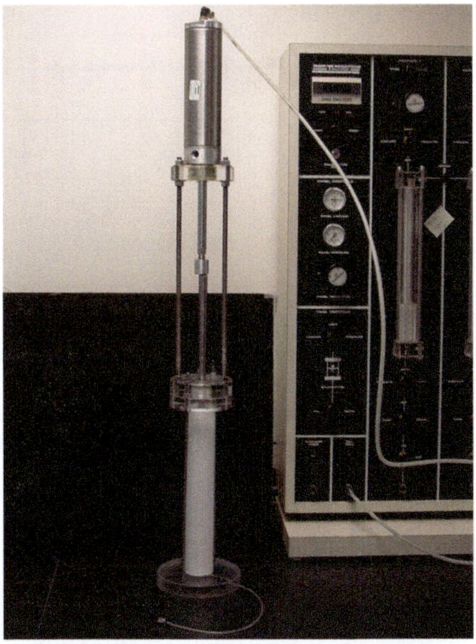

Fig. 5.3 Consolidometer for Sample Preparation

where, V_v is the total volume of voids, V_{vi} is the volume of voids inside solid aggregates, V_s is the volume of solids, γ_s is the unit weight of solids and a is the adsorption factor. The adsorption factor is defined as the volume of pore fluid absorbed per unit weight of solids, and was estimated by Mannheimer and Oswald, (1993) to be 2.1 cm^3 of pore fluid per gram of amorphous silica.

Mineral oil in the pore fluid has a high boiling point, thus it does not totally evaporate in conventional moisture (oil) content tests. The adjusted oven method was used to estimate the oil content of the specimen, as follows:

$$w_c = \frac{jW_l}{W_t - jW_l} \tag{5.2}$$

where w_c is the oil content, W_t is the total weight of the specimen, W_l is the measured weight loss due to evaporation. The correction factor, j, depends on the percentage of solvent in the mineral oil blend, and it was found to be 2 for specimens dried for 48 hours (Fig. 5.4).

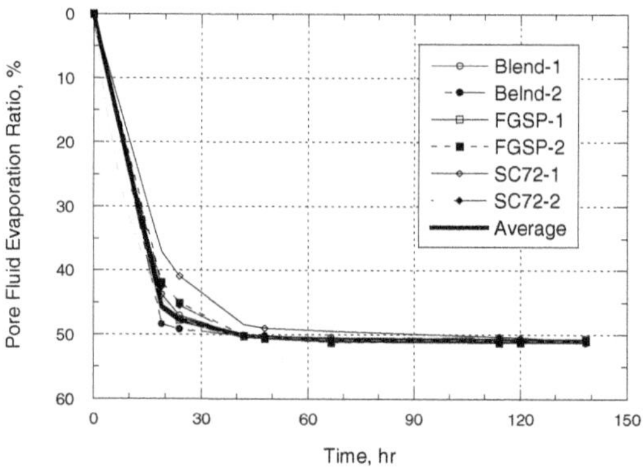

Fig. 5.4 Pore Fluid Evaporation with Time Measured by Oven Method

The pore fluid evaporation with time is shown in Fig. 5.4, where the oven temperature was 105°C. Six specimens were used; blend 1 and blend 2 were pure pore fluid. FGSP-1 and FGSP-2 were FGSP mixed with different percentage of pore fluid, so do the SC72-1 and SC72-2 specimens. A mass balance can also be used to estimate oil content assuming full saturation after sedimentation.

5.4 Undrained Triaxial Tests

Conventional consolidated undrained triaxial tests were performed on two sizes of amorphous silica: Flo-Gard SP (FGSP) and Hi-Sil T600 (HST600) according to ASTM (1993) D4767-88: Standard Test Method for Consolidated-Undrained Triaxial Compression Test on Cohesive Soils. Successful triaxial tests results are summarized in Table 5.2, where the oil content was measured by the adjusted

Table 5.2 Consolidated Undrained Triaxial Test Results on Amorphous Silica

Test No.	Oil Content	Total Void Ratio, e	Inter-aggregate Void Ratio, e_i	Sat. Unit Weight	Consol. Pressure, $\bar{\sigma}_{3c}$	OCR	Deviatory Stress at Failure, $(\sigma_1 - \sigma_3)_f$	Excess Pore Pressure at Failure, u_f
	%			kN/m^3	kPa		kPa	kPa
				Flo Gard SP (25 μm)				
1	-----	------	------	------	69	1	95.00	27.19
2	256	6.68	0.42	9.55	138	1	208.59	72.73
3	285	7.43	0.56	9.40	138	1	210.00	75.00
4	284	7.41	0.55	9.41	138	1	205.55	62.58
5	268	6.99	0.48	9.49	276	1	353.28	153.18
6	217	5.66	0.23	9.80	276	1	351.90	180.78
7	285	7.43	0.56	9.40	276	1	410.00	173.00
8	242	6.31	0.35	9.63	414	1	524.67	233.50
9	247	6.44	0.38	9.60	414	1	531.30	266.34
10	247	6.44	0.38	9.60	414	1	525.00	263.00
11	273	7.12	0.50	9.46	69	2	101.22	1.79
12	280	7.30	0.53	9.43	138	2	165.12	17.60
13	196	5.11	0.13	9.98	276	2	286.97	58.17
14	247	6.44	0.38	9.60	276	2	315.33	45.40
15	----	----	----	----	276	2.65	253.23	32.78
16	242	6.31	0.35	9.63	276	2.8	277.17	18.01
17	255	6.65	0.41	9.56	138	3.50	126.34	-6.07
18	238	6.21	0.33	9.66	414	4	398.89	14.08
19	246	6.42	0.37	9.61	276	4	206.72	11.66
20	246	6.42	0.37	9.61	276	4	266.06	1.59
21	250	6.52	0.39	9.59	276	8	232.32	-16.97
				Hi Sil T600 (1.4 μm)				
22	288	7.52	0.57	9.39	138	1	153.25	99.77
23	285	7.43	0.56	9.40	138	1	148.90	71.69
24	243	6.34	0.36	9.63	69	1	67.00	44.02
25	236	6.15	0.32	9.67	276	1	270.41	192.58
26	231	6.03	0.30	9.70	414	1	428.97	289.80

Fig. 5.5 Triaxial Test Set-up

oven method mentioned earlier. The test set-up is shown in Fig. 5.5, which consists of a load frame and two computer-controlled pumps. One is for back pressure to facilitate saturation. Another applies the confining pressure. Both pumps can keep the confining or back pressure constant by controlling the flow in or out of the pump.

Normally consolidated specimens were tested using pressures ranging from 69 kPa to 414 kPa. Specimens having overconsolidation ratios of 2–8, were tested at pressures ranging from 34.5 kPa to 138 kPa. Approximately 310 kPa of back-pressure was required for saturation. The shear rate was 0.125% per minute. Failure was defined at the maximum deviatoric shear stress.

5.4.1 Normally Consolidated Behavior

Typical stress strain and pore pressure curves of undrained normally consolidated (NC) amorphous silica powder are shown in Fig. 5.6 and 5.7. The behavior is consistent with that of natural clays. Ladd and Lambe (1963), among others, have shown that normally consolidated natural clays exhibit a unique normalized stress strain and pore pressure relationships, which was also found to hold for amorphous silica powder (Fig. 5.8). In that figure, the deviatoric shear stress, σ_1–σ_3, and the excess pore pressure, *u*, are normalized by the consolidation pressure, σ_{3c}, where σ_1 and σ_3 are the major and minor principle stresses, respectively. Normally consolidated amorphous silica powder has a higher normalized strength, but a lower modulus than natural clays (Fig. 5.9). The normalized shear stress at failure is 0.96–1.18 for HST600 and 1.27–1.45 for FGSP (Fig. 5.8). Natural clays, typically exhibit a normalized strength on the order of 0.6–1 (Table 5.3). The peak strength of transparent soils was reached at strains, which are larger than that of most natural clays. This discrepancy will not limit the use of existing transparent soils in modeling limit equilibrium and deformational problem if a scaling relationship is established.

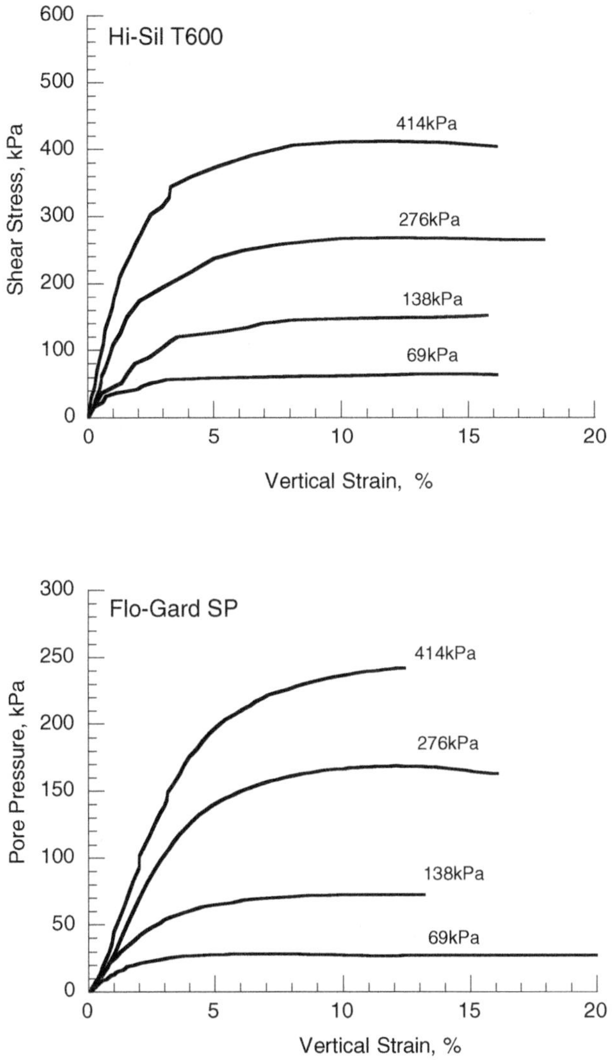

Fig. 5.6 Typical Stress Strain Diagrams of Undrained NC Amorphous Silica Powder

The pore pressure parameter, $A = u/(\sigma_1 - \sigma_3)$, also exhibited a unique relationship (Fig.5.10). The average value of pore pressure parameters at failure, A_f, is 0.35–0.7 which compares favorably with values reported by Skempton (1954) and Bjerrum (1961) for natural clays which ranged between 0.5–1.3 for normally consolidated clays, and 0.25–0.7 for lightly over-consolidated clays.

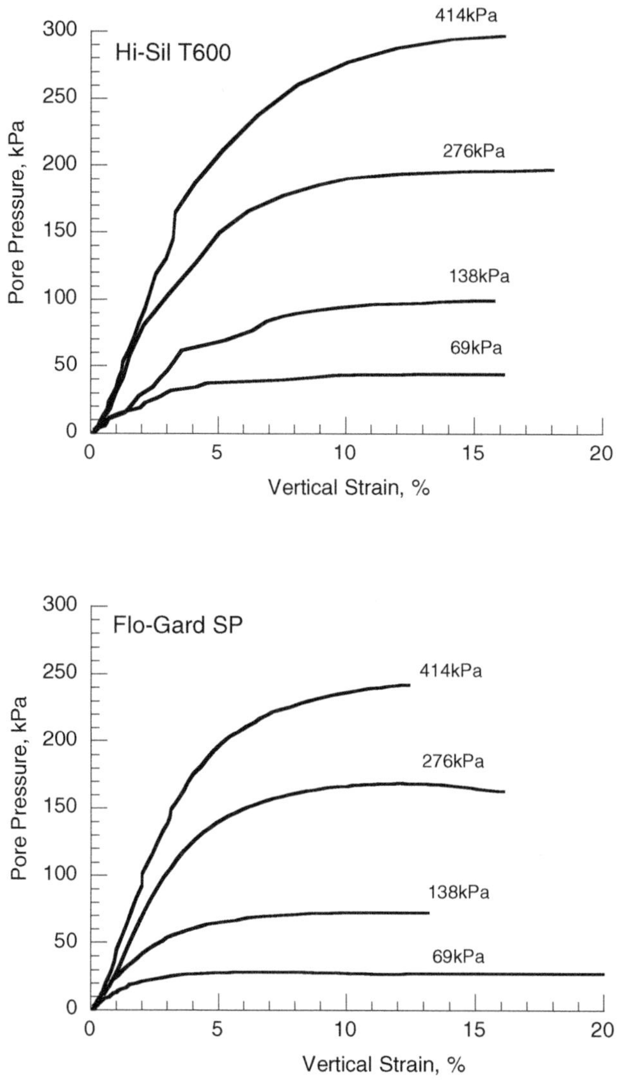

Fig. 5.7 Typical Pore Pressure Diagrams of Undrained NC Amorphous Silica Powder

The modified Mohr Coulomb failure envelope for normally consolidated undrained amorphous silica powder is shown in Fig. 5.11. The total stress parameters are c = 0–16 kPa and ϕ = 19–21°. The effective stress parameters are $\overline{c}$ =0 and $\overline{\phi}$ =36°. These parameters are higher than the strength reported for most clays. Nevertheless, they are within the range of properties reported by Kenny (1959), Bjerrum and Simons (1960), and Bishop (1971) for many clays.

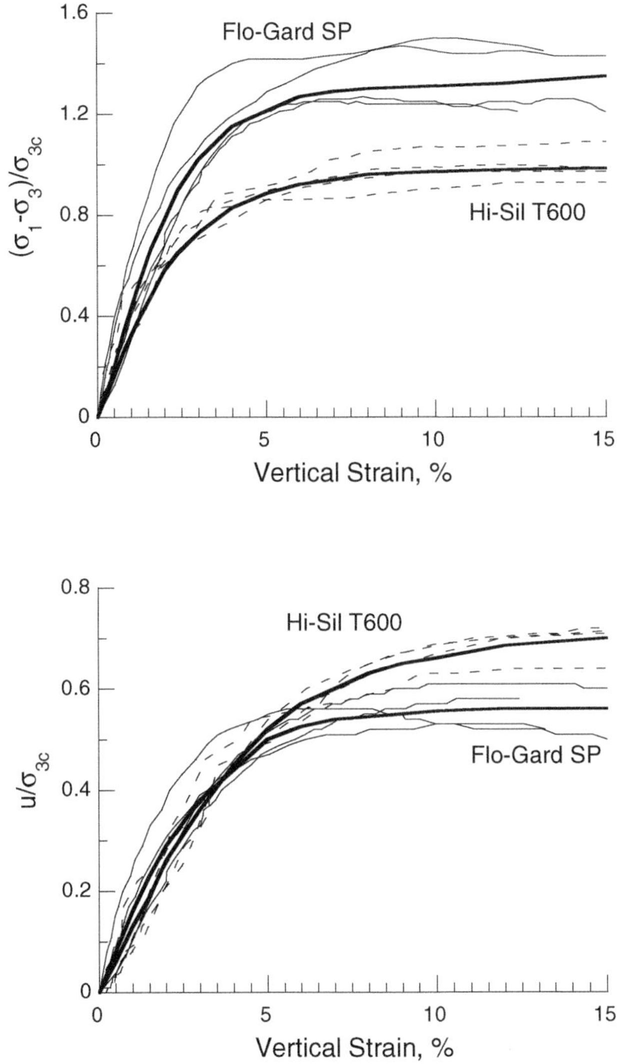

Fig. 5.8 Normalized Shear Behavior of Undrained NC Amorphous Silica Powder. (Top) Shear Stress, (Bottom) Pore Pressure. (Heavy lines are numerical averages).

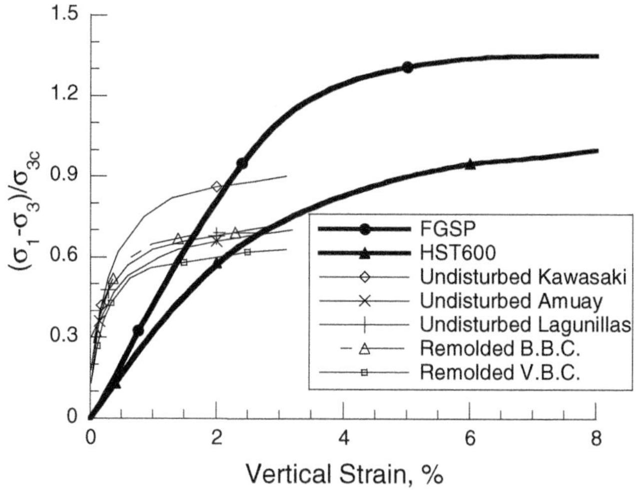

Fig. 5.9 Comparison between the Average Strength of Amorphous Silica Powder and Some Natural Clays

Table 5.3 Shear Strength of Amorphous Silica Powder & Undrained Normally Consolidated Clays

Type	Description	$(\sigma_1-\sigma_3)_f/\sigma_3$	Reference
FGSP	25 μm Amorphous Silica	1.27-1.45	----
HST600	1.4 μm Amorphous Silica	0.96-1.18	----
Boston Blue	Remolded PI = 15–21, W_L = 33–41	0.60–0.66	Ladd (1964c) Ladd et al (1977)
Lagunillas	Undisturbed PI=37, W_L =61, w_c= 60	0.80	Ladd and Lambe (1963)
Kawasaki	Undisturbed PI = 31–43	0.90–0.97	Ladd and Lambe (1963)
Drammen	Undisturbed, Soft, Silty PI = 8–29 W_L =61	0.65–0.78	Simons (1960a) Prevost (1979)
Oslo	Remolded PI=21, W_L =39, w_c=35	0.60–0.76	Simons (1960b)
Weald	Remolded PI=21, W_L =41	0.54–0.64	Ladd (1964c)
Bangkok	Undisturbed, Soft PI=43, W_L =118, w_c=112	0.60	Balasubraniam & Chaudhry (1978)

PI= Plasticity index, W_L= Liquid limit, w_c=Water Content.

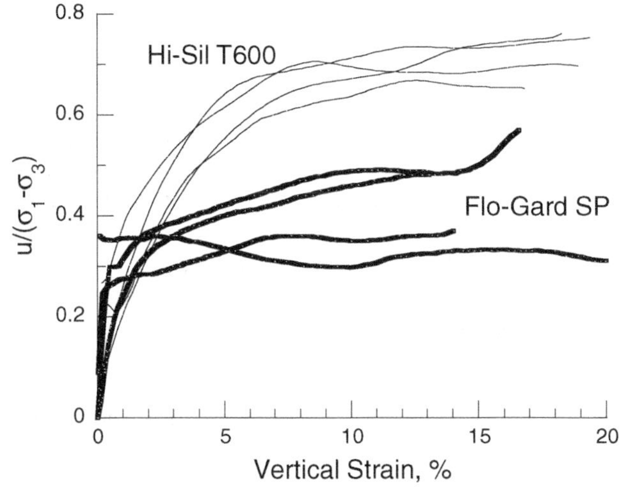

Fig. 5.10 Pore Pressure Parameter *A* from NC Undrained Triaxial Tests

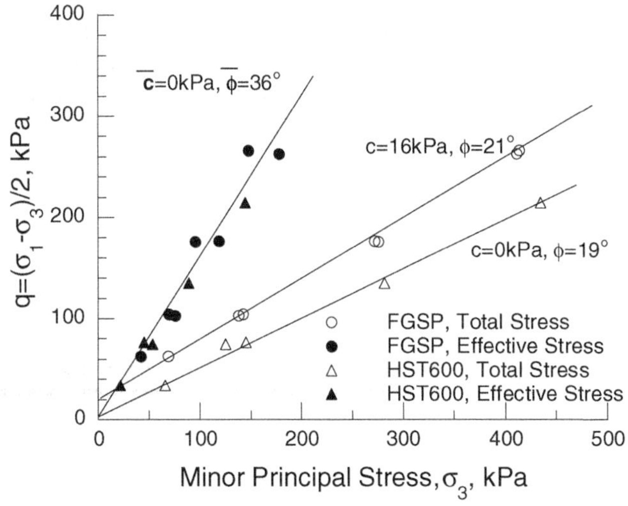

Fig. 5.11 Modified Mohr Coulomb Diagram for NC Undrained Specimens

5.4.2 Overconsolidated Behavior

Overconsolidated undrained triaxial tests were performed on specimens of FGSP (Table 5.2). Typical normalized stress and pore pressures are presented in Fig. 5.12, where the confining pressure during shear, σ_3, is used for normalization. The behavior is consistent with the reported behavior of overconsolidated natural

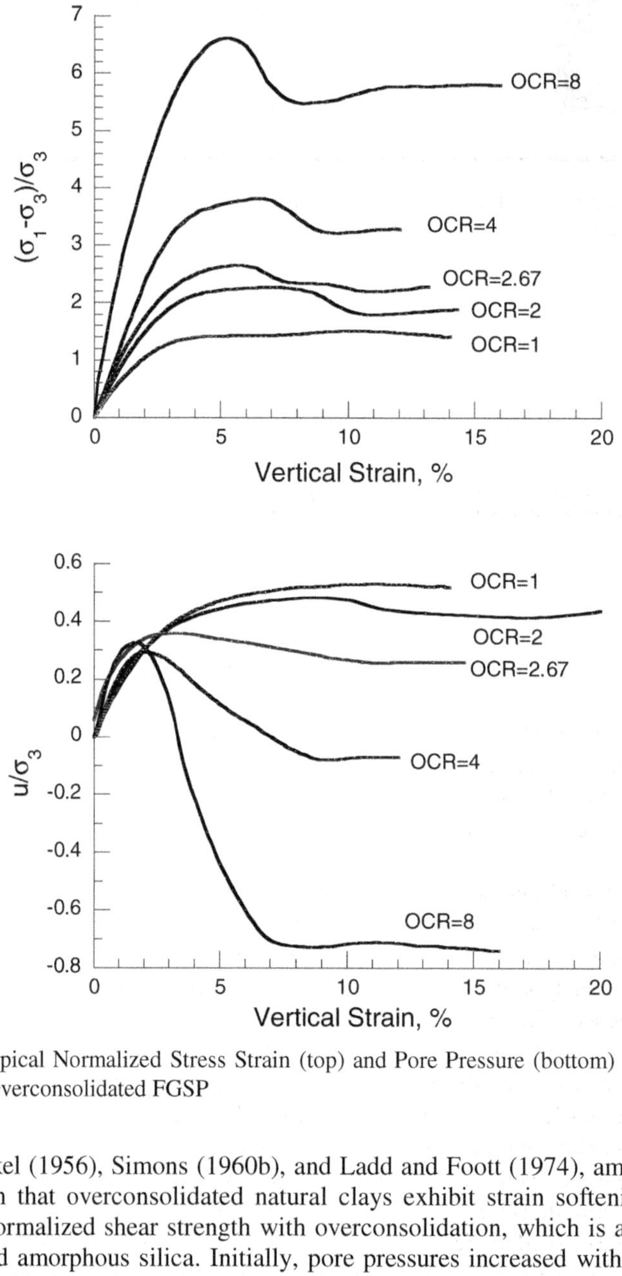

Fig. 5.12 Typical Normalized Stress Strain (top) and Pore Pressure (bottom) Diagrams of Undrained Overconsolidated FGSP

clays. Henkel (1956), Simons (1960b), and Ladd and Foott (1974), among others, have shown that overconsolidated natural clays exhibit strain softening and increase in normalized shear strength with overconsolidation, which is also evident in the tested amorphous silica. Initially, pore pressures increased with strain, and then the familiar drop in pore pressure occurred with further strain. Negative pore pressures decreased with increasing overconsolidation, which is consistent with the behavior reported by Henkel (1956) and others for overconsolidated natural clay.

The normalized strength and pore pressure of FGSP are plotted against the overconsolidation ratio (OCR), along with data from Simons (1960b) in Fig. 5.13. The strength is within the range reported for natural clay, but near the upper boundary. The normalized pore pressure at failure shows an excellent correlation with natural clays.

Table 5.4 Shear Strength of Some Undrained Overconsolidated Clays

Type	Description	OCR	$(\sigma_1-\sigma_3)_f/\sigma_3$	Reference
FGSP	25 μm Amorphous Silica	2–8	2.1–3.8	----
Boston Blue	Undisturbed PI=15–35, W_L = 42–55	Slight	0.96–1.60	Ladd (1964a)
Göta	Undisturbed PI = 29–37	High Slight	2.66 0.65–1.50	Bjerrum (1954a) Bjerrum & Wu (1960)
London	Remolded PI = 52, W_L = 78	2–16	0.8–1.8	Henkel(1956)
Oslo	Undisturbed PI = 18, W_L = 39	2–16 4–11	1.5–5.3 3.6–6.7	Simons (1960b)
Weald	Remolded PI = 25, W_L = 43	1.8–22	0.8–3.1	Henkel(1956)
Chicago	Undisturbed PI = 15, W_L = 30–42	Slight	0.5-1.4	Finno and Chung(1990)

PI= Plasticity index, W_L= Liquid limit, w_c=Water Content.

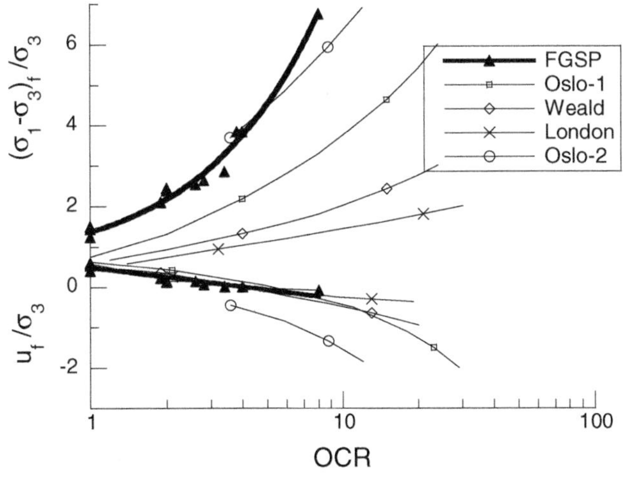

Fig. 5.13 Variation of Shear Strength and Pore Pressure at Failure with Overconsolidation for FGSP and Some Natural Clays

The pore pressure coefficient $A = u/(\sigma_1-\sigma_3)$ decreases with over-consolidation as shown in Fig. 5.14. The pore pressure coefficient at failure $A_f = u_f /(\sigma_1-\sigma_3)_f$ is plotted against the over-consolidation ratio (OCR) along with data from Henkel (1956) and Simons (1960b) in Fig. 5.15. A_f decreases from positive to negative with increasing overconsolidation. Overconsolidated amorphous silica powder exhibits A_f within the range reported for natural clays. However, A_f is lower for normally consolidated amorphous silica powder than that of natural clays.

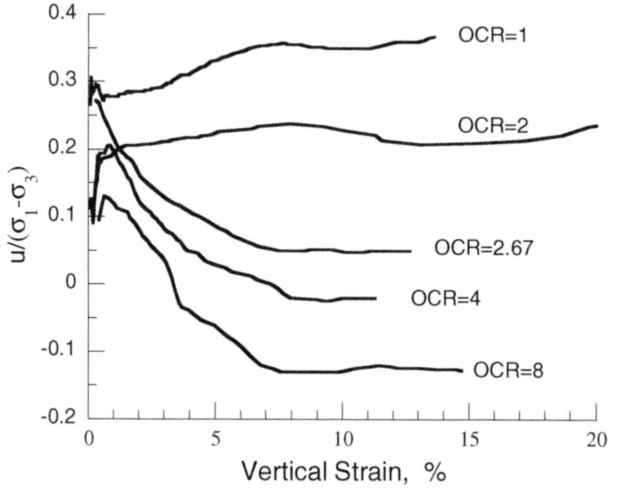

Fig. 5.14 Variation of Pore Pressure Parameter A with Overconsolidation for FGSP

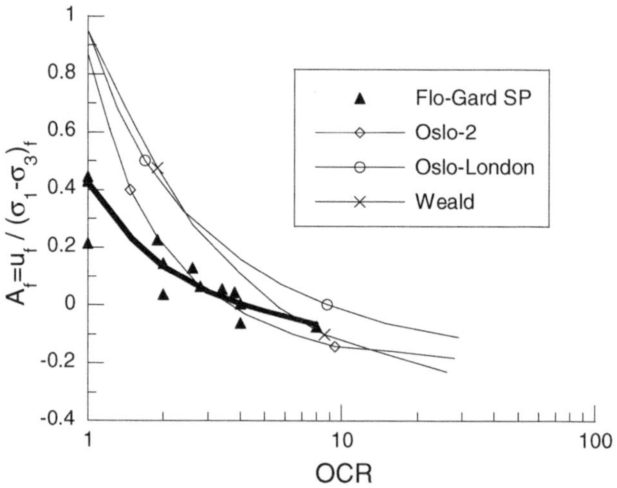

Fig. 5.15 Variation of Pore Pressure Parameter at Failure A_f with Overconsolidation for FGSP and Some Natural Clays

Typically, over-consolidation results in a decrease in the friction angle and an increase in the cohesion of natural clays. Amorphous silica powder exhibits the same property. For example in Fig. 5.16, the effective shear strength parameters were $\bar{c}$=0 kPa, $\bar{\phi}$=36° for normally consolidated FGSP and $\bar{c}$=32 kPa, $\bar{\phi}$=30° for over-consolidated FGSP (OCR = 2–4), under the same 276 kPa consolidation pressure). These values are within typical shear strength parameters shown for Boston Blue, Göta, London, Oslo, and Weald clays in Table 5.4.

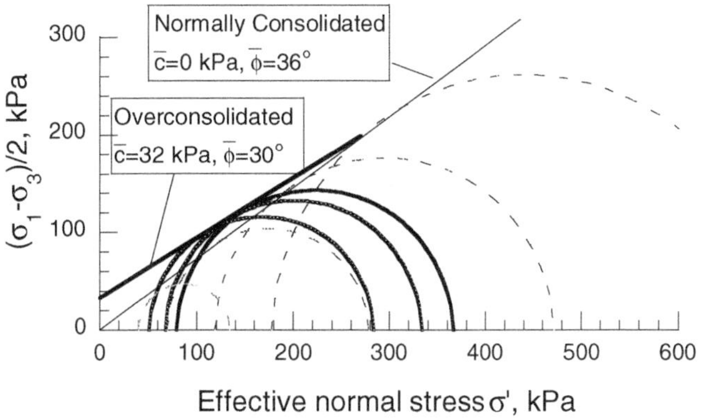

Fig. 5.16 Typical Mohr Coulomb Diagram of Normally and Overconsolidated FGSP

5.5 Drained Triaxial Tests

Conventional consolidated drained (CD) triaxial compression tests were performed following Bishop and Henkel (1962). Tests were performed on both normally and overconsolidated specimens of FGSP (Table 5.5). The two flow pumps connected to test set-up (Fig. 5.5) were used to provide pressure and measure the volume change simultaneously. One was connected to the sample to provide the back pressure and was used for measurement of the volume change of the sample. Another was connected to the chamber and was used to provide the confining pressure and to verify the sample volume change measured by the first pump. The shear rate was 0.015 % per minute.

Typical stress strain and volume change curves of drained normally and overconsolidated specimens are shown in Fig. 5.17. The behavior is consistent with that of natural clays.

In Fig. 5.17, normally consolidated tests (the heavier lines), the strength increased gradually with axial strain. Overconsolidated specimens (the lighter lines) exhibited the usual strain softening. Normally consolidated specimens exhibited increasing volume change with strain. Specimens with an OCR = 4 exhibited compressive behavior followed by dilatancy with increasing strain. The volume change in drained tests (Fig. 5.17) is similar to the pore pressure build up in

Table 5.5 Consolidated Drained Triaxial Test Results on Amorphous Silica

Test No.	Oil Content	Total Void Ratio, e	Inter-Aggregate Void Ratio, e_i	Sat. Unit Weight	Consol. Pressure $\bar{\sigma}_{3c}$	OCR	Deviatory Stress at Failure $(\sigma_1-\sigma_3)_f$	Volume Change at Failure
	%			kN/m^3	kPa		KPa	%
				Flo-Gard SP (25 μm)				
1	255	6.65	0.41	9.56	104	1	302.91	-8.42
2	268	6.99	0.48	9.49	138	1	432.70	----
3	224	5.84	0.26	9.75	414	1	965.66	-11.14
4	----	----	----	----	138	1	586.50	-9.58
8	285	7.43	0.56	9.40	276	1	900.00	-16.00
9	285	7.43	0.56	9.40	138	1	600.00	-9.00
5	---	---	----	----	138	2	242.88	-2.43
6	267	6.97	0.47	9.49	138	4	145.73	+0.70
7	240	6.26	0.34	9.65	414	4	370.25	-0.65
				Hi-Sil T600 (1.4 μm)				
8	285	7.43	0.56	9.40	35	1	200	-16.00
9	285	7.43	0.56	9.40	138	1	430	-16.20

Note: Volume change is considered to be "+" during expansion, "—" for compression.

undrained tests for both normally and overconsolidated specimens (Fig. 5.7 and 5.12).

Stress normalized by confining pressure and volumetric change are plotted against strain along with data from Lambe and Whitman (1969) in Fig. 5.18. The developed shear of FGSP is consistent with natural clays (Table 5.6). Volume change shows a remarkable agreement with natural clays in both normally and overconsolidated conditions.

The effective-stress strength parameters for the drained tests were $\bar{c}$ = 44 kPa, $\bar{\phi}$=30°, which is slightly smaller than the effective stress parameters obtained from undrained tests. Similar behavior was also found in Allschwyll and Zurich clays (Bjerrum 1954b), Kaolin and soft Bangkok clay (Balasubramaniam and Chaudhry 1978). The ratio between drained and undrained strength of FGSP ranges between 1.4 and 2.6, which is similar to values reported by Henkel, (1956) for London and Weald Clays.

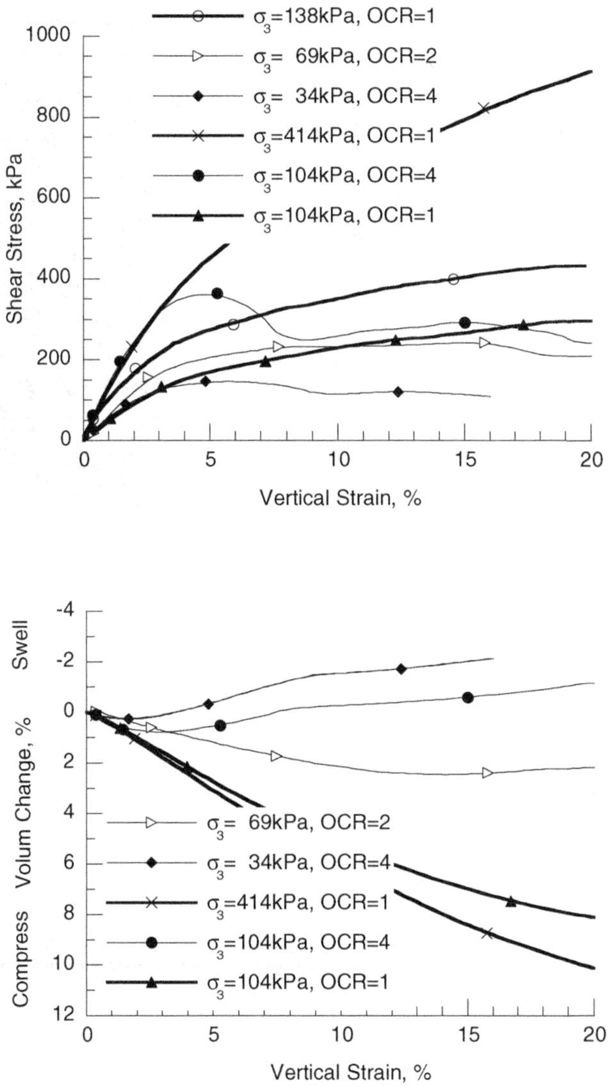

Fig. 5.17 Typical Stress Strain (top) and Volumetric Change (bottom) Diagrams of Consolidated Drained FGSP

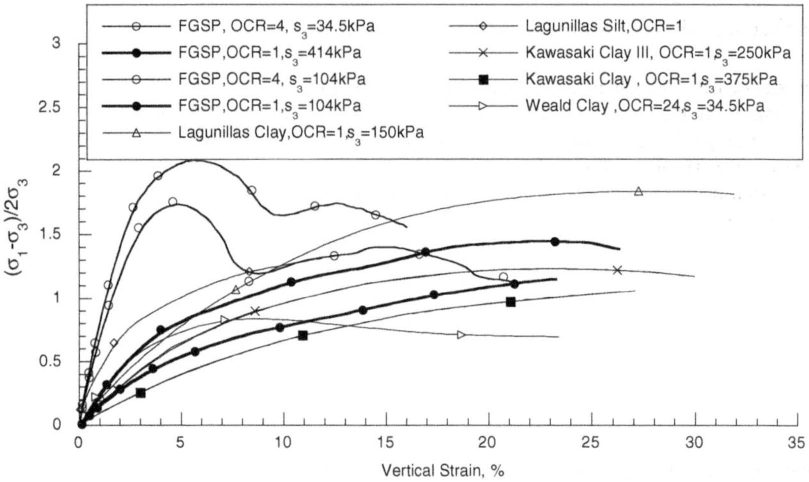

(a)

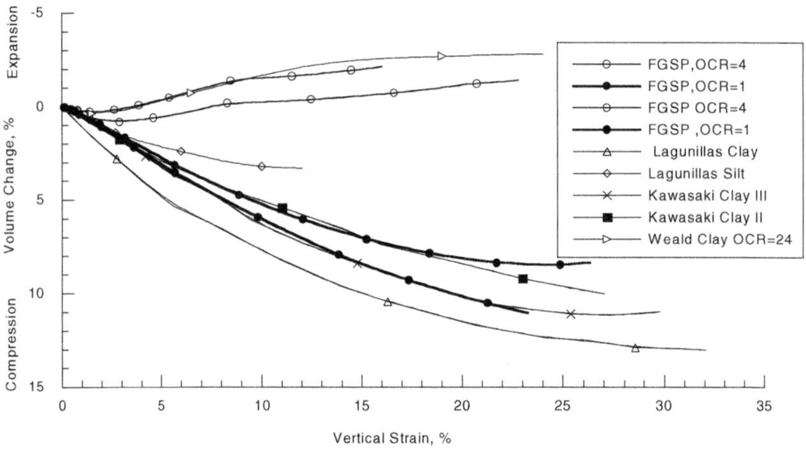

(b)

Fig. 5.18 Variation of (a) Shear Strength and (b) Volume Change of FGSP and Natural Clays

Table 5.6 Shear Strength of FGSP and Some Drained Normally Consolidated Clays

Type	Description	$(\sigma_1 - \sigma_3)_f / \sigma_3$	Reference
FGSP	25 μm Amorphous Silica	2.3–4.2	----
Drammen	Undisturbed PI ≅ 18-24	0.65–2.38	Simons(1960a)
Oslo	Remolded PI=21, W_L =39, w_c=35	1.18–1.24	Simons(1960b)
Kawasaki	Undisturbed W_L ≅ 48–93 w_c=43–75	2.20-2.50	Lambe and Whitman (1969)
Weald	Remolded PI = 25, W_L = 43	1.2	Henkel(1956)
Lagunillas	Undisturbed W_{LL} ≅ 69 w_C ≅ 63	3.70	Lambe and Whitman (1969)
London	Remolded PI = 52, W_L = 78	0.9	Henkel(1956)

PI= Plasticity index, W_L= Liquid limit, w_c=Water Content.

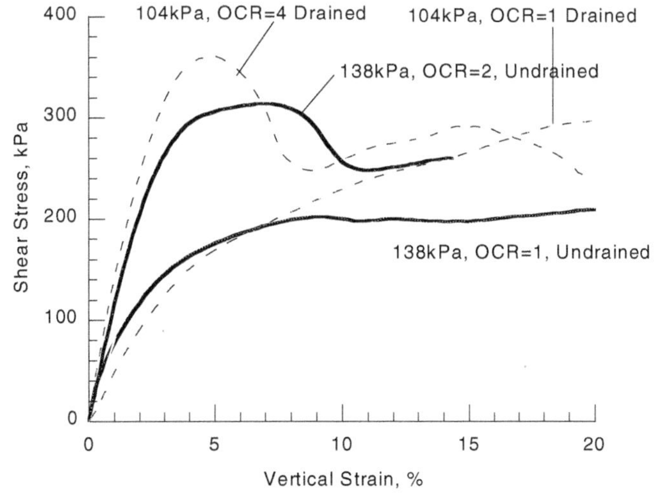

Fig. 5.19 Residual Strength of FGSP under the Same Confining Pressure

The residual shear strength of natural clays is typically independent of stress history (Skempton 1964). The stress-strain curves of normally and overconsolidated FGSP specimens tested with the same confining pressure tended to approach one another at strains exceeding 10%, for both drained and undrained tests (Fig. 5.19).

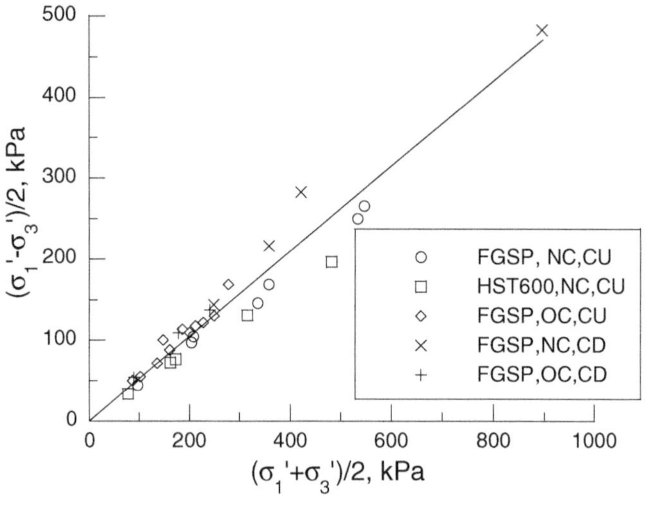

Fig. 5.20 Residual Strength for Consolidated Drained and Undrained Tests

The residual friction angle of FGSP was approximately 31°, regardless of drainage and overconsolidation conditions (Fig. 5.20), which is slightly smaller than the effective friction angle (36°), shown in Fig.5.11. This behavior is consistent with the behavior reported by Skempton (1964) for Boulder and London clays.

5.6 Elastic Properties of Amorphous Silica

The initial Young's modulus, E_i, calculated from undrained triaxial tests was 3–19 MPa for both FGSP and HST600. It was 4–15 MPa for drained tests. The secant modulus, E_s, calculated at 1/3 of the peak strength, is shown along with data from Ladd (1964b) in Fig. 5.21, where the confining pressure σ_{3c} is used for normalization. For normally consolidated undrained amorphous silica, E_s/σ_{3c} is approximately 30–60, which is smaller than natural clays. However, with overconsolidation, E_s/σ_{3c} approaches the values reported for natural clays.

Poisson's ratio, ν, was calculated by measuring the change in the volume of specimens and the corresponding axial strain during tests. For undrained tests Poisson's ratio was 0.45–0.5. For drained tests, Poisson's ratio ranged between 0.2 and 0.3.

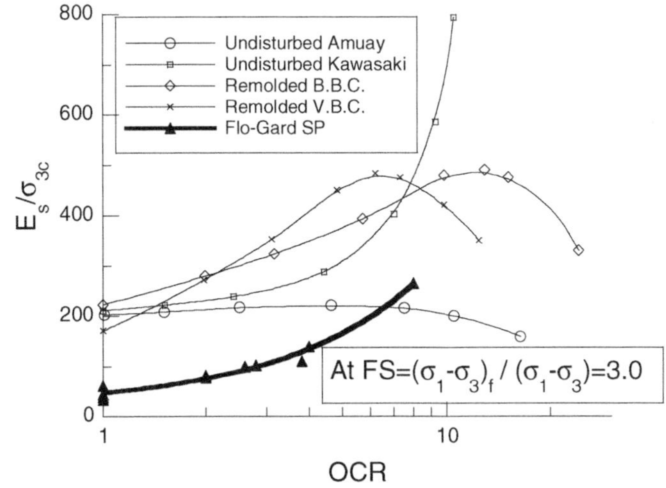

Fig. 5.21 Relation between Secant Young's Modulus and Overconsolidation Ratio

5.7 Consolidation Properties

Amorphous silica powder consists of ultra fine particles with individual diameters on the order of 0.02 μm. These particles combine to form larger aggregates, which are porous. The consolidated specimens exhibit a high apparent total void ratio, *e*, due to the internal porosity of the silica aggregates. The inter-aggregate void ratio e_i, is more representative than the total void ratio, *e*, for geotechnical purposes, since it only considers the volume in-between the aggregates. The inter-aggregate void ratio can be calculated using equation (Eq. 5.1) assuming the samples are fully saturated.

Materials used in consolidation tests are summarized in Table 5.7. One dimensional (Ko) consolidation tests were performed in accordance to ASTM D2435-90(1993): Standard Test Method For One-Dimensional Consolidation Properties of Soils. Successful tests were summarized and presented in Table 5.8. Samples were 38.1 mm (2.5 in) in diameter and 19 mm (0.75 in) high. Tests were conducted using conventional oedometer apparatus using fixed weights. Isotropic consolidation tests were also carried out in a triaxial cell using computerized apparatus, where confining pressure was applied using a computer-controlled pump, which also measured volume change with an accuracy of 0.01 ml.

Table 5.7 Some Physical Properties of Amorphous Silica Powders

Material	Median Aggregate Size (μm)	Surface Area m^2/g	Oil Absorption ml/100g	Specific Gravity	Bulk Density Kg/m^3
Hi-Sil T600 (HST600)	1.4	150	150	2.1	56
Flo-Gard FF (FGFF)	10	180	210	2.0	128
Flo-Gard SP (FGSP)	25	220	260	2.0	144
Hi-Sil SC-72 (SC72)	175	150	200-350	2.0	230

Table 5.8 Consolidation Properties of Amorphous Silica Powders

Sample No.	C_c	C_r	C_α (C_α/C_c)	Void Ratio, e	Preconsolidation Pressure, KPa
FGFF-1	1.630	0.180	0.061(0.037)	6.96	132
FGFF-2	1.800	0.199	0.064(0.036)	6.81	243
FGSP-1	2.651	0.232	0.094(0.035)	8.82	88
FGSP-2	2.582	0.227	0.079(0.030)	7.10	176
FGSP-3	2.306	0.260	0.084(0.036)	8.99	176
FGSP-4	2.020	0.162	0.087/(0.043)	8.00	138
FGSP-5‡	2.628	0.204	0.101/(0.038)	7.90	138
SC72-1	2.544	0.231	0.103(0.040)	10.55	132
SC72-2	2.933	0.300	0.073(0.025)	8.60	243
SC72-3*	2.060	----	0.105(0.051)	6.65	176

‡ Computer-controlled loading system was used.
*The pore fluid was calcium bromide brine.
Note: All tests were Ko Consolidation tests.

5.7.1 Consolidation Indices

Typical *e-log* p curves from oedometer tests are shown in Fig. 5.22 for three types of amorphous silica. Void ratios were taken 24 hours after loading because of the difficulty in identifying the end of primary consolidation. The compression and recompression indices (Cc and Cr) are 1.6–3.0 and 0.15–0.3, respectively (Table 5.8). Both indices are within the range typically reported for clay Montmorillonite minerals (Lambe and Whitman 1969). The ratio Cr/Cc is around 0.1 which is within the normal range of 0.02 to 0.2 for natural soils (Terzaghi et al 1996).

Two separate consolidation tests are presented in Fig. 5.23, where data is repeatable and compression and rebound behavior is consistent with natural clays. Conventional Casagrande method for determination of pre-consolidation pressure was found to be applicable. The measured compression index is consistent with that of natural clays and peats having water contents of 80–130% (Terzaghi et al 1996).

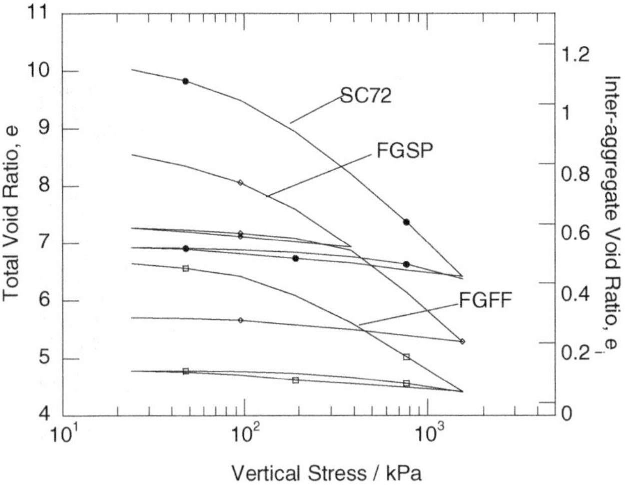

Fig. 5.22 Typical *e-log p* Curves for Amorphous Silica Powder from Oedometer Tests

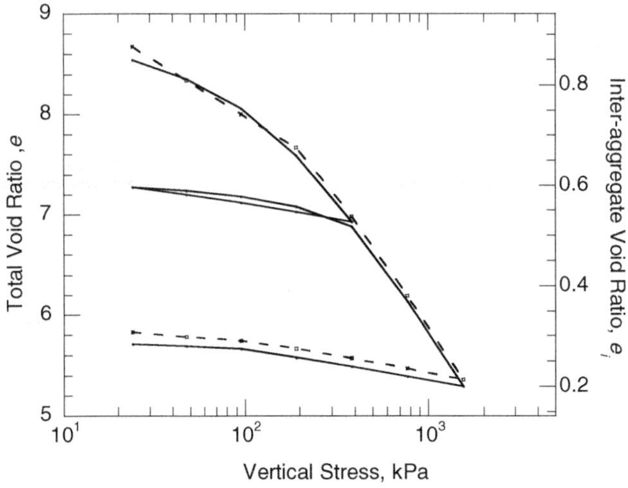

Fig. 5.23 Two Consolidation Tests on FGSP

5.7.2 Consolidation Behavior

Typical consolidation volume change versus time for both isotropic and Ko consolidation tests is shown in Fig. 5.24. The volume change from Ko consolidation was calculated based on the sample settlement. The volume changes in isotropic consolidation tests were determined by measuring the amount of oil flowing in and out of the pump connected to the sample. The volume change continued to increase with time and the end of primarily consolidation was not readily identifiable.

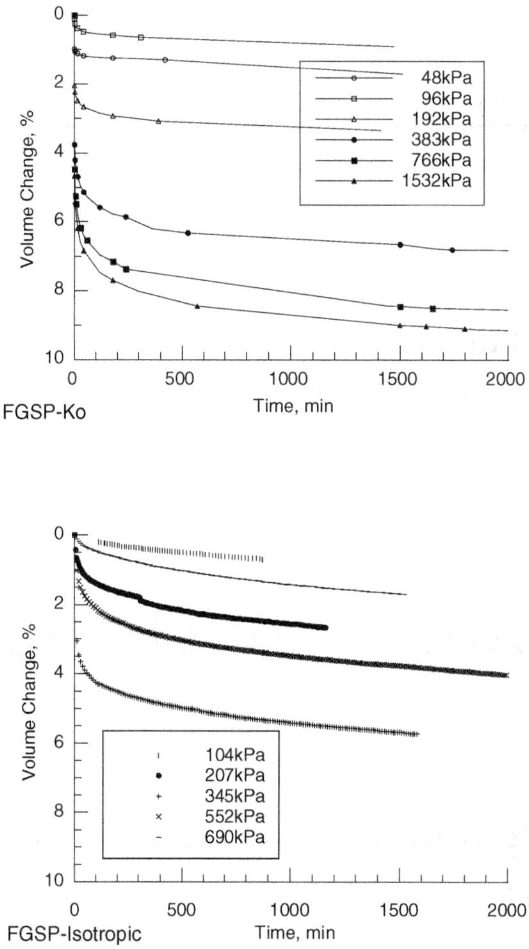

Fig. 5.24 Strain vs. Time for FGSP from (a) Ko (top) and (b) Isotropic (bottom) Tests

Conventional curve-fitting methods based on Terzaghi's theory of consolidation did not predict the observed volume change due to the occurrence of large secondary consolidation (Fig. 5.25), where the fitting points were selected within 1 minute after loads were applied to make sure the influence from secondary consolidation was limited.

The micro-pore structure of a typical amorphous silica powder is shown in Fig. 5.2. As indicated before, the change in void ratio during consolidation represents the change in both the voids within and between the silica aggregates. Two consolidation processes take place when a load is applied on a specimen. The first consolidation process involves the pores between the silica aggregates, while the second involves the pores inside the silica aggregates. Since only the first consolidation process is accounted for in Terzaghi's theory of consolidation, then, the second consolidation process appears as secondary consolidation.

Similar behaviors were reported in the consolidation of peat and organic clays (Buisman 1936, Adams 1965, Hobbs 1986, and Hanrahan 1954). Peat and organic clays have a macro-pore structure corresponding to the voids within the inorganic component in them, and a micro-pore structure corresponding to the pores within the organic matter component. When a load is applied to peat, inorganic matter consolidates first, followed by organic matter. Peat also exhibits a high void ratio typically on the order of 5 to 20 (Head 1994) like amorphous silica. Reduction in permeability due to consolidation also accounts for variation from Terzaghi's theory of consolidation.

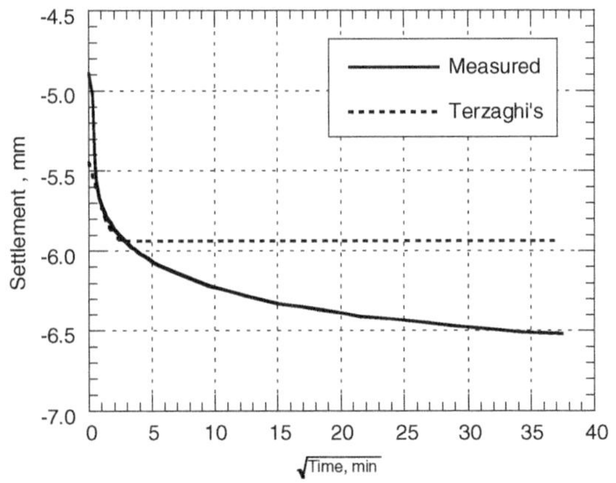

Fig. 5.25 Measured and Calculated (Terzaghi's) Settlement of FGSP

5.7.3 Settlement Components

Strain vs. logarithm of time exhibited three main components: initial strain, primary consolidation, and secondary consolidation. The initial strain was defined as the strain occurring within 0.25 minute after loading. The initial strain increased

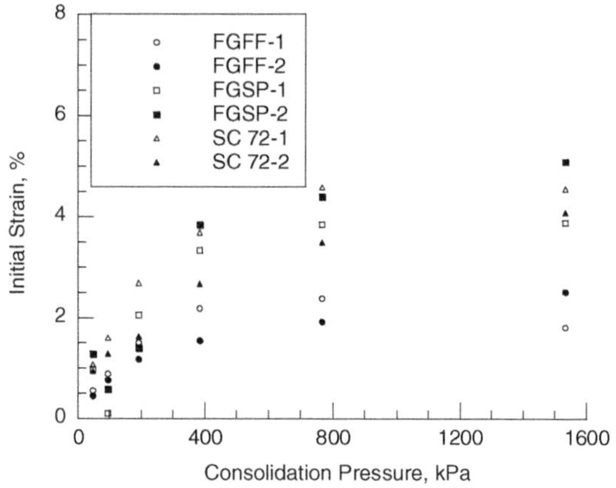

Fig. 5.26 Initial Strain vs. Pressure for Three Silicas from Oedometer Tests

with the applied load (Fig. 5.26). It accounted for approximately half the strain measured in one day.

Typical settlement vs. time curves from Ko consolidation and isotropic consolidation are presented in Fig. 5.27 and 5.28, respectively. The primary and secondary consolidation could not be clearly distinguished from both settlement curves. A sample was loaded for 78 days to investigate the secondary consolidation of transparent amorphous silica powder in an oedometer using a computer-controlled motor. The settlement vs. logarithm of time is shown in Fig. 5.29. Strain continued to increase throughout the 78 days test period. Similar behavior was reported in peats by Dhowian and Edil (1980). However, the tertiary consolidation that typically occurred in peat did not appear in transparent amorphous silica powder within the test time.

The secondary compression index, $C\alpha$, normalized by the compression index, Cc, ($C\alpha$/Cc), which is used to describe the secondary consolidation of natural soil, ranges between 0.03 and 0.05 (Table 5.8). It is consistent with the values reported by Mesri et al (1999) for 18 soft clay deposits.

5.7.4 Pore Pressure Dissipation

The definition of end of primary consolidation was the end of dissipation of built-up pore pressure. The dissipation of pore pressure during consolidation was studied using a typical specimen, FGSP-5, shown in Table 5.8. The test was performed using a computer controlled loading frame consisting of a stepper motor and sensors for measuring load, displacement, and pore pressure. Load was increased to the desired stress level in 60 seconds and maintained constant thereafter. The system allowed for maintaining the stress with an accuracy of ±0.1%. Pore pressure

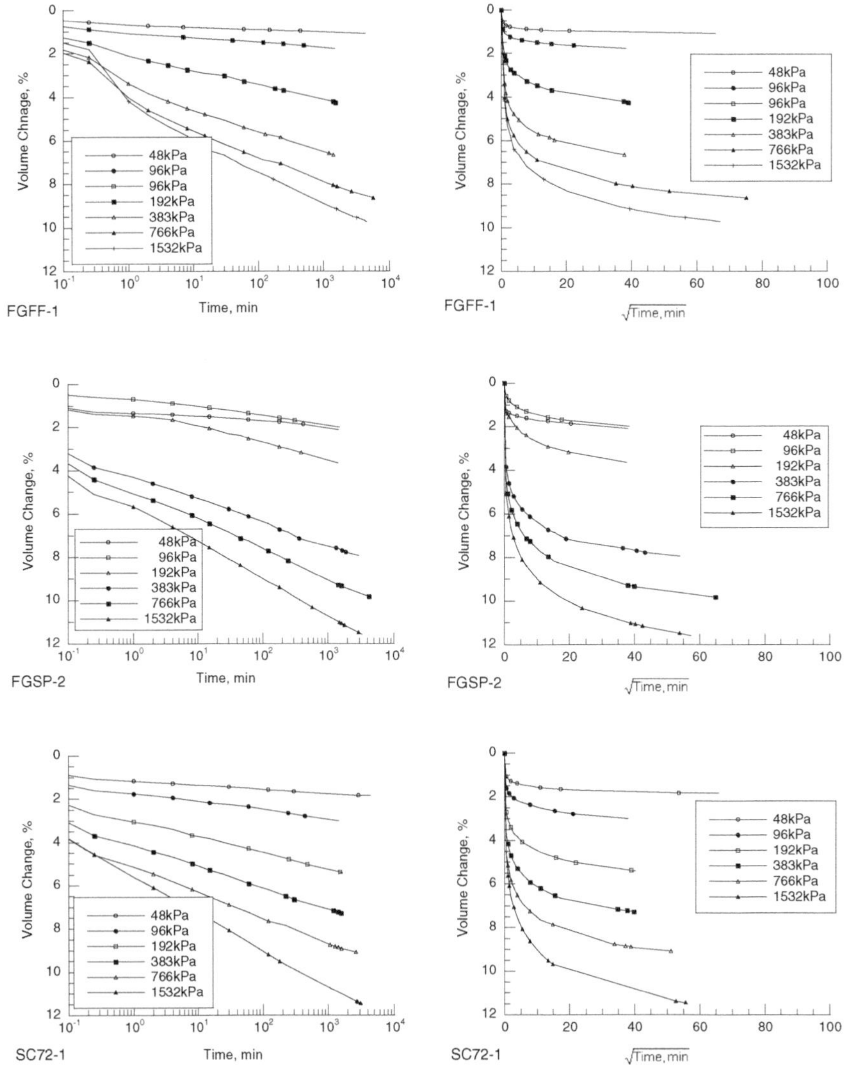

Fig. 5.27 Strain vs. Both Log and Square Root of Time from Ko Consolidation Tests

was measured at the bottom of the specimen with an accuracy of ±0.25%. Next, load was doubled and the procedure was repeated.

Typical pore pressure and volume change under load are shown in Fig. 5.30 and 5.31. It is evident that settlement continued to increase even after full pore pressure dissipation. The time for pore pressure dissipation increased with increasing vertical stress, from less than 1 minute for the 96 kPa loading stage to almost 1 day for the 6029 kPa. A similar phenomenon has been frequently observed in tests on peat (De Jong 1968, Dhowian and Edil 1980, and Lefebvre et al 1984).

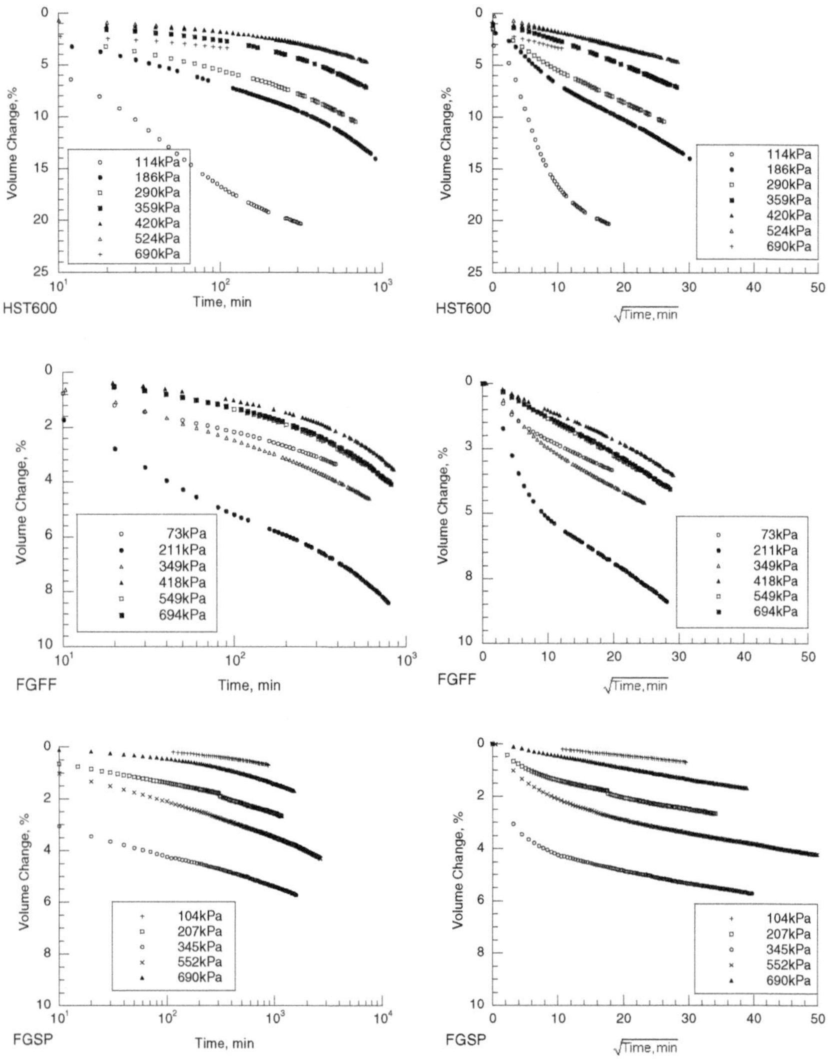

Fig. 5.28 Strain vs Both Log and Square Root of Time From Isotropic Consolidation Tests

A lag between development of pore pressure and applied load was noticed (Fig. 5.32). The lag increased with increasing applied pressure. A similar lag has been reported in tests on peat (Edil and Haan 1981 and Lefebvre et al 1984). This was probably caused by incomplete saturation. The breakage of silica powder aggregate during loading caused the volume compression, which can contribute to this phenomenon. However, the location of the pore pressure sensor at the base of the specimen (not at the center) could have also contributed to this effect (Berre and Iversen 1972).

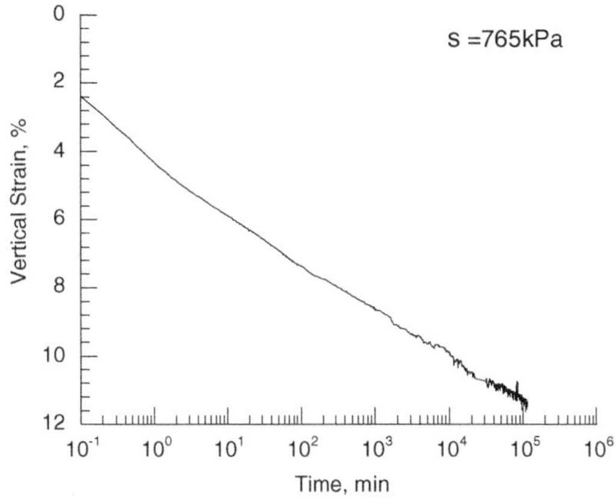

Fig. 5.29 Secondary Consolidation of FGSP (Load Kept Constant in an Oedometer)

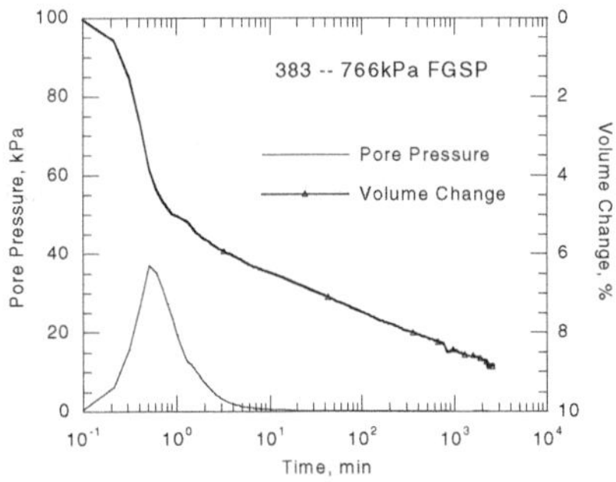

Loading from 383to 766 kPa

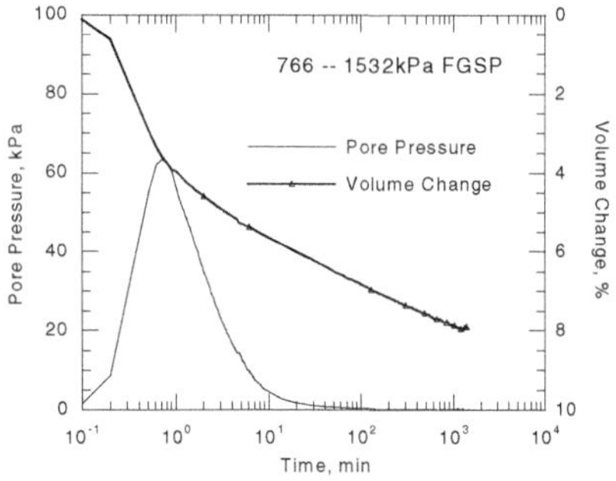

Loading from 766 to 1532 kPa

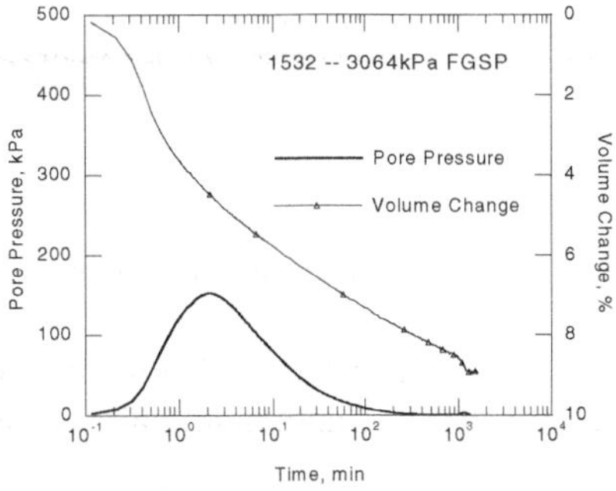

Loading from 1532to 3064 kPa

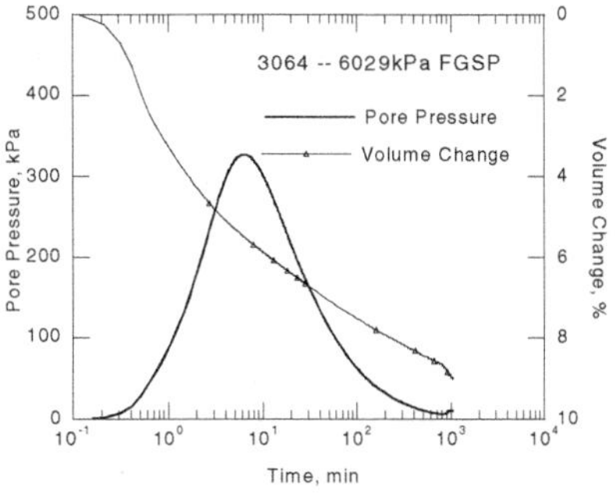

Loading from 3064 to 6029 kPa

Fig. 5.30 Volume and Pore Pressure Change in Oedometer Tests

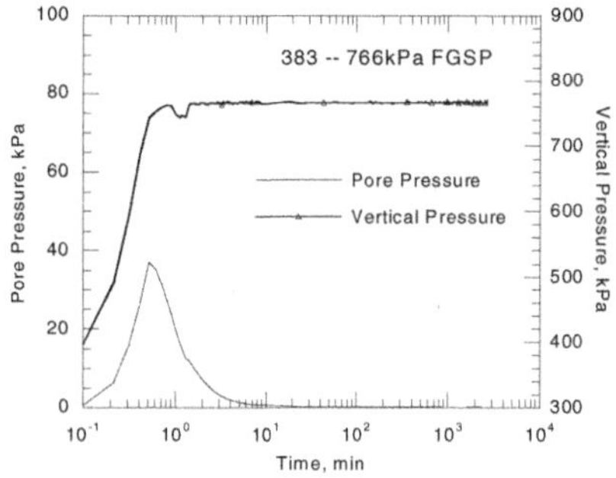

Loading from 383to 766 kPa

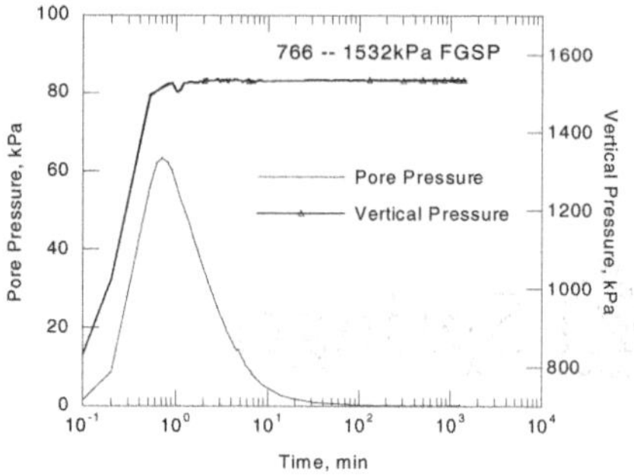

Loading from 766 to 1532 kPa

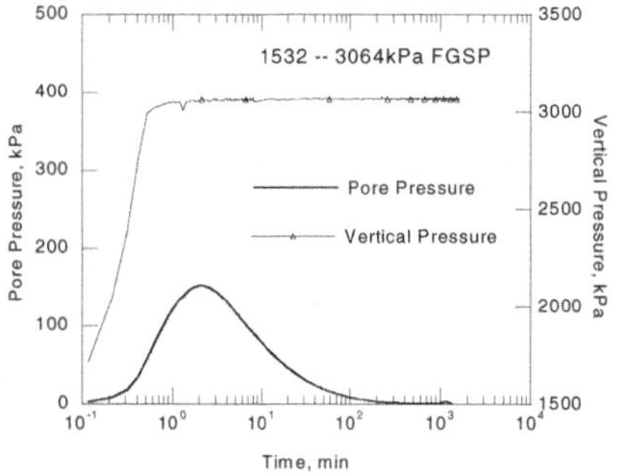

Loading from 1532to 3064 kPa

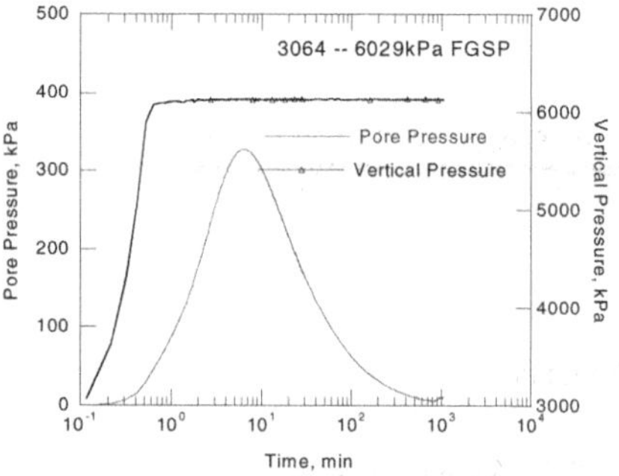

Loading from 3064 to 6029 kPa

Fig. 5.31 Pore Pressure Change in Oedometer Tests

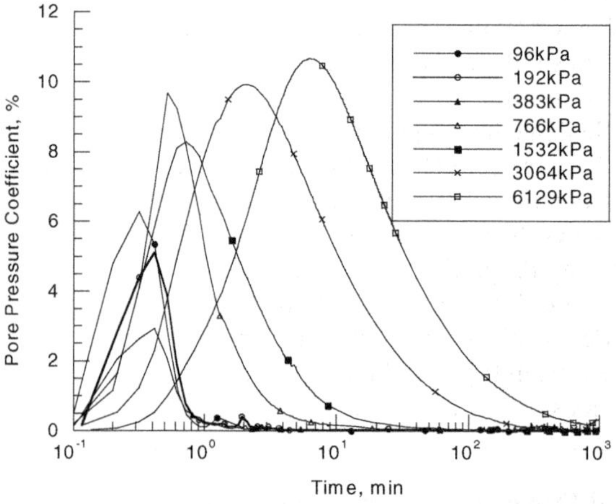

Fig. 5.32 Pore Pressure Lags the Applied Load in Oedometer Tests

5.7.5 Compression Isochrones

For materials with large secondary consolidation, such as peat, time is typically introduced in the e-log p relationship using isochrones, where the strains measured at different times are plotted against the corresponding applied total stress (Head 1994).

Hobbs (1986) reported that peat and organic clay exhibit parallel isochrones. Amorphous silica powder exhibits an increasing isochronal compression index, i.e. the slope of settlement against log pressure increases with the applied load (Fig. 5.33). For example, the isochronal compression index changed from 2.62–3.10 at 1 min to 5.44–5.87 at 1 day for the tested amorphous silicas.

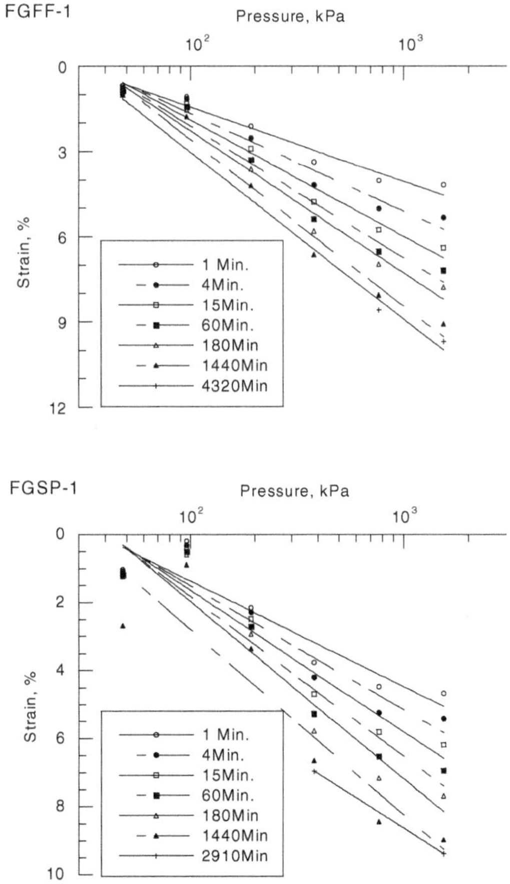

Fig. 5.33 Isochrones of Consolidation for Amorphous Silica Powder from Oedometer Tests FGFF (top) and FGSP (bottom)

5.7.6 Ko vs. Isotropic Consolidation

Volume change is plotted against logarithm of time in Fig. 5.34, for both Ko and isotropic consolidation. In Ko consolidation, the volume change increased linearly with time and the rate of settlement increased with pressure.

During isotropic consolidation, a larger non-linear increase in volume change was observed at corresponding loading stages. Similar behavior was reported by Yamaguchi et al (1985) for undisturbed fibrous peat. The relation between the slope of volume change and pressure was not apparent because of the different back pressures applied in each loading stage, which also contributed to the lower volume change in isotropic consolidation compared to that of oedometer tests for the same pressure.

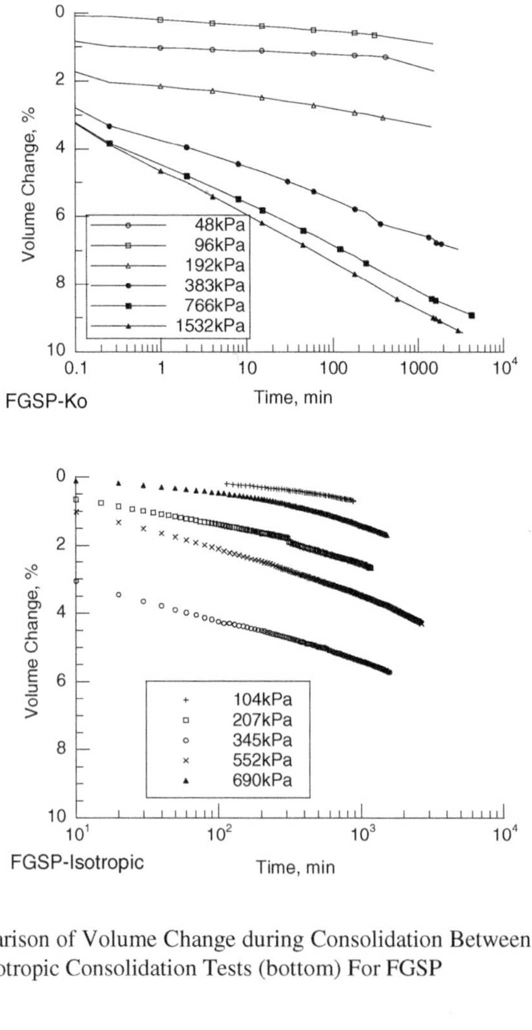

Fig. 5.34 Comparison of Volume Change during Consolidation Between (a) Ko Consolidation (top) and Isotropic Consolidation Tests (bottom) For FGSP

5.8 Permeability Properties

The permeability reported here refers to the permeability of pore fluid through transparent soil samples. All tests were carried out using a constant head setup. Flexible wall (FW) permeability tests were performed according to ASTM D5084 (1993): Standard Test Method For Measurement of Hydraulic Conductivity of Saturated Porous Materials Using a Flexible Wall Permeameter. FW tests were carried out on specimens with 37.2 mm (1.5 in) diameter and approximate 76mm (3in) in height.

A pneumatic pressure panel board was used to provide a constant head at the inflow line. A backpressure of 300–400kPa (43–58 psi) was used to facilitate saturation. Two computer-controlled pumps were used to keep the pressure head constant and measure volume change simultaneously. The first pump was connected to the outflow line. It provided backpressure and measured the volume of outflow of the specimen. The second pump was connected to the permeability cell. It controlled the cell (confining) pressure and measured the volume change of the specimen during consolidation and permeability testing. The pumps have an accuracy of 0.001 ml/min.

Hydraulic conductivities ranging between 2×10^{-8} to 1×10^{-5} m/sec were measured. These conductivities correspond to intrinsic permeability in the range of 13 to 6500 milli-darcy. Natural soils with intrinsic permeabilities in this range usually exhibit hydraulic conductivities to water at room temperature ranging between 1.3×10^{-7} and 6.5×10^{-5} m/sec.

Rigid wall (RW) permeability tests were performed on HST600 and FGSP in accordance to ASTMD2434 (1993): Standard Test Method for Permeability of Granular Soils (Constant Head). Tests were carried out on specimens 64 mm (2.5 in) in diameter and approximately 77 mm (3 in) long. After consolidation, a pneumatic pressure panel board was used to provide a constant head at the inflow line. The volume of outflow was used to compute the hydraulic conductivities. Conductivities ranging between 2.3×10^{-9} to 2.5×10^{-7} m/sec were measured in the RW tests which correspond to intrinsic permeability in the range of 1.5 to 160 milli-darcy. Natural soils with intrinsic permeability in this range usually exhibit hydraulic conductivities to water ranging between 1.5×10^{-8} to 1.6×10^{-6} m/sec.

5.8.1 Permeability with Void Ratio

The logarithm of permeability decreased linearly with void ratio as shown in Fig. 5.35. This behavior is consistent with the behavior of clay minerals reported by Mesri and Olson (1971).

5.8.2 Permeability with Vertical Pressure

The logarithm of permeability decreased with increasing pressure as shown in Fig.5.36, where RW refers to the rigid wall tests and FW refers to the flexible wall tests. The behavior was linear for RW tests but not for FW tests. The decrease in

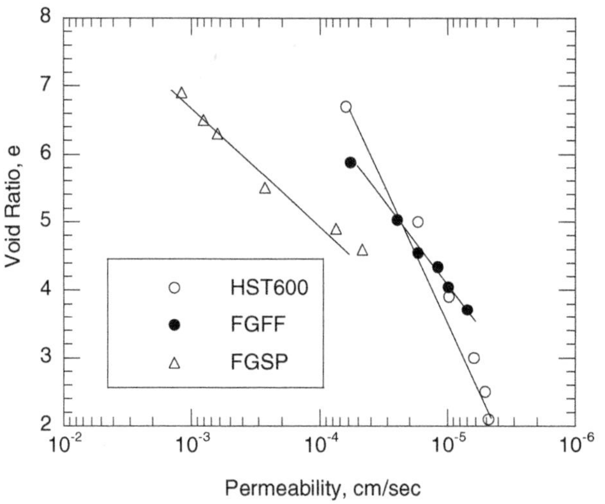

Fig. 5.35 Permeability vs. Void Ratio from Flexible Wall Permeability Test

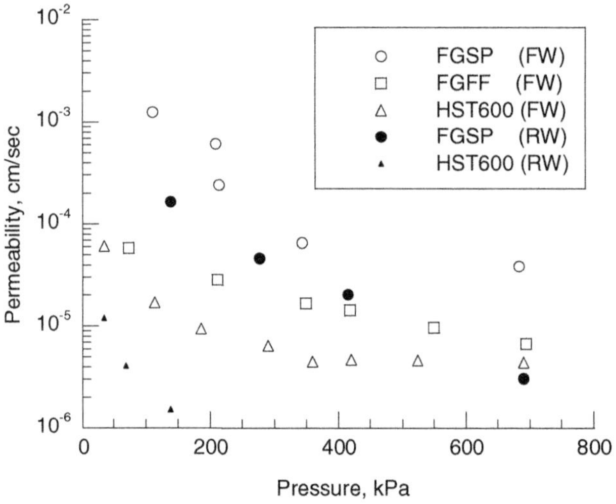

Fig. 5.36 Change in Permeability with Pressure for Both FW and RW Tests

permeability in FW tests became stable at approximately 400kPa, and no appreciable change in permeability was noticed after that.

5.8.3 Permeability with Material Type

The permeability varied from 2×10^{-8} to 1×10^{-5} m/sec in FW tests. Values ranging between 2.3×10^{-9} to 2.5×10^{-3} m/sec were measured in RW tests. The smaller the aggregate size, the lower the permeability. These permeability result range from good permeability to practically impervious permeability based on the research by Casagrande and Fadum (1940) (Terzaghi et al 1996). Soil with permeability falling within this range varies from clean sand, very fine sand, organic and inorganic silt, mixtures of sand silt and clay, glacial till, stratified clay deposit, etc. This implies that transparent amorphous silica powder can be customized to exhibit specific permeabilities representative of a wide range of natural soils.

5.9 Conclusions

Transparent amorphous silicas have shear strength, consolidation, and permeability, which are consistent with the macroscopic properties of many natural clays, but not any one clay in particular. The stress-strain behavior under both normally consolidated and over consolidated conditions are similar to those of natural clays. The shear strength parameters and permeability of this material are within the reported values for natural clays. The consolidation behavior of transparent amorphous silicas is similar to that of natural peat and organic clays, where secondary consolidation dominates the behavior.

The material can be consolidated at various pressures to meet specific model test requirements in terms of strength, deformation, and permeability. It provides an excellent media to simulate deformation and flow in clays using an optical set-up.

More research is required in a number of areas in order to prevent potential errors in extrapolating the model results from transparent soil to natural soils. Notably, the critical stress parameters of amorphous silica are also required for better modeling of geotechnical problems in natural clays. Also, the influence of particle surface charge on the geotechnical behavior should be quantified to gain a better understanding of the particle level behavior of amorphous silica.

References

Adams, J.: The engineering behavior of a Canadian Muskeg. In: Proc. 6th Int. Conf. Soil Mech. & Found. Engr., vol. 1, pp. 3–7 (1965)

ASTM D2434. Standard test method for permeability of granular soils (constant head), 1993 Annual Book of ASTM Standards, ASTM, vol. 04.08, Philadelphia, PA, pp. 308–312 (1993)

ASTM D2435. Standard test method for one-dimensional consolidation properties of soils, 1999 Annual Book of ASTM Standards, ASTM, vol. 04.08, Philadelphia, PA, pp. 210–219 (1999)

ASTM D4767. Standard test method for consolidated-undrained triaxial compression test on cohesive soils, 1999 Annual Book of ASTM Standards, ASTM, vol. 04.08, Philadelphia, PA, pp. 860–869 (1999)

ASTM D5084. Standard test method for measurement of hydraulic conductivity of saturated porous materials using a flexible wall permeameter, 1993 Annual Book of ASTM Standards, ASTM, vol. 04.08, Philadelphia, PA, pp. 1199–1206 (1993)

Balasubramaniam, A., Chaudhry, A.: Deformation and strength characteristics of soft Bangkok clay. J. Geotech. Eng. Div., ASCE 104(GT9), 1153–1167 (1978)

Berre, T., Iversen, K.: Oedometer tests with different specimen heights on a clay exhibiting large secondary compression. Geotechnique 22(1), 53–70 (1972)

Bishop, A., Henkel, D.: The measurement of soil properties in the triaxial test, 2nd edn. Edward Arnold Publishers, London (1962)

Bishop, A.: Shear strength parameters for undisturbed and remolded soil specimens. In: Parry, R., (ed.) Proc. Roscoe Memorial Symposium: Stress-Strain Behavior of Soils, Cambridge University, G.T. Foulis, Henley-on-Thames, England, pp. 3–58 (1971)

Bjerrum, L.: Geotechnical properties of Norwegian marine clays. Geotechnique 4(2), 49–69 (1954a)

Bjerrum, L.: Theoretical and experimental investigations on the shear strength of soils, Publication No. 5, Norwegian Geotechnical Institute, p. 111 (1954b)

Bjerrum, L.: The effective shear strength parameters of sensitive clays. In: Proc. 5th Int. Conf. on Soil Mech. & Found. Engr., Paris, France, vol. 1, pp. 23–28 (1961)

Bjerrum, L., Simons, N.: Comparison of shear strength characteristics of normally consolidated clays. In: Proc. Res. Conf. Shear Strength of Cohesive Soils, ASCE, pp. 711–726 (1960)

Bjerrum, L., Wu, T.: Fundamental shear strength properties of Lilla Edet clay. Geotechnique 10(3), 101–109 (1960)

Buisman, A.: Results of Long Duration Settlement Tests. In: Proc. 1st Int. Conf. Soil Mech. & Found. Eng., vol. 1, pp. 103–105 (1936)

Casagrande, A., Fadum, R.: Notes on soil testing for engineering purposes. Harvard Univ. Grad. School of Engineering, Publ. 268 (1940)

De Jong, D.: Consolidation models consisting of an assembly of viscous elements of a cavity channel network. Geotechnique 18(2), 195–228 (1968)

Dhowian, A., Edil, T.: Consolidation behavior of peats. ASTM Geotechnical Testing Journal, GTJODJ 3(3), 105–114 (1980)

Edil, T., Haan, E.: Settlement of peats and organic soils. In: Proc. of Settlement 1994, Vertical and Horizontal Deformation of Foundations and Embankments, ASCE, pp. 1543–1572 (1994)

Finno, R., Chung, C.: Stress-strain-strength response of compressible Chicago glacial clays. J. Geotech. Eng., ASCE 118(10), 1607–1625 (1992)

Gill, D.: Experimental and theoretical investigations of pile and penetrometer installation in clay, Ph.D. Thesis, Trinity College, Dublin, UK (1999)

Hanrahan, E.: An investigation of some physical properties of Peat. Geotechnique 4, 108–123 (1954)

Head, K.: Manual of Soil Laboratory Testing, 2nd edn. Permeability, Shear Strength and Compressibility Tests, vol. 2. Pentech Press (1994)

Henkel, D.: The effect of overconsolidation on the behavior of clays during shear. Geotechnique 6, 139–150 (1956)

Hobbs, N.: Mire morphology and the properties and behavior of some British and foreign Peats. Quarterly Journal of Engineering Geology, London 19(1), 7–80 (1986)

Holtz, R., Kovacs, W.: An introduction to Geotechnical Engineering. Prentice Hall, Englewood Cliffs (1981)

Iler, R.: The colloid chemistry of Silica and Silicates. Cornell University Press, Ithaca (1955)

Iskander, M.: A transparent material to model the geotechnical properties of soils. In: Proceedings, ICSMFE, Hamburg, Germany, vol. 1, pp. 315–319 (1997)

Iskander, M.: Transparent soils to image 3D flow & deformation. In: Proc. 2nd Int. Conf. on Imaging Technologies: Techniques and Applications in Civil Engineering, ASCE, pp. 255–264 (1998)

Iskander, M., Lai, J., Oswald, C., Mannheimer, R.: Development of a transparent material to model the geotechnical properties of soils. ASTM Geotechnical Testing J. 17(4), 425–433 (1994)

Kenny, T.: Discussion. J. Soil Mechanics and Foundations Div., ASCE 85(SM3), 67–79, ASCE (1959)

Ladd, C.: Stress-strain behavior of saturated clay and basic strength principles. Research Report R64-17, Dept. Civil Eng., MIT (1964a)

Ladd, C.: Stress-strain modulus of clay from undrained triaxial tests. J. Soil Machanics and Foundation div. ASCE 90(SM5) (1964b)

Ladd, C.: Stress-strain behavior of anisotropically consolidated clays during undrained shear. In: Proc. 6th Int. Conf. of Soil Mech. and Found. Engr., Available as Professional Paper No. P64-21, Soils Publication No. 160, Dept. Civil Engr., M.I.T (1964c)

Ladd, C., Foott, R.: New design procedure for stability of soft clays. J. Geotech. Eng. Div., ASCE 100(GT7), 763–786 (1974)

Ladd, C., Lambe, T.: The strength of undisturbed clay determined from undrained tests. Laboratory Shear Testing of Soils, ASTM STP No. 361, pp. 342–371 (1963)

Ladd, C., Foott, R., Ishihara, K., Schlosser, F., Poulos, H.: Stress-deformation and strength characteristics," State of the Art Report. In: Proc. 9th Int. Conf. on Soil Mech. & Foundation Engr., Tokyo, Japan, vol. 2, pp. 421–494 (1977)

Lambe, T., Whitman, R.: Soil Mechanics. Wiley, New York (1969)

Lefebvre, G., Langlois, P., Lupien, C.: Laboratory testing and in-situ behavior of peat as embankment foundation. Canadian Geotechnical Journal 21, 322–337 (1984)

Levine Era, N.: Physical Chemistry, 4th edn. McGraw-Hill, Inc., New York (1995)

Mannheimer, R., Oswald, C.: Development of transparent porous media with permeabilities and porosities comparable to soils, aquifers, and petroleum reservoirs. Ground Water 31(5), 781–788 (1993)

Mesri, G., Olson, R.: Mechanism controlling the permeability of clays. Clays and Clay Minerals 19(3), 151–158 (1971)

Mesri, G., Feng, T., Shahien, M.: Coefficient of consolidation by inflection point method. ASCE J. of Geotechnical and Geoenvironmental Engineering 125(8), 716–718 (1999)

Mitchell, J.: Fundamentals of Soil Behavior, 2nd edn. Wiley, New York (1993)

Ni, Q., Hird, C., Guymer, I.: Physical modeling of pile penetration in Clay using trapnsparent soil and particle image velocimetry. Geotechnique 60(2), 121–132 (2010)

PPG Industries. Product Specification Data, PPG Industries Inc., One PPG Place, Pittsburgh, PA 15272 (1996)

PPG Industries Inc., Silica Products, One PPG Place, Pittsburgh, PA 15272 (2000)

Prevost, J.: Undrained shear test on clays. J. Geot. Eng. Div., ASCE 105(GT1), 49–61 (1979)

Sadek, S., Iskander, M., Liu, J.: Geotechnical properties of transparent silica. Canadian Geotechnical Journal 39(1), 111–124 (2001)

Simons, N.: Comprehensive investigation of the shear strength of an undisturbed Drammen clay. In: Proc. Res. Conf. Shear Strength of Cohesive Soils, ASCE, pp. 727–746 (1960a)

Simons, N.: The effect of overconsolidation on the shear strength characteristics of an undisturbed Oslo clay. In: Proc. Res. Conf. Shear Strength of Cohesive Soils, ASCE, pp. 747–763 (1960b)

Skempton, A.: The pore-pressure coefficients A and B. Geotechnique 4, 143–147 (1954)

Skempton, A.: Long-term stability of clay slopes. Fourth Rankine Lecture, Geotechnique 14(2), 77–102 (1964)

Song, Z., Hu, Y., O'Loughlin, C., Randolph, M.: Loss of Anchor Embedment during Plate Anchor Keying in Clay. J. Geotechnical and Geoenvironmental Engr. 135(10), 1475–1485 (2009)

Terzaghi, K., Peck, R., Mesri, G.: Soil Mechanics in Engineering Practice, 3rd edn. Wiley, New York (1996)

Tomlinson, M.: Foundation Design and Construction, 4th edn. Pitman, London (1986)

Weast, R.: Handbook of Chemistry & Physics, 66th edn. CRC Press, Boca Raton (1986)

Welker, A., Bowders, J., Gilbert, R.: Applied research using transparent material with hydraulic properties similar to soil. Geotechnical Testing Journal 22(3), 266–270 (1999)

Wong, R.: Mobilized strength components of Athabasca oil sand in triaxial compression. Canadian Geotechnical J. 36(4), 718–735 (1999)

Yamaguchi, H., Ohira, Y., Kogure, K.: Volume change characteristics of undisturbed fibrous Peat. Soils and Foundations 25(2), 119–134 (1985)

Chapter 6
Geotechnical Properties of Silica Gels

Magued Iskander, Samer Sadek, and Louis Ge

Abstract. Silica gel is a colloidal form of silica, which resembles coarse white sand. It is primarily used as a desiccant for packaging. When mixed with a pore fluid having a matched refractive index it resembles transparent sand. In this chapter, the physical, geotechnical, and dynamic properties of some silica gels are discussed in detail.

6.1 Background

The amorphous silica family has two forms: powders and gels. Both shapes (or families) have the same refractive index. The geotechnical properties of the powders are presented in chapter five. The geotechnical properties of the solids are presented in this chapter. Commercial silica gels showed an excellent transparency in the same pore fluids used for silica powders. The granular appearance of silica gel particles is suitable for modeling sands. Many manufacturers were contacted for supplying samples of their product and many samples were investigated. The following sections explain details of the performed investigation.

6.2 What Is Silica Gel?

Silica gel is a colloidal form of silica, which resembles coarse white sand. It has been known since the 17th century but it was first used as a powerful adsorbing agent in gas masks during World War I.

Silica gel has great adsorptive power due to its many tiny pores. As a result, it is used as a drying agent, as a catalyst or catalyst carrier, and in purifying various substances. It is available commercially in many sizes and shapes. The most common size ranges are from 0.5mm to 5mm. The common shapes are rounded beads and angular particles.

Five manufacturers were contacted and fourteen different gels were identified as potential modeling materials (Table 6.1). The identification process involved (1) assessing clarity by matching the refractive index of the silica particles and pore fluid (Fig. 6.1), (2) identifying the grain sizes and shapes, which are

M. Iskander: Modeling with Transparent Soils, SSGG, pp. 85–115.
springerlink.com

Table 6.1 Some Identified Transparent Silica Gels

Supplier	Particles Size Range	Particle Shape	Clarity
	1-3mm	Rounded	Good
	0.5-1mm	Rounded	Good
	2-3mm	Rounded	Good
AGM Container Controls	2-5mm	Rounded	Good
	1.4-3.5mm	Rounded	Good
Eagle Chemical Co., Inc	1.4-2.4mm	Angular	Fair
	1.4-3.5mm	Angular	Fair
	2.5-6.7mm	Angular	Fair
Sphinx Adsorbents, Inc	1.4-3.4	Angular	Fair
	1.7-3.4mm	Angular	Fair
	0.85-2.4mm	Angular	Fair
Multsorb Technologies Inc.	0.5-1.5mm	Angular	Excellent
	2-5mm	Rounded	Excellent
Zeochem®	2-5mm	Rounded	Excellent

Fig. 6.1 Samples of Silica Gels Mixed with Matching Oil (Some turned yellow with time)

Fig. 6.2 Fine Dry Silica Gel Particles (SG1 on Left and SG2 on Right)

representative of natural sands, and (3) stability of the colorless appearance with time.

The two silica gels (shown in Fig. 6.2) were selected for detailed geotechnical investigation. The first is angular, 0.5–1.5 mm in size, and is referred to throughout the chapter as SG1. The second is round, 2–5 mm in size, and is referred to throughout the chapter as SG2. These gels were selected for two reasons. First, the materials were the most transparent of the identified gels. Second, these sizes are believed to encompass the range of available gels, which can be used for modeling medium to coarse sands. So far, we have not been able to identify a transparent silica gel representative of fine sands.

In this chapter, the physical and geotechnical properties of the identified silica gels are discussed in details. The chemical properties of silica gels are also briefly introduced.

6.3 Basic Chemical Properties and Chemical Preparation

Silica Gel is an amorphous, highly porous, partially hydrated[1] form of silica, which is a substance made from the 2 most abundant elements in the earth's crust, silicon and oxygen. Silica (silicon dioxide) and silicates form together 55% of the earth crust. Silica gel is inert and non-toxic.

[1] Hydrated = chemically combined with water in its molecular form.

The discovery of the preparation method for silica gel is usually credited to Sir Thomas Graham in 1861 (Scott 1993). He prepared the gel by dialyzing[2] dilute silica sols obtained from mixing a dilute aqueous solution of sodium silicate with hydrochloric acid. The same method is still used in manufacturing silica gel today. Silica gel is used in industry as a filter, catalyst, catalyst carrier, adsorbent, dehydrating agent, buffing media in toothpaste and in chromatography.

In general, silica gel manufacturing is carried out in three steps. The first step is production of sodium silicate from crushed quartz or sand. The second step is the release of silicic acid from the sodium silicate by means of a suitable acid followed by the conversion of the hydrogel into xerogel (totally dried gel). The third and final stage is the grinding and grading of the xerogel to particles of the correct shape and size for whatever purpose the material is to be used for.

Spherical silica gel preparation is a little bit different in the preparation details, as the final particles are not grinded. Usually it is prepared by spraying a neutralized silicate solution into fine droplets before gelling. Subsequently drying the droplets in a stream of hot air results in the required spherical solid shape of silica gel.

6.4 Physical Properties

The internal structure of silica gel particles is very different from sands. It contains an enormous number of micro-pores. This fact affects all the physical properties that are discussed in the following sections.

6.4.1 Particle Structure

As discussed previously, silica gel particles are shaped grinded from "xerogel" mass formed from the mentioned chemical reaction. This mass consist of a three dimensional vast networks made up of chains of particles. These particles may range in size from 3-30 nanometer (Iler 1955). This vast network has inter-connected spherical pores having an average diameter of approximately 21 $\mathring{A}$, and totaling 0.43 cm^3 per gram of silica (Sphinx 1993).

6.4.2 Specific Gravity and Unit Weight

The specific gravity of silica gels is 2.2 (Weast 1986), which is approximately 80% of the specific gravity of natural silicate sands. The dry unit weight is 6–9 kN/m^3 (40–60 pcf), which depends on its packing, shape and grain size. The saturated unit weight depends on the pore fluid used, and is 11-14 kN/m^3 (70-90 pcf), for the mineral oil and solvent pore fluid used in this research.

---

[2] Dialyses = The separation of smaller molecules from larger molecules or of dissolved substances from colloidal particles in a solution by selective diffusion through a semi permeable membrane (dictionary.com).

6.4.3 Void Ratio

Void ratio calculation for silica gel is different from natural soils, as silica gel particles have pores inside the particle. These in-particle pores should not be considered when calculating the void ratio for geotechnical engineering purposes. In that respect, silica gel is similar to silica powder. A complete discussion of void ratio calculation is presented in section 5.3. Equation 5.1 estimates the inter-aggregate void ratio based on phase diagram and assuming a predefined constant "adsorption factor". The adsorption factor for silica gel is assumed to be 0.43 cm^3 per gram of silica for silica gel (Sphinx 1993).

Using Eq. 5.1 with a= 0.43, Gs = 2.1 a relationship between change in both void ratio and inter-aggregate void ratio with the change in dry unit weight is plotted in Fig. 6.3.

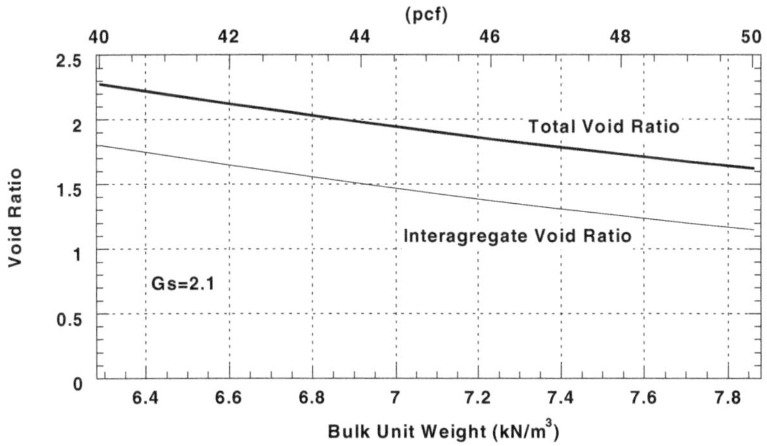

Fig. 6.3 Change in Void Ratio with Unit Weight of Silica Gel

6.4.4 Particle Size Distribution and Uniformity

Commercial silica gels are usually sold in uniform size ranges, *i.e.* the purchased silica will have a pre-defined maximum particle size and minimum particle size. Sieve analysis of as received sample (Fig. 6.4) shows this size ranges clearly.

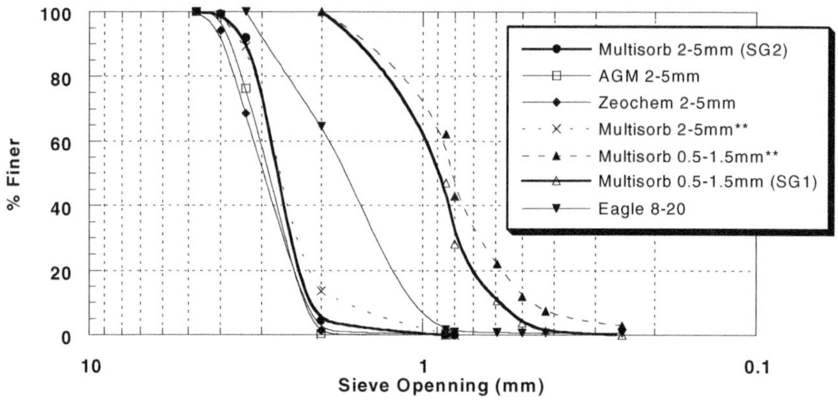

Fig. 6.4 Grain Size Distribution Curves of Silica Gels (Solid lines are for as received samples and dashed lines are for samples after triaxial testing)

6.5 Static Geotechnical Properties of Silica Gel

6.5.1 Shear Strength

6.5.1.1 Triaxial Tests

Consolidated drained triaxial compression tests were performed on dry and saturated specimens (Table 6.2 and 6.3).

6.5.1.1.1 Sample Preparation and Testing Procedure

Bishop & Henkel (1962) were followed whenever possible. However, testing silica gels, with a matched refractive index pore fluid poses a number of peculiarities, which require deviation in sample preparation from traditional procedures. In particular, the pore fluid, which consists of mineral oil and solvent, degrades latex membranes. Therefore, limiting exposure of the latex membrane to the pore fluid during test set-up is crucial. In addition, saturation of the voids inside the silica particles was time consuming, and was therefore performed prior to forming the triaxial test specimens. Saturation of the particles was accomplished by soaking and mixing the particles in pore fluid under vacuum for several hours. One advantage of "transparent soils," is that saturation can be readily viewed when the particles become transparent (compare Fig. 6.5 and Fig. 6.6). Excess oil was drained after saturation of the particles.

Both dry and saturated particles were formed by tamping to the desired dry density in a commercial sand mould. Compaction effort was maintained a constant, because of the difficulty in estimating the dry unit weight of specimens made by

Table 6.2 Summary of Consolidated Drained Triaxial Test Results on SG1 (Angular particles, 0.5-1.5mm in size)

Sample ID	Testing Condition	Initial Dry Unit Wt. (kN/m^3)	Initial Inter-aggregate Void Ratio	Initial Total Void Ratio	Confining pressure (psi)	ϕ**
SG0.5 SL-1	Sat. Loose	6.85	1.21	2.15	10	36°
SG0.5 SL-2	Sat. Loose	7.84	0.81	1.75	20	34°
SG0.5 SL-3	Sat. Loose	6.77	1.24	2.19	40	34°
SG0.5 SL-5	Sat. Loose	7.55	0.91	1.86	60	31°
SG0.5 SD-4	Sat. Dense	9.17	0.41	1.35	10	42°
SG0.5 SD-3	Sat. Dense	8.24	0.67	1.62	20	39°
SG0.5 SD-1	Sat. Dense	7.55	0.91	1.86	40	36°
SG0.5 DL-1	Dry Loose	6.71	1.27	2.22	10	36°
SG0.5 DL-2	Dry Loose	6.79	1.23	2.18	20	35°
SG0.5 DL-3	Dry Loose	6.72	1.27	2.21	40	31°
SG0.5 DL-4	Dry Loose	6.50	1.37	2.32	60	29°
SG0.5 DD-1	Dry Dense	7.16	1.07	2.01	10	40°
SG0.5 DD-2	Dry Dense	7.35	0.99	1.94	20	37°
SG0.5 DD-3	Dry Dense	6.97	1.15	2.09	40	33°
SG0.5 DD-4	Dry Dense	6.86	1.20	2.15	60	31°
SG0.5 SMD-3	Sat. Med. Dense	8.24	0.67	1.62	10	40°
SG0.5 SMD-1	Sat. Med. Dense	8.24	0.67	1.62	20	39°
SG0.5 SMD-2	Sat. Med. Dense	7.36	0.99	1.93	40	34°
SG0.5 SMD-4	Sat. Med. Dense	8.89	0.48	1.43	60	33°

**Based on one test only and assuming no cohesion.

tamping oil-filled particles. A different compaction effort was used for dense, loose, and medium-dense specimens. The existence of pore fluid in oil-filled particles resulted in a higher dry unit weight for saturated specimens (Table 6.4), which is also observed in compaction of natural sand.

The dry unit weight of saturated samples is calculated taking into consideration the absorption of pore fluid into the particles. Thus, the values are highly dependent on the estimated absorption amount. Further research is required to investigate the exact absorbed quantities for better estimation of dry unit weight.

Table 6.3 Summary of Consolidated Drained Triaxial Test Results on SG2 (Round Particles 2-5mm)

Sample ID	Testing Condition	Initial Dry Unit Wt. (kNm^3)	Initial Inter-aggregate Void Ratio	Initial Total Void Ratio	Confining Pressure (psi)	φ**
SG25 SL-1	Sat. Loose	6.91	1.18	2.13	10	30°
SG25 SL-2	Sat. Loose	7.46	0.95	1.89	20	29°
SG25 SL-3	Sat. Loose	7.36	0.99	1.93	40	32°
SG25 SD-1	Sat. Dense	7.97	0.76	1.71	10	36°
SG25 SD-2	Sat. Dense	7.89	0.79	1.74	20	33°
SG25 SD-3	Sat. Dense	7.50	0.93	1.88	40	34°
SG25 DL-1	Dry Loose	7.75	0.84	1.78	10	37°
SG25 DL-2	Dry Loose	7.73	0.85	1.79	20	31°
SG25 DL-3	Dry Loose	7.75	0.84	1.78	40	33°
SG25 DD-1	Dry Dense	7.89	0.79	1.74	10	34°
SG25 DD-2	Dry Dense	7.91	0.78	1.73	20	34°
SG25 DD-3	Dry Dense	8.03	0.74	1.69	40	35°
SG25 DD-4	Dry Dense	8.64	0.55	1.50	60	33°

**Based on one test only and assuming no cohesion.

Table 6.4 Summary of Triaxial Test Results for SG1

Testing Condition	Average Unit weight (kN/m^3)	Angle of Internal Friction Range	Angle of Friction (Failure Envelope)
Saturated Loose	7.25	31°-36°	33°
Sat. Dense	8.53	35°-42°	36°
Dry Loose	6.68	36°-39°	30°
Dry Dense	7.08	31°-40°	32°
Sat. Med. Dense	8.18	33°-40°	34°

All tests were performed with a shearing rate of 0.5mm/min. SG1 Samples were 38 mm (1.5 in) in diameter, while SG2 were 70 mm (2.8 in) in diameter. A larger diameter was used for SG2 in order to (1) obtain a better particle to sample diameter ratio, and (2) reduce membrane failure rate. One 0.3mm thick membrane was used to setup the SG1 Specimens, and two 0.4mm thick membranes were used for SG2 specimens. Nevertheless, 40% of SG2 tests incurred membrane failures particularly in saturated cases under high confining pressures.

For tests on saturated specimens, a vacuum source was connected to the top and pore fluid was introduced at the base of the specimen in order to saturate the inter-particle void space. Complete saturation can be inspected visually as samples turn transparent.

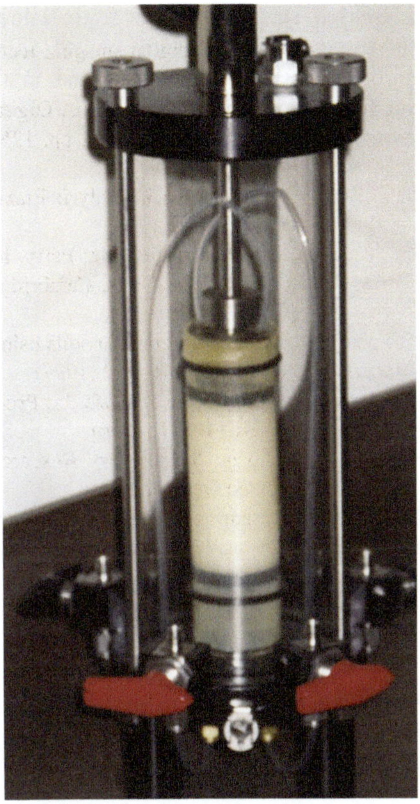

Fig. 6.5. Fig. Prepared Dry Sample of Silica Gel (Fine Particles)

Fig. 6.6. Same Sample shown in Fig 6.5 after Saturation

The testing program was designed to investigate the shear strength properties of dense and loose silica gel samples in both cases dry and saturated for both coarse and fine silica gels.

6.5.1.1.2 Triaxial Tests Results

The stress strain behavior of SG1 silica (Fig. 6.7 – 6.11) is consistent with typical stress strain behavior of sand, for both dense and loose conditions. Peak strength was reached at higher strains than normally expected for dense sands, due to the compressibility of the silica gel particles. Dense specimens exhibited typical strain softening behavior, particularly at low confining pressures. Saturation resulted in higher strength, possibly due to the increase in unit weight of saturated specimens, or because pore fluid acted as an anti-lubricant. Water is reported as an anti-lubricant of siliceous sand (Horn and Deere 1962).

The stress strain behavior of SG2 silica (Fig.6.12 – Fig.6.15) is also consistent with typical behavior of coarse sand. Peak strength occurred at lower strain values than the finer SG1 specimens, which is more consistent with typical sand behavior. A large fluctuation in stress-strain behaviors is observed post-peak in all tests, due to particle breakage. Saturation had a negligible effect on strength. Strain softening was also observed in most specimens.

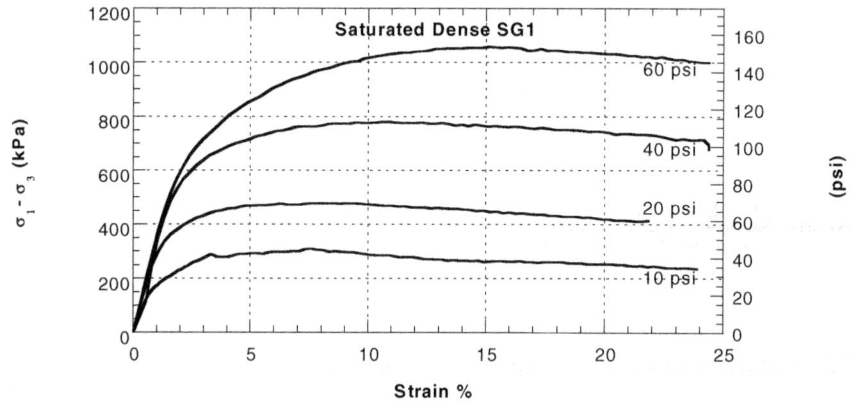

Fig. 6.7 Stress Strain Curves for CD Triaxial Test of Saturated Dense SG1

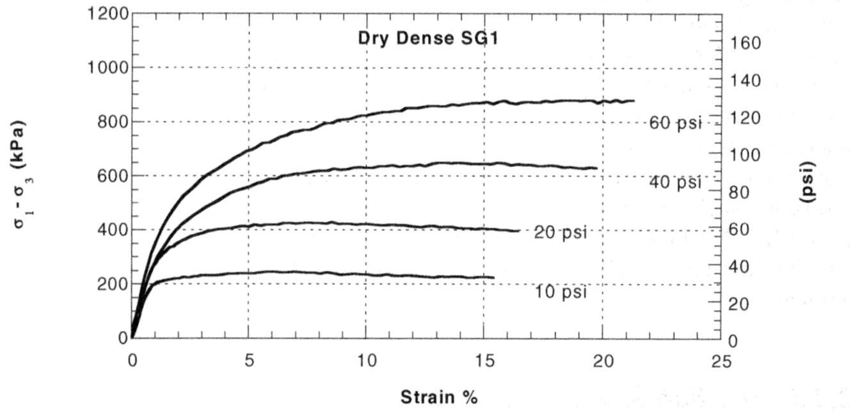

Fig. 6.8 Stress Strain Curves for CD Triaxial Test of Dry Dense SG1

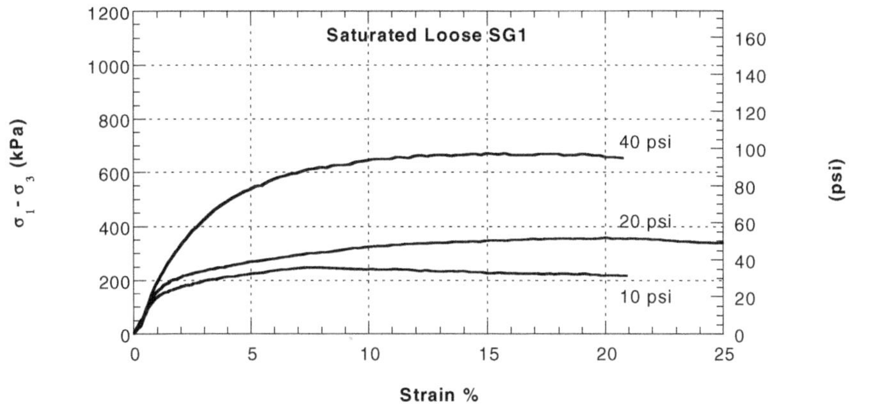

Fig. 6.9 Stress Strain Curves for CD Triaxial Test of Saturated Loose SG1

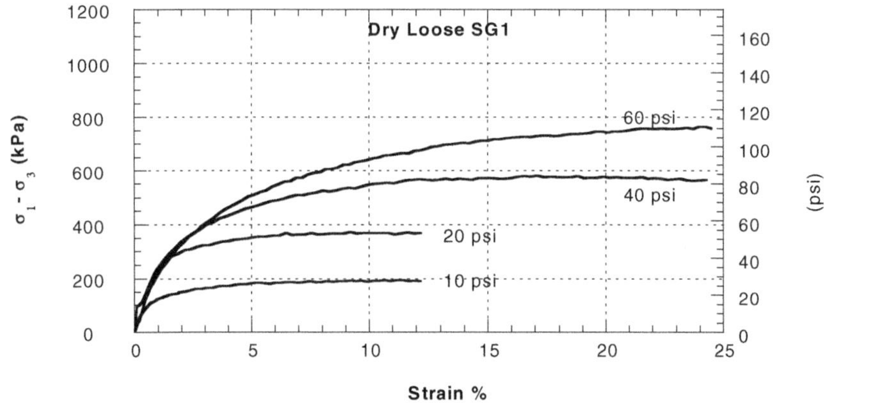

Fig. 6.10 Stress Strain Curves for CD Triaxial Test of Dry Loose SG1

6.5.1.1.3 Angle of Internal Friction

The angles of internal friction, ϕ, calculated for each test, assuming no cohesion, ranged between 29° and 42° (Table 6.2), which is consistent with medium to coarse sands. The angle of internal friction decreases with increasing confining pressure, resulting in a curved failure envelope (Fig. 6.16). This phenomenon appears in sands, but at much higher confining pressure (Holtz & Kovacs, 1981). For example curvature begins at approximately 0.2 MPa (29 psi) in silica gels (Fig. 6.16), but was reported to begin at approximately 2 MPa (290 psi) in Sacramento River Sand (Lee & Seed 1967).

The P-Q failure envelopes shown in Fig. 6.17 and Fig. 6.18 were used to calculate the friction angles, for each tested category. No cohesion was observed. The angles of friction of the Mohr-Coulomb failure envelopes were 30°–36°, for SG1 and 31°–34° for SG2 envelopes (Table 6.4, 6.5).

6.5.1.2 Direct Shear Testing Results

Direct shear tests were performed on dry samples according to ASTM test method D3080. The stress strain behavior of SG1 silica (Fig. 6.19 and Fig. 6.20) is consistent with typical behavior of sand, for both dense and loose conditions. Dense samples show a defined peak and exhibit a tendency to compress prior to expanding, like natural sands. Loose sands exhibit strain hardening and contractive behavior, particularly at higher normal pressures. The peak strength coincides with peak volume change for both dense and loose samples. The residual strength of dense sands approaches the peak strength of loose samples at the same normal pressure, thus validating the critical void ratio concept (Casagrande, 1936, 1975) for silica gels.

SG2 specimens exhibited a higher strength and volume change than SG1 specimens (Fig. 6.21 and Fig. 6.22). A large fluctuation in stress-strain behaviors was observed post-peak in all tests, due to particle breakage. All specimens exhibited strain softening. However, the results could have been influenced by the large diameter of the SG2 particles (2–5 mm) relative to the height of the direct shear specimen (20 mm).

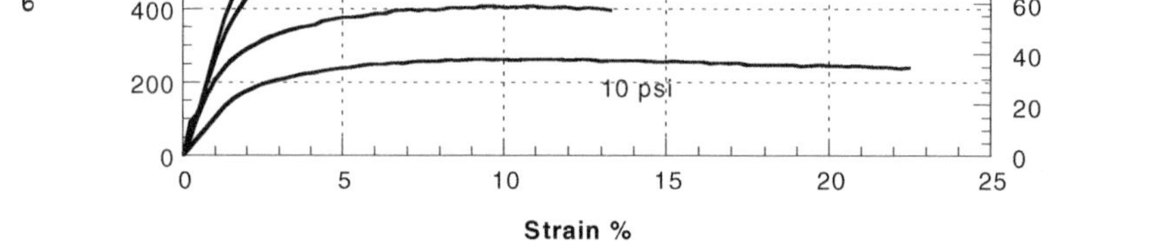

Fig. 6.11 Stress Strain Curves for CD Triaxial Test of Saturated Med. Dense SG1

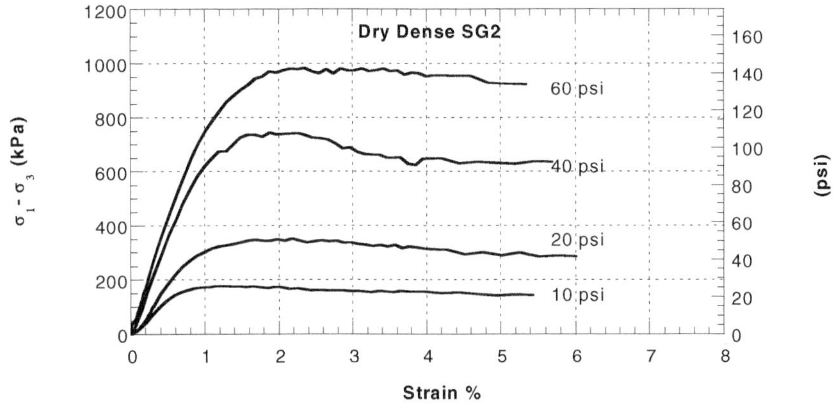

Fig. 6.12 Stress Strain Curves for CD Triaxial Test of Dry Dense SG2

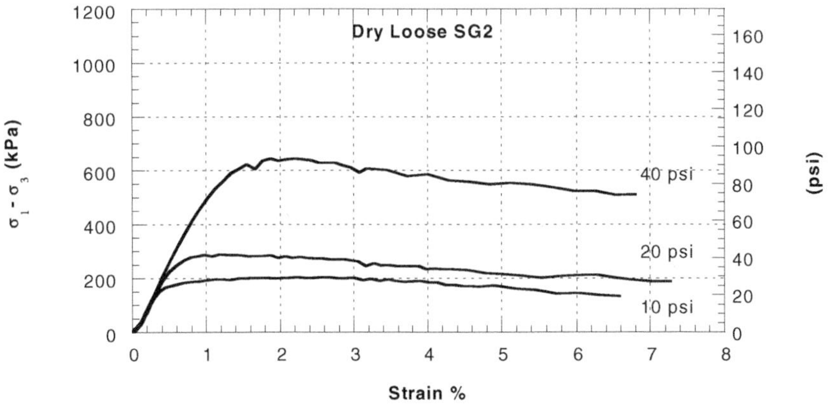

Fig. 6.13 Stress Strain Curves for CD Triaxial Test of Dry Loose SG2

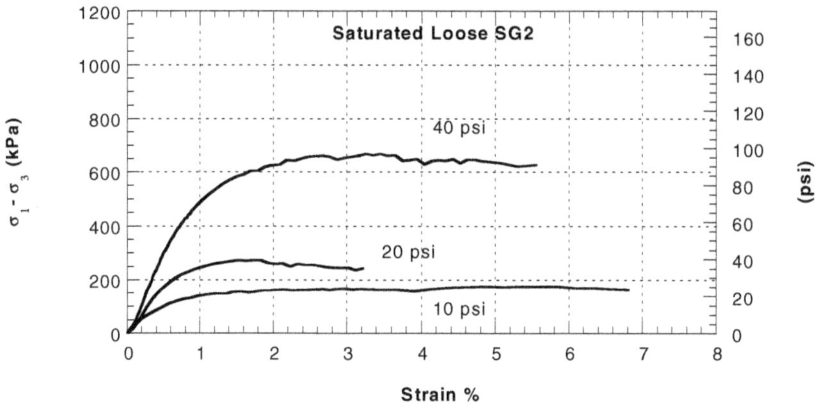

Fig. 6.14 Stress Strain Curves for CD Triaxial Test of Saturated Loose SG2

Angles of friction ranged between 34° and 50° decreasing with the increase in normal pressure (Table 6.6). The angles of friction of the Mohr-Coulomb failure envelopes were 36–40° for SG1 and 38–42° for SG2 envelopes, which is consistent with most natural sands and the triaxial test results.

6.5.2 Modulus of Elasticity

Young's Modulus for all SG1 and SG2 Triaxial tests was calculated based on stress associated with 50% of failure strain. Table 6.7 shows a detailed Young's Modulus calculation for all performed triaxial tests. For SG1 Young's modulus was found to be 15-22 MPa (2-3 ksi) for loose samples and 26-32 MPa (3-5 ksi) for dense samples. For SG2 it was found to be 24-52 MPa (3-8 ksi) and 36-84 MPa (5-12 ksi) for loose and dense respectively. These values are similar to the reported values for sand (Hunt 1984).

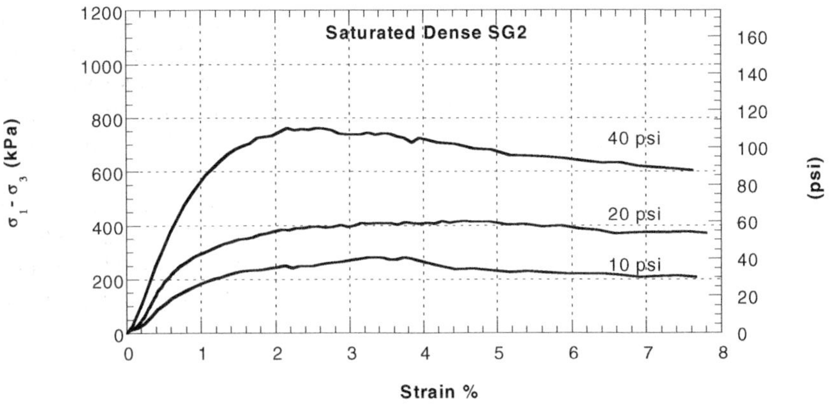

Fig. 6.15 Stress Strain Curves for CD Triaxial Test of Saturated Dense SG2

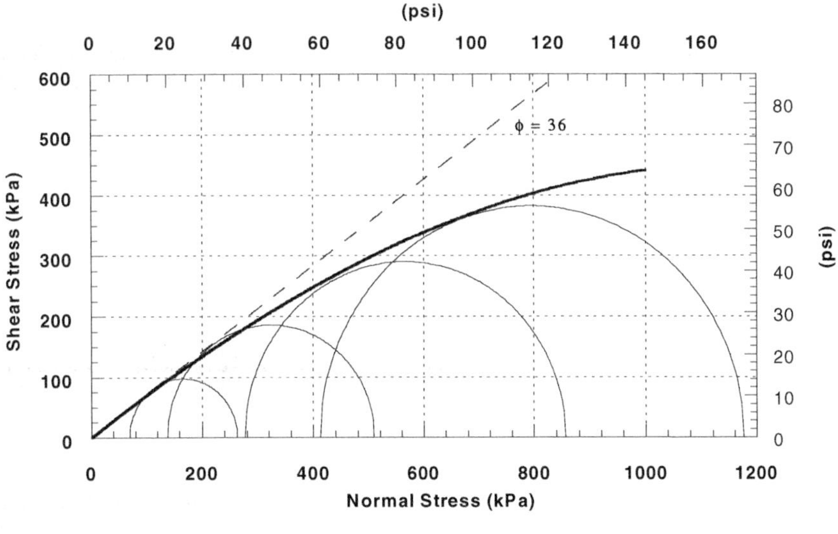

Fig. 6.16 Typical Mohr Coulomb Envelope of Silica Gels Tested in Triaxial Compression (Dry Loose SG1)

Table 6.5 Summary of Triaxial Test Results for SG2

Testing Condition	Average Unit weight (kN/m^3)	Angle of Internal Friction Range	Angle of Friction of Failure Envelope
Saturated Loose	7.24	29°-32°	31°
Sat. Dense	7.79	33°-36°	34°
Dry Loose	7.74	31°-37°	33°
Dry Dense	8.12	33°-35°	34°

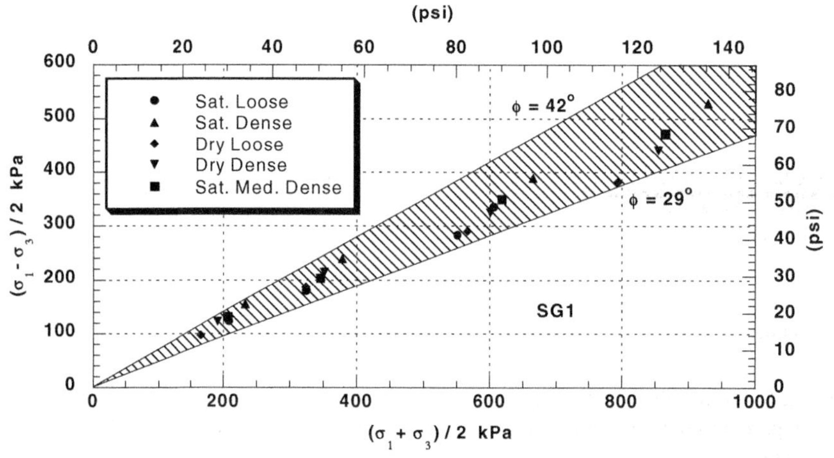

Fig. 6.17 P-Q Plot for all Triaxial Test Results of SG1

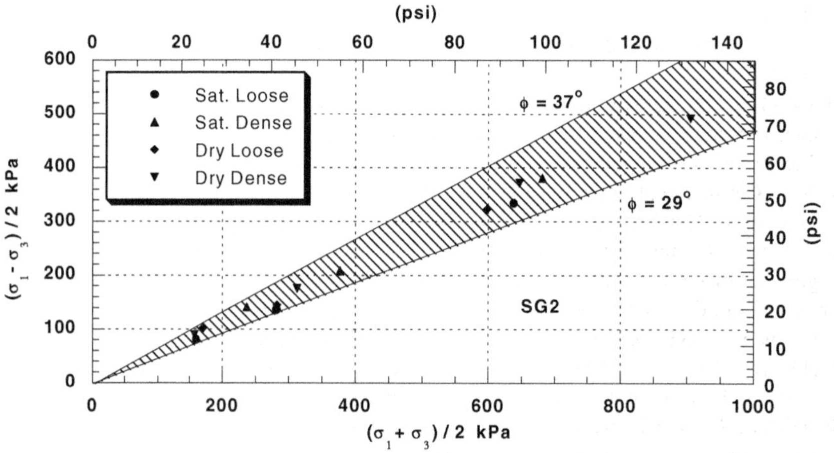

Fig. 6.18 P-Q Plot for all Triaxial Test Results of SG2

Table 6.6 Summary of Direct Shear Test Results on Dry Silica Gel

Particles Shape	Testing Condition	Average Initial γ_d, (kN/m^3)	Initial inter-aggregate void ratio "e_i"	Initial total void ratio "e"	"ϕ" at Low Normal Pressure*	"ϕ" at High Normal Pressure**	Overall Failure Angle
Angular 0.5-1.5mm	Dry Loose	7.17	1.06	2.01	44°	34°	36°
Angular 0.5-1.5mm	Dry Dense	8.42	0.62	1.56	47°	38°	40°
Round 2-5mm	Dry Loose	7.89	0.79	1.73	42°	37°	38°
Round 2-5mm	Dry Dense	8.65	0.55	1.49	50°	40°	42°

* Less than 140 kPa (20 psi); ** More than 140 kPa (20 psi).

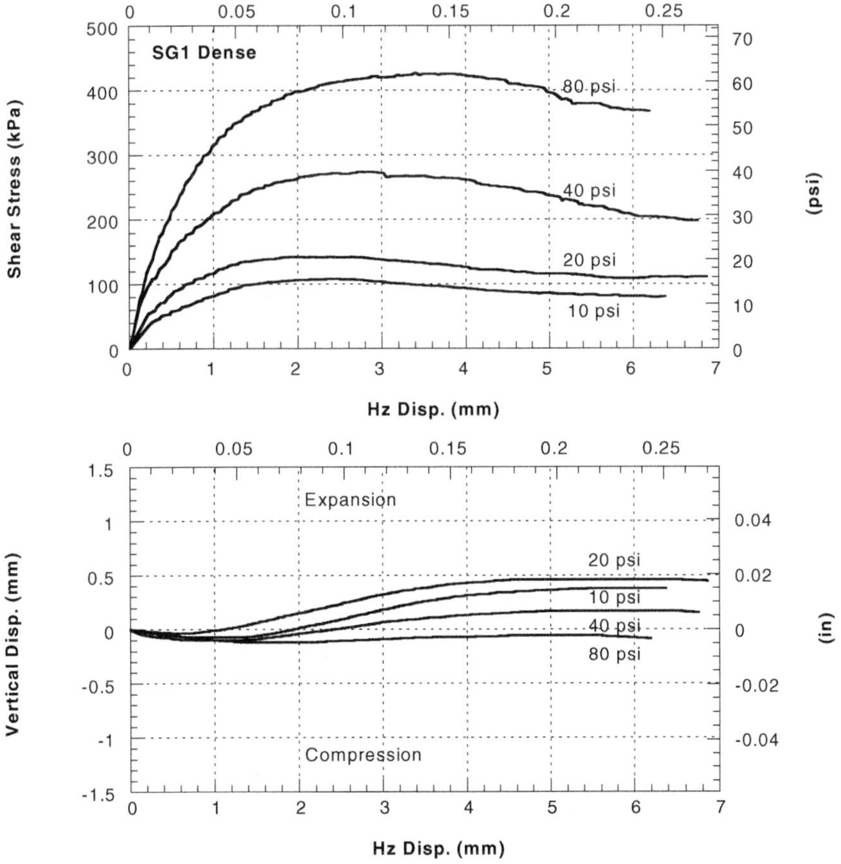

Fig. 6.19 Direct Shear Test Results for Dense SG1

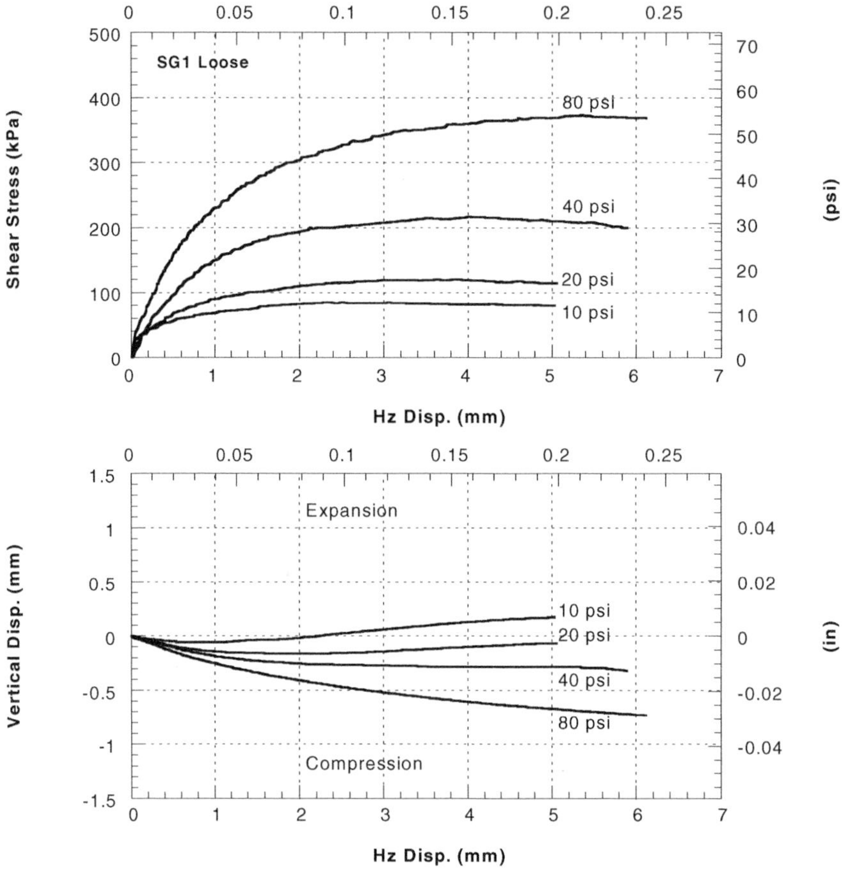

Fig. 6.20 Direct Shear Test Results for Loose SG1

6.5.3 Compressibility

Confined compression odometer tests were performed on both SG1 and SG2 specimens according to ASTM test method D2439-96, for One Dimensional Consolidation of Soils. The results of the tests are shown in Fig 6.23 along with data from Lamb & Whitman (1969) for Ottawa and Calcareous sands. Both SG1 and SG2 show nonlinear stress-strain behavior like most sands (Hendron 1963). However, Ottawa sand is far less compressible than SG1, SG2, and Calcareous sand.

The "virgin" compressibility of SG1 is larger than that of calcareous sands. SG1 exhibited large plastic deformation, probably due to particle breakage. This is also supported by the sieve analysis results for post-tested Triaxial samples (Fig. 6.3). Particle breakage is reported in many sands at different normal pressures, and can be audibly heard during compression tests at high normal pressures (Lee 1965). In our tests, particle breakage was audible at approximately 250 kPa (36 psi). The high non-linear elastic compressibility of SG1 explains the lack of a defined peak strength in the triaxial compression tests shown in Fig. 6.7 and Fig 6.8. This also explains why the peak strength of dense SG1 specimens was reached at a strain of 8–15%, which is relatively high. As expected, dense samples are less compressible than loose ones.

Although SG2 specimens exhibit the same virgin compression like calcareous soils, their elastic rebound is considerably larger. It is therefore believed that elastic compression of SG2 specimens is the dominant compressive mechanism, which is probably due to the relatively small odometer sample height (20 mm) relative to the 2–5 mm particles mean diameter.

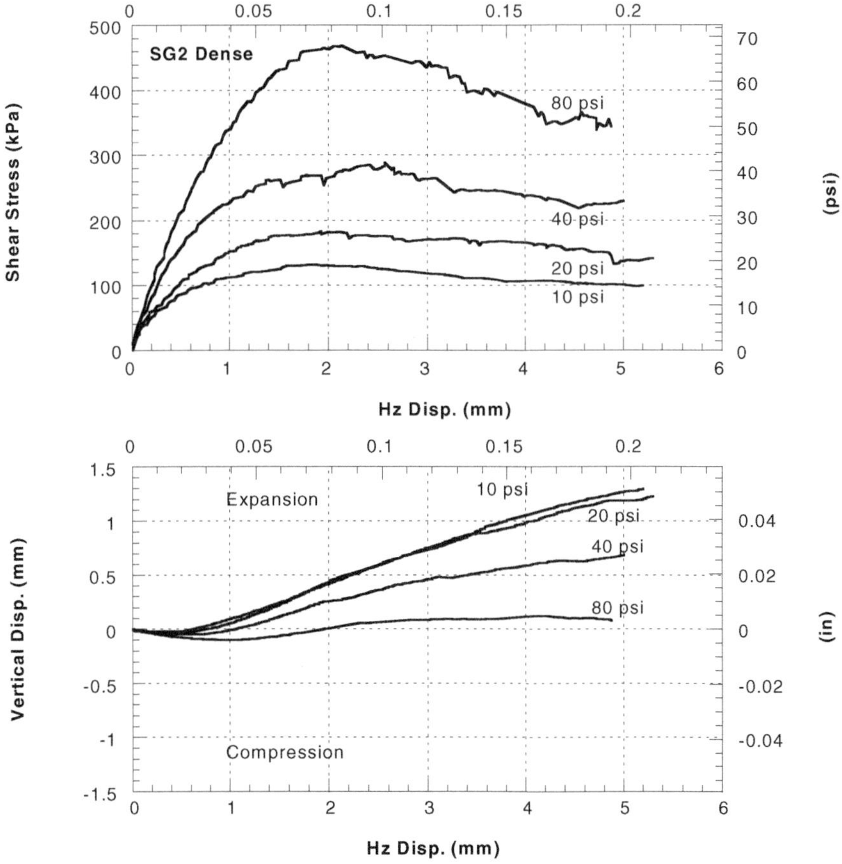

Fig. 6.21 Direct Shear Test Results for Dense SG2

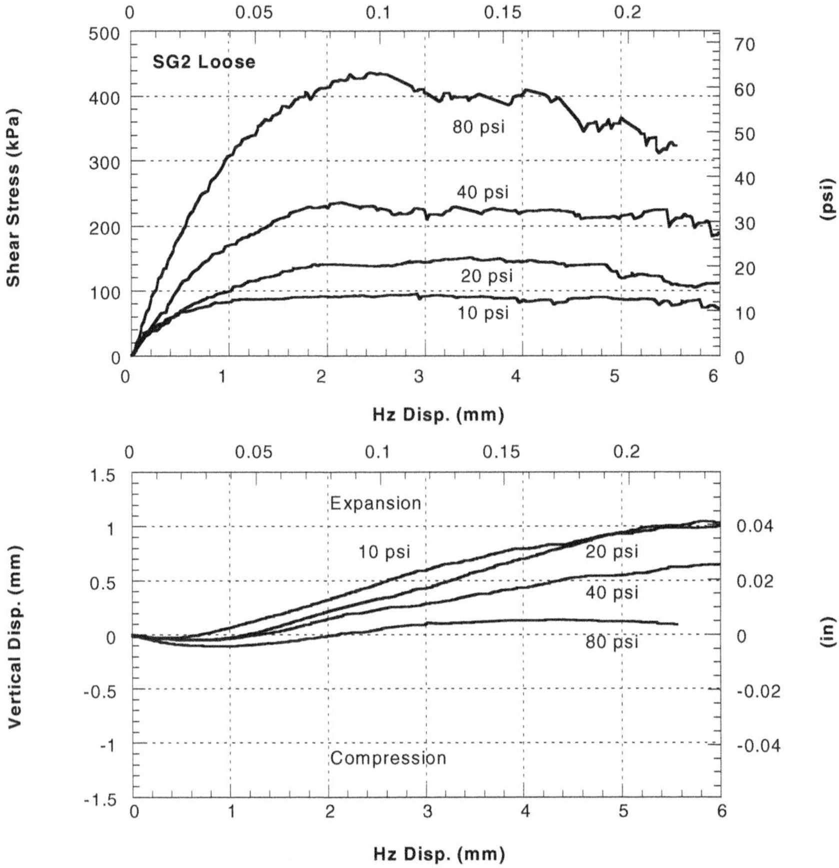

Fig. 6.22 Direct Shear Test Results for Loose SG2

6.5.4 Hydraulic Conductivity

Flexible wall permeability tests (Fig. 6.24) were performed on SG1 & SG2 specimens according to ASTM test method D5084 for Measurement of Hydraulic Conductivity of Saturated Porous Materials. The samples were permeated with the mineral oil & solvent mixture during triaxial tests. Hydraulic conductivities of $1.5*10^{-4}$ cm/sec for SG1 and $7*10^{-3}$ cm/sec for SG2 were measured. These conductivities correspond to intrinsic permeabilities of 1 and 45 darcys, respectively. Natural soils with intrinsic permeabilities in this range usually exhibit a hydraulic conductivity to water at room temperature ranging between $5*10^{-2}$ and $1*10^{-3}$ cm/sec. These permeabilities are in the range typically reported for sands (Hunt, 1984).

Table 6.7 Young's Modulus Calculated for all Triaxial Tests

Silica Gel Used	Sample Number	Confining Pressure, psi	Max. Deviatoric Stress, psi	Strain% at 50% of Failure	Stress Associated with Strain	Young's Modulus (psi)	Young's Modulus (MPa)
SG 25	3	10	29.56	0.24	14.73	6117	42
Dry	4	20	41.96	0.31	20.42	6547	45
Loose	5	40	93.52	0.60	45.55	7579	52
SG 25	1	10	25.76	0.30	10.59	3501	24
Dry	2	20	50.79	0.41	21.69	5331	37
Dense	5	40	107.77	0.51	52.91	10342	71
	4	60	142.53	0.55	66.81	12241	84
SG 25	6	40	97.01	0.57	47.81	8421	58
Saturated	4	20	39.54	0.36	19.18	5289	36
Loose	2	10	23.40	0.29	9.83	3442	24
SG 25	1	40	110.53	0.58	54.06	9283	64
Saturated	2	20	60.40	0.56	29.30	5238	36
Dense	3	10	40.86	0.61	18.76	3076	21
SG 0.5	1	10	28.12	0.54	13.55	2532	17
Dry	2	20	53.92	0.62	24.35	3926	27
Loose	3	40	84.20	1.18	38.18	3222	22
	4	59.7	110.89	2.61	53.87	2067	14
SG 0.5	1	10	35.62	0.45	16.58	3704	26
Dry	2	20	62.13	0.63	29.35	4627	32
Dense	3	40	94.28	1.11	43.47	3905	27
	4	60	127.86	1.29	58.94	4571	32
SG 0.5	5	60	82.10	1.40	38.78	2762	19
Dry	3	40	97.23	1.97	47.53	2408	17
Dense	2	20	51.97	1.16	24.83	2137	15
	1	10	35.91	0.82	17.62	2142	15
SG 0.5	5	60	153.45	1.61	75.10	4651	32
Dry	1	40	113.05	1.07	50.35	4702	32
Dense	3	20	69.33	0.55	24.95	4561	31
	4	10	44.94	0.77	21.85	2842	20

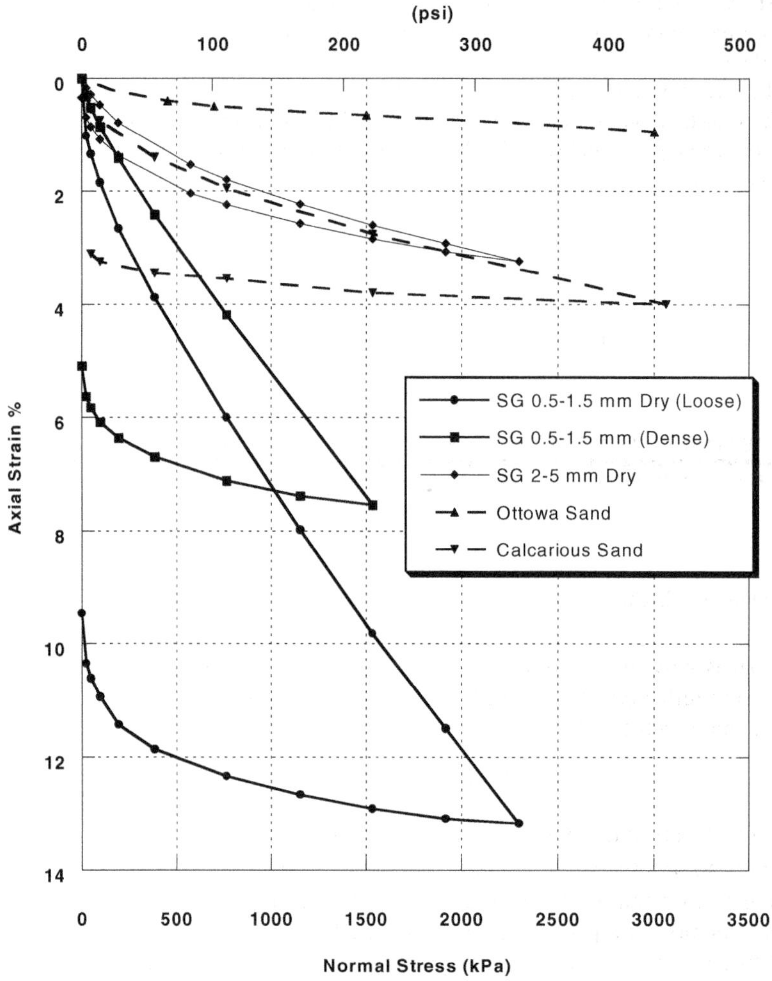

Fig. 6.23 Compressibility of SG1, SG2, Ottawa Sand, and Calcareous Sand from Odometer Tests.

Fig. 6.24 Transparent Silica Gel in a Triaxial Chamber During Permeability Testing

6.6 Dynamic Properties of Silica Gel

6.6.1 Testing Program and Sample Preparation

Resonant column tests were performed by Zhao (2007). A torsional-resonant column apparatus from Geotechnical Consulting and Testing Systems (GCTS) was used for the tests (Fig 6.25). It has a fixed bottom and a rotating top. A torsional drive is used to vibrate the top of the soil specimen at first-mode resonance at frequencies up to 250 Hz while the bottom is fixed. Non-contacting rotational sensor and a torque transducers attached to the top platen monitor the

Fig. 6.25 GCTS Resonant Column Apparatus

torsional motion and torque directly. The non-contacting sensor is located on extended arms to amplify the rotational deformation providing a shear strain resolution of 10^{-6}. The input voltage was adjusted to maintain consistent strain values between tests.

Hardin and Drnevich (1972) summarized that the shear strain amplitude, effective stress level, and void ratio are the primary factors that affect the shear modulus in clean sands. The grain size, grain shape, and degree of saturation have a secondary influence on the shear modulus of sands. The damping ratio was also considered to be affected by these factors. The effective confining stress, void ratio, and strain amplitude were selected as the controlling factors in setting up the test program.

ASTM D 4015 specification was followed as the operating and data reduction procedure. The shear strain level was in the order of 10^{-5} to 10^{-3}. Four gradations of silica gel were tested within the range of 2-5 mm, 1-3 mm, 0.5-1 mm, and 0.5-5 mm. For each gradation, silica gel specimens at two void ratios were prepared, representing the dense and loose states. The confining pressure for each specimen varied from 50 kPa to 100 kPa, 200 kPa, 300 kPa, and 400 kPa. The air pluviation method was used to prepare the loose specimens, where silica gel was poured from a nozzle while maintaining a constant drop-height (4 cm) through the preparation. The confining pressure was applied to the specimen and was raised to the desired level. For dense specimens, the entire specimen was evenly divided into five layers, and each layer was compacted by a miniature tamper. The mass of silica gel for each layer was determined by trial and error. The average specimen diameter and height were 70 mm and 158 mm.

Typical shear modulus degradation with shear strain for tests conducted by Zhao (2007) for loose and dense silica gel under different confining pressures are shown in Fig. 6.26-6.27. The reduction in the shear modulus varied significantly throughout a wide range of shear strains tested. Zhao (2007) reports that the silica gel specimen with a smaller void ratio had a higher shear modulus than did silica gel with a large void ratio under the same confining pressure. Within the linear shear strain range, the shear modulus of silica gel increased with an increase in the confining pressure. However, beyond the threshold shear strain value, the silica gel samples tended to have the same values of shear modulus irrespective of the initial void ratio and confining pressure of the silica gel.

6.6.2 *Shear Modulus of Silica Gel*

The normalized shear modulus (G/G_{max}) is shown along with data from Seed and Idriss (1970) and Rollins et al (1998) in Fig. 6.28. Since the lowest strain level of the silica gel resonant column tests was 10^{-5}, the maximum shear modulus was determined analytically through the equation (Borden et al, 1996)

$$\frac{G}{G_{\max}} = \frac{1}{\left(1 + m\gamma^{n}\right)^{p}} \tag{6.1}$$

where *m, n,* and *p* are model parameters. It is found that the maximum shear modulus for silica gel is nearly identical to the shear modulus at strain level of 10^{-5}. The normalized shear modulus values of silica gel are above the range of values for sand. The normalized moduli fall into a narrow band as typically found in cohesionless soils. Also, the threshold strain level is found to be very consistent for all four gradations and is ranging between 10^{-4} and 2×10^{-4}. Silica gel exhibit a typical linear elastic behavior followed by nonlinear elastic range.

The maximum shear modulus G_{max}, at strain level of 10^{-6}, and four different confining pressures is presented Fig. 6.29. The higher confining pressure, the greater the maximum shear modulus is for each gradation of silica gel. The solid lines in Fig. 6.29 are the best fit curves of describing the relationship between Gmax and confining pressure (σ_m) through the equation (Borden et al 1996)

$$G_{max} = a\sigma_m^{\ b} \tag{6.2}$$

where *a* and *b* are the model parameters. It can be seen that the parameter *a* ranges between 4 and 9 with the mean value about 5, while the parameter *b* ranges between 0.3 and 0.5 with the mean value around 0.4.

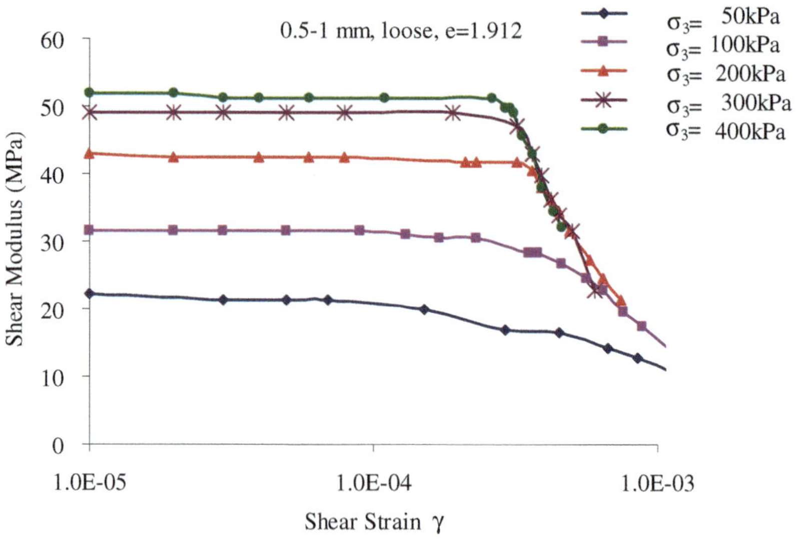

Fig. 6.26 Typical Shear Modulus Degradation with Shear Strain for a 0.5-1 mm Loose Silica Gel Sample under Different Confining Pressures (Zhao 2007)

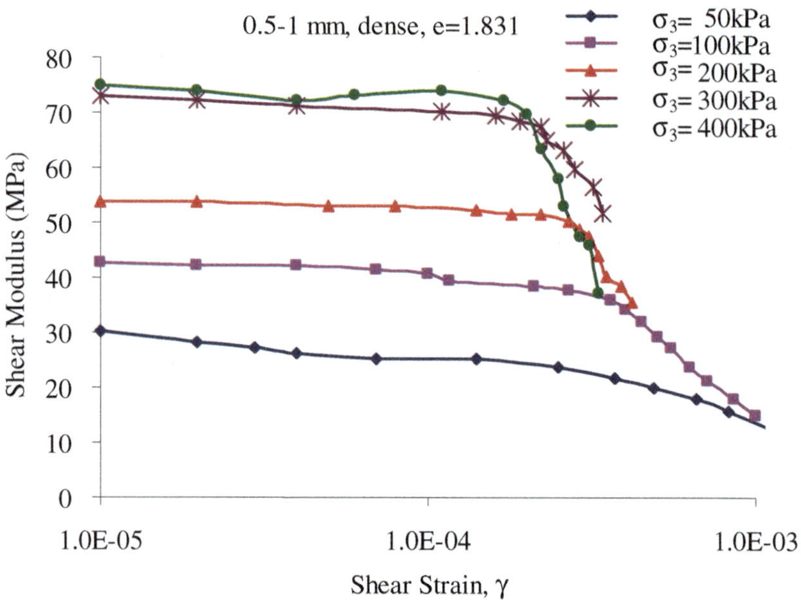

Fig. 6.27 Shear Modulus Degradation with Shear Strain for a 0.5-1 mm Dense Silica Gel Sample under Different Confining Pressures (Zhao 2007)

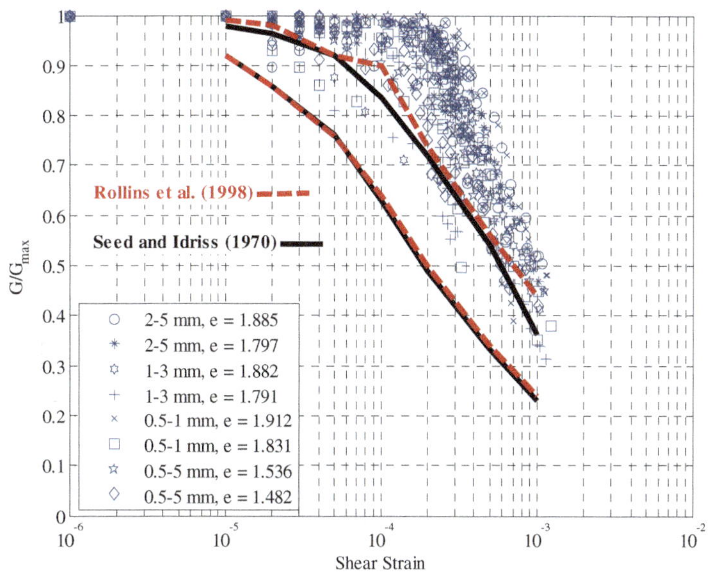

Fig. 6.28 Variation of Shear Modulus for Sand and Silica Gel with Shear Strain at the Confining Pressure of 100 kPa (Zhao 2007)

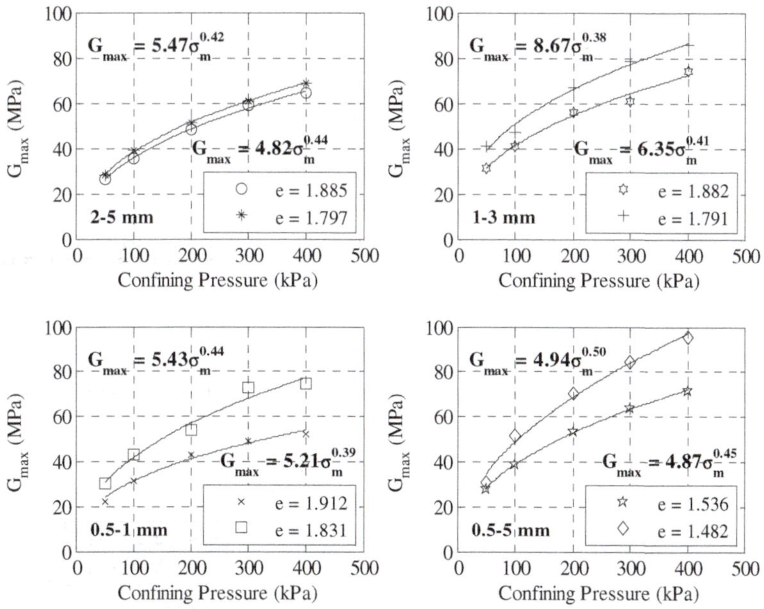

Fig. 6.29 Variation of G_{max} with Confining Pressure (Zhao et al 2010)

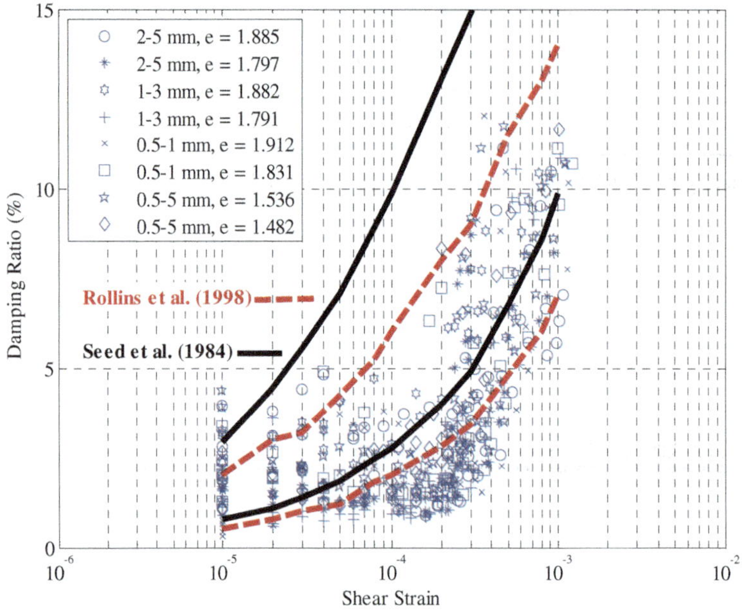

Fig. 6.30 Typical Damping Behavior of Silica GEL (Zhao 2007)

6.6.3 Damping Ratio of Silica Gel

The variation of damping ratio variation with the shear strain for tests conducted by Zhao (2007) is shown in Fig 6.30. Even though the data are scattered, a general trend can still be observed, the damping ratios increases with increase in shear strain amplitude. For both loose and dense silica gel specimens of different grain sizes, the damping ratio increases slowly at a low strain level and then increases quickly with increasing strain amplitude. Investigations carried out by Seed and Idriss (1970), Hardin and Drnevich (1972), and others have shown that the shear strain level and confining pressure are the main factors affecting the damping ratio behavior of sand. The same conclusions can be drawn for silica gel. For sand at the same strain level, increasing the confining pressure causes the damping ratio to decrease.

Damping is induced in sand due to the friction between the sand grains, viscous drag between the pore fluid and sand grains, and plastic deformation of the sand grains. The sand grains are very stiff and tend to keep their elasticity at a high confining pressure, and thus very little energy dissipates through plastic deformation. However, the silica gel grains are much easier to yield compared with sand grains. Under low confining stress, silica gel grains experience very small deformation; energy dissipates primarily through friction between the silica gel grains, and thus it displays similar damping behavior as sand. When the confining pressure becomes large, yielding of silica gel is the dominant energy dissipation mechanism, so silica gel exhibits a different damping behavior from sands. Dense specimens are more difficult to yield than the loose ones. Thus, an increase of the damping ratio for dense specimens was not as quick as that for the loose ones.

6.6.4 Comparison with Results of Sands and Gravels

The relationship between the normalized shear modulus and shear strain and the relationship between the damping ratio and shear strain of silica gel were found to follow the general trend in sands and gravels; however, under a given tested shear strain level, the normalized shear modulus of silica gel tends to be greater and the damping ratio is lower than the typical value of sands and gravel.

With the test findings, silica gel could be used as transparent soil. It also should be born in mind that proper scaling rules need to be defined when carrying out small scale dynamic model tests.

6.7 Modeling Capabilities of Transparent Soils

The main advantage of silica gels described in this chapter and amorphous silica powders described in chapter 5 is that they are the only system which is capable of resembling soils having a wide range of grain sizes including sand and clay using the same pore fluid. For example stratified soil models can be built easily. The transparent soil model shown in Fig. 6.31 was prepared by separately mixing fine silica gels and amorphous silica (Flo Gard SP) with the mineral oil-solvent blend. Both mixes were de-aired under vacuum. The de-aired amorphous silica mix was carefully placed in a transparent container. The amorphous silica mix was consolidated in-place, with vertical drainage allowed at the top only. After consolidation, the silica gel mix was placed carefully, in the pore fluid, on the top of the consolidated amorphous silica. Silica gel was carefully compacted to form the second layer and excess pore fluid was drained.

6.8 Recommended Future Work

Silica gel is produced by a large number of factories. However to date only a limited number of manufacturers were contacted and the geotechnical properties of some of identified materials were fully investigated. This research can be extended for more materials from the same family. For example, the best clarity in all investigated silica was fused silica particles (Fig. 6.32). Unfortunately during our commercial survey for silica products we did not succeed in obtaining more than 1 pound of a sufficiently clean fused silica to allow for producing a large models. The advantage of fused silica is not only its clarity but also that particles are solid thus can be more representative of sand.

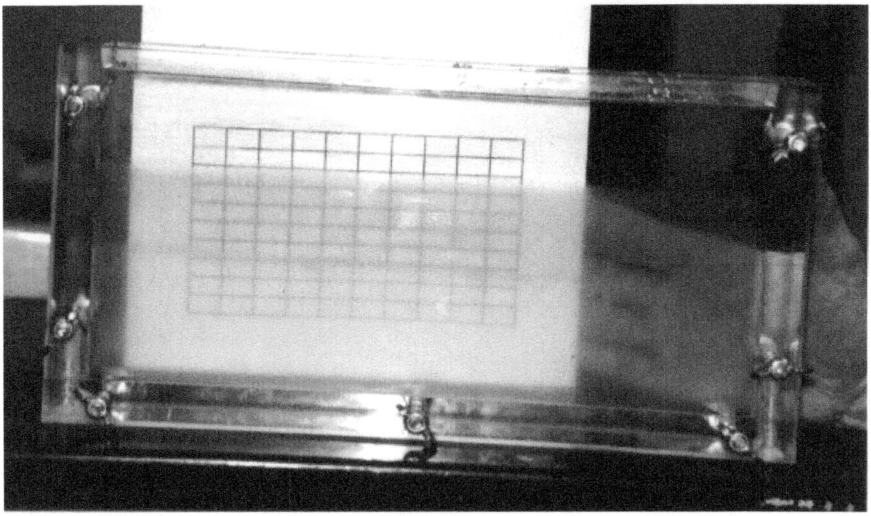

Fig. 6.31 Layered Model Made from Both Silica Gel and Amorphous Silica Powders

Fig. 6.32 Fused Silica Particles (Particle size 2-5 mm)

References

ASTM, Standard D 4015, Standard test methods for modulus and damping of soils by resonant-column method. Annual Book of ASTM Standards, ASTM International, West Conshohocken, PA

Bishop, A., Henkel, D.: The Measurement of Soil Properties in the Triaxial Test, 2nd edn. Edward Arnold Publishers, London (1962)

Borden, R., Shao, L., Gupta, A.: Dynamic properties of piedmont residual soils. Journal of Geotechnical Engineering 122(10), 813–821 (1996)

Casagrande, A.: Characteristics of cohesionless soils affecting the stability of slopes and earth fills. Journal of Boston Soc. of Civil Engineers (1936) (Reprinted in Contributions to Soil Mechanics 1925-1940, Boston Soc. of Civil Engineers)

Casagrande, A.: Liquefaction and cyclic deformation of sands, a critical review. In: Proceedings of the Fifth Pan-American Conference on Soil Mechanics and Foundation Engineering, Buenos Aires, Argentina, January 1976. Harvard Soil Mechanics Series, vol. 88. Prentice Hall, Englewood Cliffs (1975)

Hardin, B., Drnevich, P.: Shear modulus and damping in soils: measurement and parameter effects. J. of Soil Mechanics and Foundations Division, ASCE 98(6), 603–624 (1972)

Hendron, A.: The Behavior of Sand in One-Dimensional Compression. Ph.D. Thesis, The University of Illinois at Urbana-Champaign (1963)

Holtz, R., Kovacs, W.: An Introduction to Geotechnical Engineering. Prentice Hall, Englewood Cliffs (1981)

Horn, H., Deere, D.: Frictional characteristics of minerals. Geotechnique 7, 319–335 (1962)

Hunt, R.: Geotechnical Engineering Investigation Manual. McGraw-Hill, New York (1984)

Iler, R.: The colloid chemistry of silica and silicates. Cornell University Press, New York (1955)

Lamb, T., Whitman, R.: Soil Mechanics. Wiley, New York (1969)

Lee, K.: Triaxial Compressive Strength of Saturated Sand Under Seismic Loading Conditions. Ph.D. Dissertation, The University of California, Berkeley (1965)

Lee, K., Seed, H.: Drained strength characteristics of sands. Journal of The Soil Mechanics and Foundation Division, ASCE 93(SM6), 117–141 (1967)

Mannheimer, R., Oswald, C.: Development of transparent porous media with permeabilities and porosities comparable to soils, aquifers, and petroleum reservoirs. Ground Water 31(5), 781–788 (1993)

Mannheimer, R.: Slurries you can see through. Technology Today, p. 2 (March 1990)

Rollins, K., Evans, M., Diehl, N., Daily, W.: Shear modulus and damping relationships for gravels. J. of Geotechnical and Geoenvironmental Engr. 124(5), 396–405 (1998)

Scott, R.: Silica Gel and Boded Phases. Wiley, New York (1993)

Seed, H., Idriss, I.: Soil moduli and damping factors for dynamic response analysis. Report No. EERC 70-10, Earthquake Engr. Research Ctr., University of California, Berkeley (1970)

Sphinx Adsorbent Inc., Granular silica gel data sheet, 53 Progress Ave. Springfield, MA 01104 (1993)

Weast, R.: Handbook of Chemistry & Physics, 66th edn. CRC Press, Boca Raton (1986)

Zhao, H.: Development and Assessment of Transparent Soil and Particle Image Velocimetry in Dynamic Soil-Structure Interaction. Dissertation, University of Missouri-Rolla (2007)

Chapter 7
Geotechnical Properties of Aquabeads

Magued Iskander and Kazunori Tabe

Abstract. This chapter presents the geotechnical properties of a water-based transparent synthetic polymer, named *Aquabeads*, suitable for modeling some natural soils. A comprehensive study of the geotechnical properties of Aquabeads including permeability, compressibility, and yield stress was performed in order to understand its modeling capabilities. Aquabeads is suitable for representing soils with hydraulic conductivities in the 10^{-2} to 10^{-5} cm/sec range. The strength and compressibility are suitable for representing very weak sediments. Aquabeads has a number of advantages over available transparent synthetic soils for modeling flow. First, it is water based. Second, it possesses similar hydraulic characteristics to natural soils. Third, it is compatible with water, oil, selected surfactants, and alcohols used for modeling multiphase flow. Finally, it does not suffer from transparency degradation in large-scale models. These advantages make Aquabeads a practical alternative for simulating flow in educational settings including undergraduate labs and demonstrations in high schools.

7.1 Introduction

In recent years, several experimental studies have been performed using transparent synthetic soils. Glass beads have been widely used in research of 2D flow problems (Corapcioglu and Fedirchuck 1999, Huang et al 2002, Lunati et al 2003, Theodoropoulou et al 2003, and Gaganis et al 2005). Amorphous silica gels and powders is another good tool for studying flow characteristics and soil structure interaction as described elsewhere in this book.

Although glass beads and amorphous silica possess potential advantages in modeling many flow and soil structure interaction situations, they still have some limitations. The pore sizes, porosities, and permeability of glass beads are not representative of natural soils and engender difficulties in simulating flow transport process. Large scaled models using glass beads are also limited by their poor transparency. In addition, in order to achieve transparency the refractive index of the solids and pore fluids must be matched using synthetic pore fluids such as mixtures of mineral oil and paraffinic solvents, alcohols, or calcium bromide brine. These pore fluids are difficult to handle and use and often present difficulties in long term tests due to change in optical properties with time. For

M. Iskander: Modeling with Transparent Soils, SSGG, pp. 117–135.
springerlink.com

example, Welker et al (1999) demonstrated that the degradation of transparency can be a problem of large amorphous silica powder samples.

Developing a new water based transparent soil is desirable to overcome the limitations of glass beads and available transparent soils made of amorphous silica powders and precipitated silica gels.

7.2 What Is Aquabeads?

Aquabeads is a strong water adsorption polymer sold under the commercial name KI-GEL 201K. It has the same refractive index like water and is produced by *Kuraray Chemical Company, Osaka, Japan.* Aquabeads is composed of isobutylene and maleic anhydride copolymer, and is delivered as light yellow spherical particles, having a density of 980 g/l, water content of 7% or less, and a pH of 9-10. Although the dry appearance of Aquabeads is yellowish they become very clear (transparent) after absorbing water (Fig.7. 1).

According to Kuraray (1998), Aquabeads was originally designed for ground improvement purposes such as stabilization of soft soils or mud by mixing with in-situ soils that contain high water contents during excavation. In addition, this resin is utilized to manufacture chemical pocket heaters, and perfume carriers.

Aquabeads can absorb water up to 200 times their own weight (Tabe 2009). They have good durability and stable water absorption under different temperature conditions (Table 7.1). Once Aquabeads absorbed water, it releases minimal water even under a pressurized condition, as long as the ambient stress remains constant. For example, Kuraray (1998) reports that adsorption increased by only 2.5% after immersion in water for 899 days. In this study, water absorption was controlled to achieve (1) stable volume, (2) full saturation, and (3) extreme transparency.

Fig. 7.1 No Smoking sign viewed through 2.8 cm thick Aquabeads 200 in a Plexiglas mold with internal dimensions = 40x2.8x14.7 cm

Table 7.1 Durability of fine Aquabeads (Kuraray 1998)

Test	Test conditions	Test period	Change of absorbency (ml/g)
Stability at room temperature	10 to 30 °C	2 years	198 • 201
	75 to 85 % humidity	13 months	190 • 197
Immersion in liquids	Cool water	899 days	194 • 199
	70 °C hot water	30 days	193 • 207
	100 °C boiling	3 days	200 • 218
	30% NaOH at 80 °C	10 days	14 • 14
Heating	140 °C	24 hours	200 • 190
	180 °C	24 hours	200 • 180

Research using two types of Aquabeads (Table 7.2) is reported in this study, but other types are available, as follows:

1. ***Aquabeads M*** having a dry diameter of 0.6-2 mm. When water is added to Aquabeads M the particles reach a size of approximately 5-7mm (Fig 7.2). These particles were sometimes crushed after being fully cured in order to reduce their particle size. The crushed particles were oven dried at 100 °C for 24 hours and the resulting blueish material was re-hydrated with water. This was done to produce a material with a lower permeability. The uncrushed Aquabeads are referred to as *NC100.* Two types of crushed Aquabeads are also discussed, *C50* and *C75.*
2. ***Aquabeads 200*** is delivered as a powder passing #200 sieve. It becomes a transparent gel when it is mixed with water. The percentage of Aquabeads (by weight) in the mixture is indicated as either 200-0.5% or 200-1% in Table 7.2.

Table 7.2 Aquabeads used in this Study

Designation	Type	% Aquabeads in Mixture by weight	Crushing
NC100	M	0.5%	No
C50	M	1%	Yes
C75	M	0.66%	Yes
200-0.5%	200	0.5%	No
200-1%	200	1%	No

Fig. 7.2 Types of Aquabeads – (Top) As-delivered Specimens; (Bottom) Water Hydrated Specimens (From left to right: C50, 200-1%, and NC100)

7.3 Grain Size Distribution

Sieve analysis of dry and hydrated Aquabeads is shown in Fig.7.3. Tests were performed using the dry method for as delivered Aquabeads (ASTM D1921 - 06 Standard Test Methods for Particle Size (Sieve Analysis) of Plastic Materials), and the wet method (ASTM C92-95 (2005) Standard Test Method for Sieve Analysis and Water Content of Refractory Materials) for hydrated Aquabeads. The grain size distributions of both dry and hydrated Aquabeads exhibit poorly graded or uniform behavior.

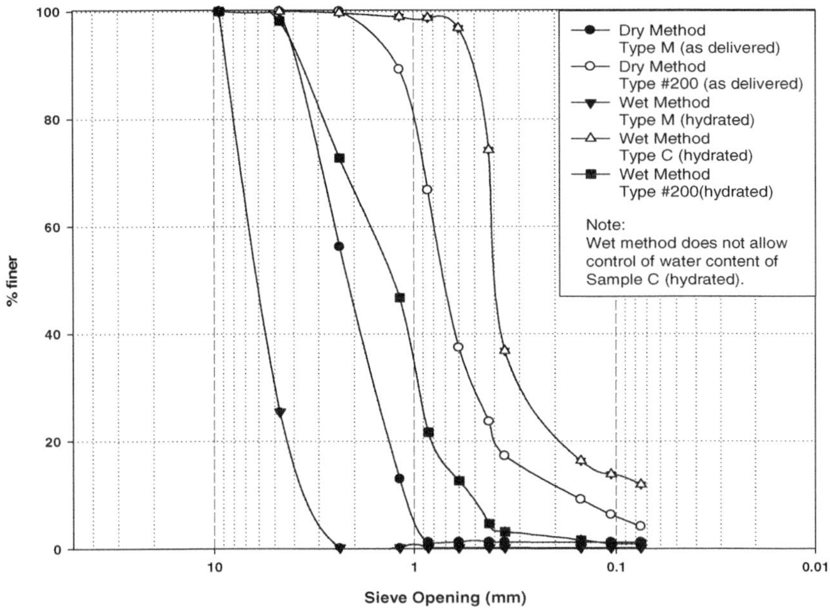

Fig. 7.3 Grain size distribution of Aquabeads

7.4 Hydraulic Conductivity of Aquabeads

The hydraulic characteristics of various types of Aquabeads were investigated (Tabe 2009). Fixed wall permeability tests indicate that the hydraulic conductivity of different types of Aquabeads ranges from 7×10^{-2} to $2\text{x}10^{-5}$ cm/sec (Table 7.3). Flexible wall hydraulic conductivity tests were also performed to study the variation of hydraulic conductivity of Aquabeads after being consolidated at different confining pressures (Fig 7.4). Since Aquabeads are made of a water based soft polymer, they undergo volume change due to small variations in confining pressure. For example, when a 20 kPa (2.9 psi) confining pressure was applied, specimens of Aquabeads experienced a volume reduction of 6.9% for NC100, 8.6% for C50, and 10.5% for #200–0.5%. Once the volume is decreased, the hydraulic conductivity is reduced. The consolidation behavior of Aquabeads is important to research in future studies.

The modeling conditions in this study are similar to fixed wall tests without a confining pressure. In comparison to the conductivity of natural soils (Table 7.4), the Aquabeads reported in this study are suitable for representing the hydraulic conductivities of sands and silts. Consolidated Aquabeads are suitable for modeling clays. Fine Aquabeads which may also be suitable for representing clays are available, but were difficult to de-air and make transparent in our initial studies.

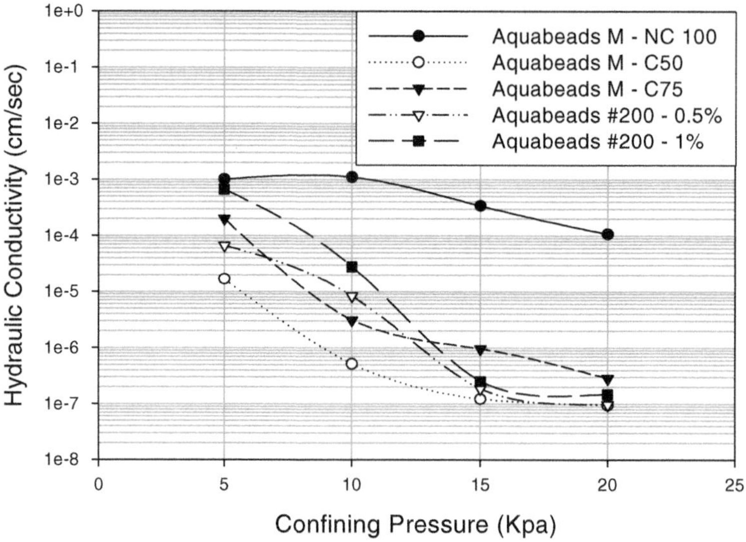

Fig. 7.4 Results of Flexible Wall Hydraulic Conductivity Test

Table 7.3 Summary of Fixed Wall Hydraulic Conductivity Tests

Types of Aquabeads	Hydraulic Conductivity (cm/sec)		Suitable for Representing
	Constant Head Test	Falling Head Test	
NC100	2.81×10^{-2}	7.34×10^{-2}	Sand
C50	1.02×10^{-5}	2.44×10^{-5}	Silt
C75	1.45×10^{-4}	1.85×10^{-4}	Silt
#200 – 0.5 %	5.61×10^{-4}	4.18×10^{-4}	Silt
#200 – 1 %	2.68×10^{-4}	2.54×10^{-4}	Silt

Table 7.4 Typical Range of Hydraulic Characteristics for Various Natural Soils

Types of Soils	Hydraulic Conductivity (cm/sec)	Intrinsic Permeability (Darcy)	Porosity
Clay	$10^{-9} \sim 10^{-6}$	$10^{-6} \sim 10^{-3}$	0.35 ~ 0.65
Silt	$10^{-6} \sim 10^{-4}$	$10^{-3} \sim 10^{-1}$	0.35 ~ 0.60
Sand	$10^{-3} \sim 10^{-1}$	$1 \sim 10^{2}$	0.25 ~ 0.55
Gravel	$10^{-2} \sim 1$	$10 \sim 10^{3}$	0.20 ~ 0.40

7.5 Compressibility of Aquabeads

Aquabeads changes its volume by approximately 200 times from its dried particle size when it absorbs water. Aquabeads particles are porous. The consolidated specimens exhibit a high apparent total void ratio, *e*, due to the internal porosity of the Aquabeads.

Hydrated Aquabeads samples were used in one dimensional consolidation tests in accordance to ASTM D2435-90(1993): Standard Test Method for One-Dimensional Consolidation Properties of Soils. Samples were 38.1 mm (2.5 in) in diameter and 19 mm (0.75 in) high. Tests were conducted using conventional oedometer apparatus using fixed weights. The results of these tests are summarized in Table 7.5.

Table 7.5 Consolidation Properties of Aquabeads

Sample No.	C_c	C_r	C_r/C_c	Total Void Ratio, e	Inter-aggregate Void Ratio, e_i	Preconsolidation Pressure, psf
NC100	0.15	0.03	0.20	197.48	0.26	103
C50	0.10	0.02	0.20	127.8	0.19	132

7.5.1 Void Ratio

Aquabeads exhibit a high apparent total void ratio, *e*, due to the internal porosity of the Aquabeads aggregates. The inter-aggregate void ratio, e_i, is more representative than the total void ratio, *e*, for geotechnical purposes, since it only considers the volume in-between the aggregates. The inter-aggregate void ratio (effective porosity) can be computed using Eq. 5.1.

7.5.2 Consolidation Behavior

Typical consolidation volume change versus time is shown in Fig. 7.5. The volume change was calculated based on the sample settlement. Volume change continued to increase with time and the end of primarily consolidation was not readily identifiable. This phenomena may be explained with the existence of two types of voids within and between the Aquabeads. The first consolidation process involves the pores between the Aquabeads, while the second involves the pores inside the Aquabeads. Peat and organic clays have a macro-pore structure corresponding to the voids within the inorganic component in them, and a micro-pore structure corresponding to the pores within the organic matter component (Buisman 1936, Adams 1965, Hobbs 1986, and Hanrahan 1954). Similar behaviors were reported in those soils.

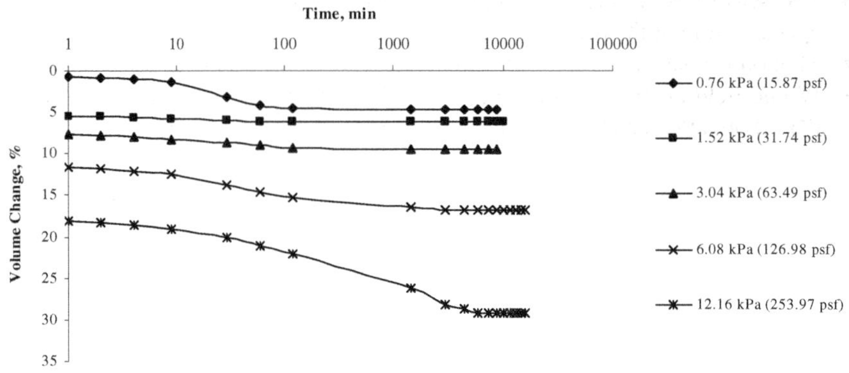

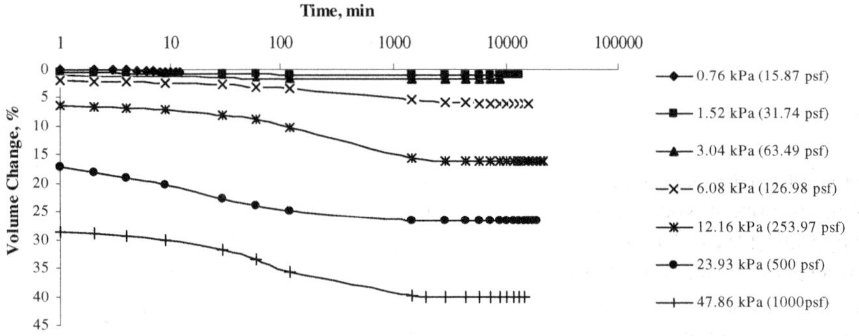

Fig. 7.5 Volume Change vs. Time for (Top) NC100 and (Bottom) C50

7.5.3 Consolidation Indices

Typical e-log p curves from oedometer tests are shown in Fig. 7.6 for two types of Aquabeads. Void ratios were taken 24 hours after loading because of the difficulty in identifying the end of primary consolidation. The compression and recompression indices (*Cc* and *Cr*) are 0.1-0.15 and 0.02-0.03, respectively (Table 7.5). Both indices are within the range typically reported such as Ft. Gordon Clay, Georgia: *Cc*=0.12; New Orleans clay: *Cr*=0.05; Montana clay: *Cr*=0.05 (Das 1998), and Montmorillonite minerals (Lambe and Whitman 1969). The ratio *Cr/Cc* is 0.2 which is within the normal range of 0.02 to 0.2 for natural soils (Terzaghi et al 1996).

The coefficient of consolidation, *Cv*, for NC100 and C50 is plotted versus vertical stress normalized by pre-consolidation stress in Fig. 7.7. This figure also shows the relationship in some inorganic and organic soils. Change in *Cv* is observed in both inorganic and organic soil data in the normally consolidated (NC) region (v/p=1). This change in *Cv* around NC region can also be seen in NC100 and C50. For Middletown peat and Celery Bog organic soil, *Cv* markedly decreases with increasing vertical stress, however, Boston Blue Clay gradually increases with increasing vertical stress. The change in *Cv* after NC region for Aquabeads samples cannot be clearly identified but represents intermediate characteristics between inorganic and organic soils.

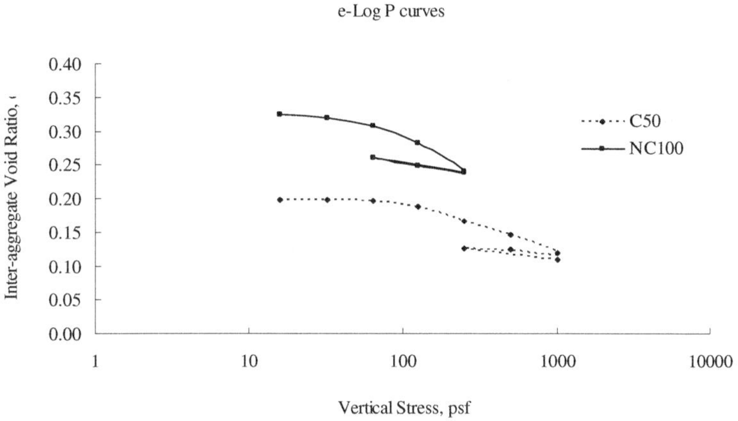

Fig. 7.6 Typical e-log p Curves for Aquabeads from Oedometer Tests

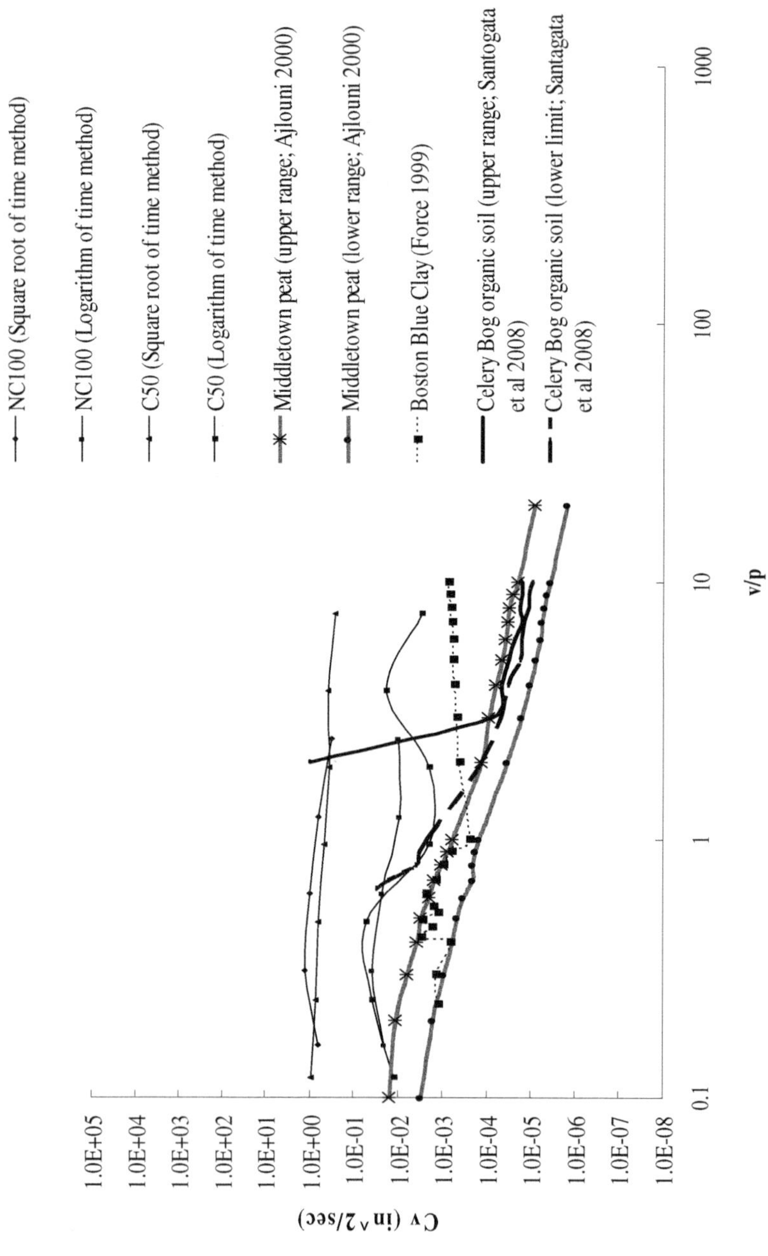

Fig. 7.7 Variation of the Coefficient of Consolidation with Stress Level for Aquabeads and some Natural Soils

7.6 Strength of Aquabeads

7.6.1 Yield Stress

Aquabeads are very weak. They exhibit solid-like behavior at low stresses and liquid-like behavior at high stresses. It was not possible to form specimens for conventional triaxial measurement of shear strength. Therefore, the yield stress was employed as a measure of strength. The yield stress of a material is defined as the minimum shear stress that must be applied to the material to induce flow (Liddell and Boger 1996).

Generally, elasticity is associated with solid-like behavior where the material strains to a fixed value when subjected to a certain stress, and there is complete strain recovery upon removal of that stress. On the other hand, viscosity is associated with liquid-like behavior such that the material strains continuously when subjected to a certain stress, and there is no strain recovery upon removal of that stress. The yield stress is generally regarded as the transition stress between elastic solid-like behavior and viscous liquid-like behavior.

The most common example of materials which exhibit yield stresses is a suspension in which solid particles are dispersed in a liquid medium. The particles in the suspension interact with each other to create a continuous, three-dimensional network, a flocculated structures, which can preclude flow at low stresses. The suspension flows only when the stress is large enough to break down the flocculated structure. Based on this phenomenon, the yield stress is related to the strength of the coherent network structure as the force per unit area required to break down the flocculated structure. Examples of yield stress materials include cement, soils, paints, pastes, printing inks, greases, pharmaceutical creams and ointments, and a large variety of food products such as dressings, sauces and spreads.

Fakher et al (1999) defined super soft clay as insensitive clays with a very high water content and very low yield stress that neither behaves like a liquid nor like a solid. Measurement of yield stress has many practical applications. For example, Yano et al (1985) reported that in land-reclamation projects, an established technique is to introduce a primary construction stage that is used as a platform to provide access for conventional ground improvement techniques. A working platform can be produced by laying a layer of geosynthetic reinforcement over the super soft clay and covering it with a layer of cohesionless fill. According to Inoue (1990), formation of the sand layer over super soft clay, with or without the aid of a geosynthetic layer, can be difficult and depends on the yield stress of the clay. The yield stress of the clay is also an important factor in the formation of mud waves in construction processes involving the placement of fills on super soft soils (Broms 1987).

7.6.2 Yield Stress Measurement

There are only a limited number of methods that may be used to measure the yield stress of super soft clays and Aquabeads. The yield stress can be directly measured using a laboratory vane. A typical vane consists of four thin blades arranged at equal angles around a small cylindrical shaft. After the vane is immersed in the soft clay, it is rotated at a constant rotational speed. The resulting stress it measured as a function of time. The yield stress is then determined as the maximum stress in the stress-time profile.

The measurement of yield stress has been driven by the vane geometry. The vane geometry offers some advantage in the measurement of the yield stress (Liddell and Boger 1996). The vane circumscribes a localized cylindrical surface along which yielding occurs. So the material yields within itself, so all the problems and errors associated with slip flow on smooth surfaces are absent. Another advantage of the vane geometry is that the introduction of the vane into the suspension does not cause significant disturbance to the sample prior to measurement, which is particularly important for thixotropic suspensions.

In general, the larger the diameter of the outer cylinder the more accurate the results because a larger torque is measured. Whorlow (1992) mentioned the following assumptions are made or implied when using a rotary viscometer:

1. The cylinders are long enough to ensure that end effects due to turbulence can be ignored.
2. There is no slip between the slurry and the cylinders.
3. Each fluid particle on a cylindrical surface of radius, *r*, moves at a constant angular velocity, which is a function of r.

7.6.3 Yield Stress Calculation

When the vane is set in motion at a constant speed, the total torque (T) experienced by the shaft is the sum of shearing on the side (T_s) and the two ends ($2T_e$) of the cylindrical shear surface. Thus,

$$T = \left(\tfrac{\pi}{2} D^2 H\right)\tau_s + 2\left(2\pi \int_0^{D/2} \tau_e r^2 dr\right) \tag{7.1}$$

where τ_s is the constant shear stress along the side of the sheared cylinder, and τ_e is the shear along the ends. At the yielding moment, when $T=T_m$, (maximum torque) and the shear stresses are typically assumed to be $\tau_s = \tau_y$.

Two assumptions are employed for conventional yield stress calculation when the torque is at maximum (T_m). The first assumption is that the material yields along a cylindrical surface having an area where D and H are diameter and length of the vane, respectively. The second assumption is that along the ends τ_e increases linearly from zero at the center to a value τ_y at the edge (r=D/2), as follows:

$$T_m = \frac{\pi D^3}{2}\left(\frac{H}{D} + \frac{1}{3}\right)\tau_y \tag{7.2}$$

Dzuy and Boger (1985) proposed two methods of yield stress analysis. The first proposal (Method A) is based on the assumption that the end shear stress is distributed. The second proposal (Method B) assumes that only the contribution of the wall shear stress to the total stress is considered. In Method A τ_e may be arbitrarily assumed to vary with radial position (r) according to a power relationship:

$$\tau_e(r) = \left(\frac{2r}{D}\right)^m \tau_s, \quad 0 \le r \le \frac{D}{2} \tag{7.3}$$

where m is a constant describing the radial distribution function of τ_e. At the time of yielding, the solution of Eq. 7.1 becomes:

$$T_m = \frac{\pi D^3}{2}\left(\frac{H}{D} + \frac{1}{m+3}\right)\tau_y \tag{7.4}$$

To determine both τ_y and m simultaneously from experimental data, Eq. 7.4 may be rearranged in the following form:

$$\frac{2T_m}{\pi D^3} = \tau_y \frac{H}{D} + \frac{\tau_y}{m+3} \tag{7.5}$$

Method B makes use of the general torque balance in Eq. 7.1 at yielding. Thus, for vanes of the same diameter but different lengths, the second term in Eq. 7.5 should be constant, and an experimental plot of T_m versus H would be linear for a given suspension. From the slope of this line, the yield stress can be readily calculated, regardless of the nature of the end shear distribution.

7.6.4 Yield Strength of Aquabeads

The measurement of yield stress of Aquabeads was performed using a Wykeham Farrance WF23500 Laboratory Vane (Wykeham 1999). The vane is connected to the motor via a spring (Fig 7.8). When rotation of the motor is engaged the spring winds up, thereby exerting an increasing torque on the vane. Prior to yielding of the material, the spring begins to unwind causing the vane to rotate within the material. The maximum torque is related to the maximum rotation of the spring. this is marked with a mechanical spindle.

The rotational speed of the vane was fixed to 0.0029 rad/sec (0.03%/sec). Two vane sizes were used to measure the stress. The first vane was 12.7 mm diameter x 12.7 mm height (D/H=1; Vane No. A) The second was 12.7 mm diameter x 25.4 mm height (D/H=2; Vane No. B).

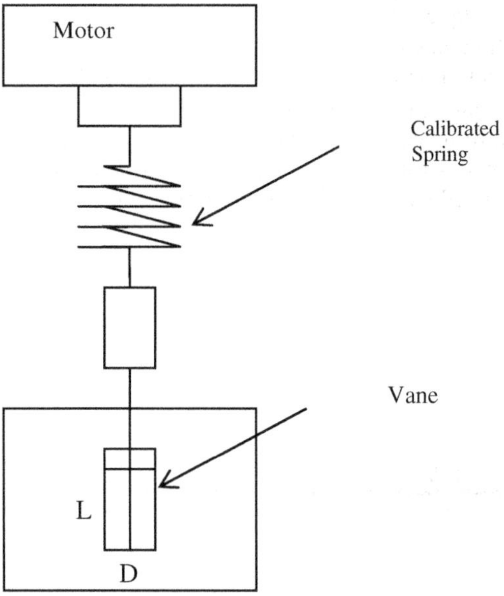

Fig. 7.8 The mechanism of vane test equipment

To make test samples, dried Aquabeads and de-aired water were mixed with the appropriate ratio in a glass beaker whose height and diameter meet criteria established for the satisfactory measurements with the vane method. Samples were stored 24 hours under vacuum, to remove all entrapped air. After observing no entrapped air in the samples, samples were placed in the viscometer (vane device). The height of test samples was constant in all tests.

Tabe (2009) measured the yield stress of many varieties of Aquabeads (Table 7.6). The yield stress measurement of the each sample was performed 20 times to average out mechanical and sample preparation errors (Fig. 7.9). Aquabeads samples generally behave as an elastic solid up to 70 sec. Yielding of the samples generally occurs at 450 to 1000 sec, where torque reaches a maximum value. The torque declines slowly or remains constant after yielding. As expected, the peak stress becomes higher with lower water absorption for all specimen types (Fig. 7.10).

The yield stress results from the conventional method are generally larger than the yield stress obtained from both Method A and B. The maximum differences of values are recorded for #200-1.0% and #200-1.5%, respectively. The empirical index, *m*, calculated in Method A varies widely from -2.098 to 10.674. According to Dzuy and Boger (1985), a zero-value for m suggests a uniform distribution of the stress on the two end surfaces and the side of the yield cylinder. These results, however, indicates that the usual assumption of a uniform stress distribution everywhere on the sheared surface may not be reasonable, at least at the moment of yielding.

Table 7.6 Summary of Yield Stress Calculations (Tabe 2009)

Sample #	Method A		Method B	Conventional Method	
	τy (Pa)	m	τy (Pa)	τy (Pa)	Standard Deviation (Pa)
C25	7.59	-1.465	7.73	9.0	0.54
C50	7.83	0.902	7.73	7.5	0.15
C75	7.35	10.674	7.73	6.2	0.45
#200-0.5%	4.9	-1.888	3.87	6.2	0.67
#200 -1.0%	4.9	-2.098	3.87	7.1	0.86
#200 -1.5%	5.63	-1.915	3.87	7.6	0.75
#200 -2.0%	8.57	-0.965	7.73	9.4	0.3

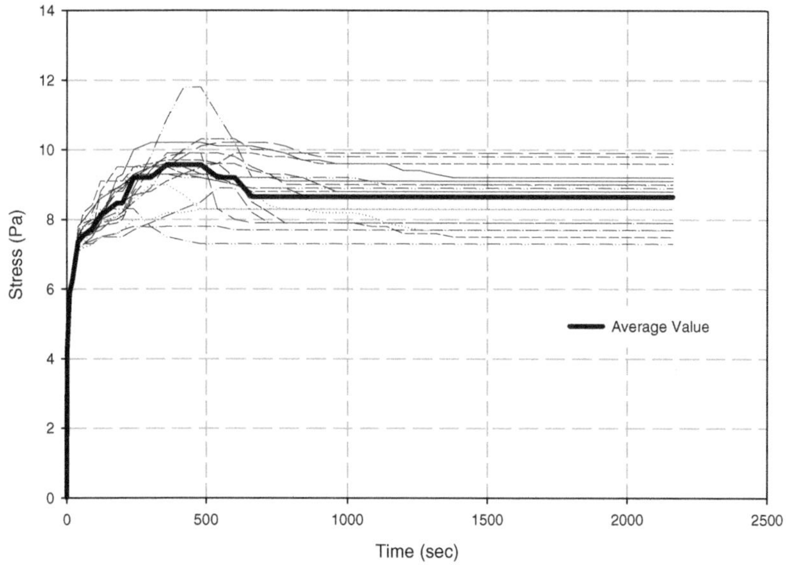

Fig. 7.9 Typical Repeatability of Vane Tests on Aquabeads (#200-2%)

The results calculated using Method B appear to be the correct way for calculating the yield stress from vane data since the end shear stress is avoided in the analysis. However, the procedure involved is tedious because more than two vanes of equal diameter are needed to construct a plot of T_m versus H. So, the accuracy achieved is not greatly improved when compared with the conventional method which requires only a single vane test.

The conventional method is thus more preferable for practical work because it represents a simple, quick, and direct method for yield stress measurement.

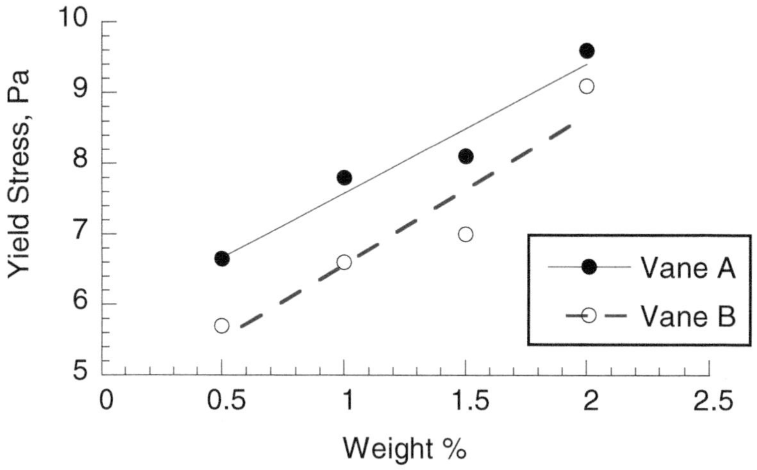

Fig. 7.10 Typical Relationship between Yield Stress and Weight % of Aquabeads

7.6.5 Comparison to the Yield Stresses of Natural Soils

Most research on super soft clays (Tsuruya et al 1986, Inoue 1990) involves a series of shear stress measurements, at low strain rates. A plot of shear stress versus strain rate shows a linear relationship between shear stress and strain rate. Fakher et al (1999) conducted a series of shear stress measurements at different small strain rates (<7%/sec) to estimate the yield stress using an assumed model.

The relationship between the measured shear stress and strain rate for several types of super soft clays is plotted along with measured shear stress of Aquabeads in Fig. 7.11. There is a scatter of super soft clay data when measuring shear stress at different strain rates. The scatter seemed to reduce as the water contents of natural soils reduced because the values of shear stress were higher. All measured yield (shear) stresses of Aquabeads were plotted at 0.03%/sec because the strain rate was fixed in this study. Nevertheless, the range of measured shear stresses of Aquabeads is within the range of measured shear stress for natural super soft clays, except for Brown Till (w=70%).

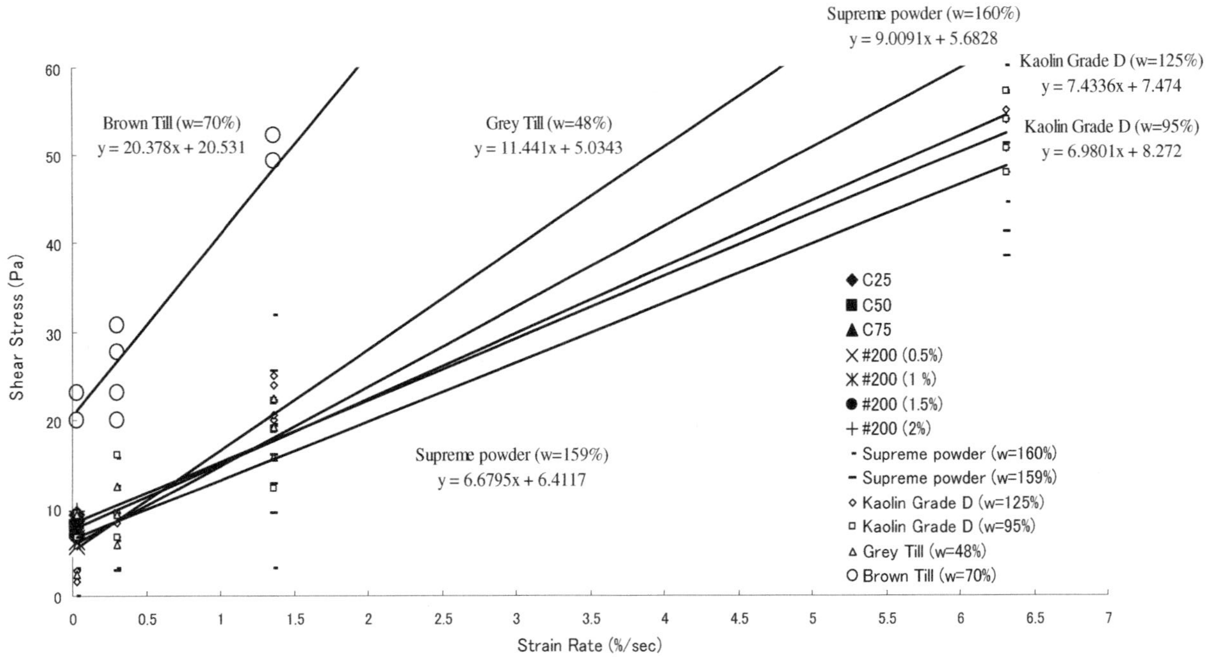

Fig. 7.11. Shear Stress Measurement on Aquabeads Compared with Data for Natural Soils from Fakher et al (1999)

7.7 Conclusions

Although available transparent soils made of silica gels and silica powder have been implemented with good results in modeling 2D flow and soil structure interaction, some difficulties with their application exist. Aquabeads were developed as a new family of water-hydrated transparent synthetic soils. Aquabeads have four advantages which make them easier to use in educational settings such as undergraduate labs and high schools. First, they are more transparent after absorbing water than silica gel and silica powder. Second, they have similar hydraulic characteristics to natural sands and silts. Third, they are compatible with water, oil, surfactants, and alcohols used for modeling. Finally, they do not exhibit transparency degradation in large scale models.

Permeability test results indicate that the hydraulic conductivities of the tested Aquabeads are consistent with fine sand to silty clay.

The compression and recompression indices of tested Aquabeads (Cc and Cr) are 0.1-0.15 and 0.02-0.03, respectively, which is within the range reported for natural soils. Volume change continued to increase with time and the end of primary consolidation was not readily identifiable due to the existence of a dual porosity structure, like many peats and organic clays.

The yield stress of Aquabeads was obtained using a laboratory vane. The yield stress ranged from 4 to 9 Pa which is similar to values reported for super soft natural clays.

In summary, Aquabeads are suitable for representing soils with hydraulic conductivities in the 10^{-2} to 10^{-5} cm/sec range. The strength and compressibility are suitable for representing very weak sediments.

References

Ajlouni, M.: Geotechnical properties of peat and related engineering problems, Ph.D. Dissertation, University of Illinois at Urbana-Champaign (2000)

Adams, J.: The engineering behavior of a Canadian Muskeg. In: Proc. 6th Int. Conf. Soil Mech. & Found. Eng., vol. 1, pp. 3–7 (1965)

Broms, B.: Stabilization of very soft clay using geofabric. Geotextiles and Geomembranes 5(1), 17–28 (1987)

Buisman, A.: Results of long duration settlement tests. In: Proc. 1st Int. Conf. Soil Mech. & Found. Eng., vol. 1, pp. 103–105 (1936)

Corapcioglu, M., Fedirchuk, P.: Glass bead micromodel study of solute transport. Contaminant Hydrology 36(3-4), 209–230 (1999)

Das, B.: Principles of Geotechnical Engineering, 4th edn. PWS Publishing Company, MA (1998)

Dzuy, N., Boger, D.: Direct yield measurement with the vane method. Journal of Rheology 29(3), 335–347 (1985)

Fakher, A., Jones, C., Clarke, G.: Yield stress of soft clays. J. of Geotechnical and Geoenvironmental Engineering, 499–509 (1999)

Force, E.: Factors controlling pore-pressure generating during Ko-consolidation of laboratory tests, MS Thesis, Dept. of Civil and Environmental Engineering, MIT (1998)

Gaganis, P., Skouras, E., Theodoropoulou, M., Tsakiroglou, C., Burganos, V.: On the evaluation of dispersion coefficients from visualization experiments in artificial porous media. J. of Hydrology 307(1-4), 79–91 (2005)

Hanrahan, E.: An investigation of some physical properties of peat. Geotechnique 4(3), 108–123 (1954)

Hobbs, N.: Mire morphology and the properties and behavior of some British and foreign peats. Quarterly J. of Engineering Geology, London 19(1), 7–80 (1986)

Huang, W., Smith, C., Lerner, D., Thornton, S., Oram, A.: Physical modeling of solute transport in porous media: evaluation of an imaging technique using UV excited fluorescent dye. Water Research 36(7), 1843–1953 (2002)

Inoue, T.: Behavior and treatment of dredged clay, Ph.D. Thesis, National University of Singapore, Singapore (1990)

Kuraray America Inc., Water absorbent resin KI-GEL 201K-F2, Technical document, Kuraray America, Inc., 200 Park Avenue, New York, NY10166 (1998)

Lambe, T., Whitman, R.: Soil Mechanics. Wiley, New York (1969)

Liddell, P., Boger, D.: Yield stress measurements with the vane. J. of Non-Newtonian Fluid Mechanics 63(2-3), 235–261 (1996)

Lunati, I., Kinzelbach, W., Sorensen, I.: Effect of pore volume-transmissivity correlation on transport phenomena. Contaminant Hydrology 67(1-4), 195–217 (2003)

Santogata, M., Bobet, A., Johnston, C., Hwang, J.: One-dimensional compression behavior of a soil with high organic matter content. J. Geotech. and Geoenvir. Engrg. 134(1), 1–13 (2008)

Tabe, K.: Aquabeads to model the geotechnical behavior of natural soils, Ph.D. Dissertation, Polytechnic Institute of NYU, NY, USA (2009)

Terzaghi, K., Peck, R., Mesri, G.: Soil mechanics in engineering practice, 3rd edn. John Wiley and Sons, New York (1996)

Theodoropoulou, M., Karoutsos, V., Kaspiris, C., Tsakiroglou, C.: A new visualization technique for the study of solute dispersion in model porous media. J. of Hydrology 274(1-4), 176–197 (2003)

Tsuruya, H., Nakano, S., Tanaka, J.: Investigation in rheological properties of soft muds with a rotary viscometer, Res. Rep. No. 566, Port and Harbor Research Institute, Ministry of Transport, Japan (1986)

Welker, A.J., Bowder, J., Gilbert, R.: Applied research using transparent material with hydraulic properties similar to soil. Geotechnical Testing J., GTJODI 22(3), 266–270 (1999)

Wykeham, WF23500 Laboratory Vane Equipment Handbook, Wykeham Farrance International (1999)

Whorlow, R.: Rheological techniques, p. 460. Ellis Horwood, Chichester (1992)

Yano, K., Watari, Y., Yamanouchi, T.: Earthwork on soft clay grounds using rope-netted fabrics. In: Proc. Recent Developments in Ground Improvement Techniques, pp. 225–237 (1985)

Chapter 8
Digital Image Correlation

Jinyuan Liu and Magued Iskander

Abstract. Digital image correlation (DIC) is a classic pattern recognition technique that is widely used for experimental measurements, mainly for: (1) surface displacement analysis in solid mechanics, and (2) particle image velocimetry (PIV) in fluid mechanics. DIC is based on using a correlation function to locate the best matching position of two images thus predicting movements. An advanced form of DIC called adaptive cross correlation (ACC) has been adapted for this research. In this chapter, the basis of the technique and its application are explained. The anticipated errors of using DIC and ACC are also evaluated.

8.1 Introduction

Motion detection from a sequence of images is of interest to many fields, such as, tracking of targets in the military field (Markandey et al 1996), heart motion estimation from magnetic resonance images in the medical field (Loncaric and Majcenic 2000), traffic monitoring in transportation engineering (Leuck and Nagel 1999), estimation of the movement of crowds in traffic engineering (Bouchafa et al 1997), obstacle detection in robot development (Willersinn and Enkelmann 1997), and estimation of the movement and intensity of cyclones in meteorology (Hossain et al 1999). Digital image cross-correlation (DIC) has been adapted for use in this research to measure spatial deformations in transparent soil models.

8.2 Digital Imaging

8.2.1 Digital Image Format

The digital image format is illustrated in Fig. 8.1. This example image is a grayscale image of the Moon, where 0 is black, 255 is white, and the intermediate values are shades of gray. The image shown is represented by 192,246 samples arranged in a two-dimensional array of 358 columns by 537 rows. In image processing, each sample is called a pixel. Each pixel in this grayscale image is an integer number between 0 and 255. This number is related to the amount of energy received from the sensor during digital photography. It represents the light intensity change in the object.

M. Iskander: Modeling with Transparent Soils, SSGG, pp. 137–164.
springerlink.com

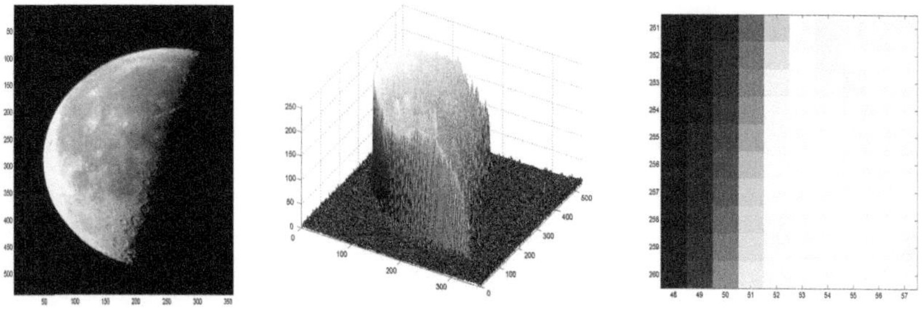

(a) (b) (c)

Column Number

Row Number	48				52					57
251	24	29	36	98	205	249	246	249	249	248
	22	29	46	117	226	250	247	246	248	248
	19	24	50	131	243	250	247	246	248	248
	23	30	58	148	247	250	248	247	248	248
255	26	37	70	173	250	249	249	248	248	247
	27	39	78	194	250	249	249	249	247	247
	26	41	88	208	250	249	250	249	247	248
	28	40	103	218	249	248	249	249	248	248
	32	50	121	226	248	248	249	250	249	246
260	34	57	140	234	248	248	249	249	249	247

(d)

Fig. 8.1 Illustration of Digital Image Format (a) Grayscale Image of Moon, (b) Image Values Shown in 3D, (c) Image Segment (Column 48-57; Row 251-260) (d) Cell Value of Image(c)

8.2.2 *Digital Image Resolution*

Until recently, a "typical" digital image was composed of about 480 rows by 640 columns. This is the image quality encountered in television and general scientific research. Images with more than 1000 by 1000 pixels are considered good quality. In the last few years, cameras have evolved dramatically; many digital cameras can capture 3872x2592 pixels (10 mega pixel). Full high definition video (1080p) projects images at a resolution of 1920x1080 pixels.

One of the most difficult problems in image processing is managing massive amounts of data. It is common for 256 gray levels to be used in image processing of technical images. For example, the 537 by 358 pixel Moon image will occupy 0.19 megabyte with the density of a byte per pixel assuming no compression. In the 8-bit color image format, each pixel will have three registers corresponding to red, green, and blue components. Thus, a color image of the Moon in the same image format will be 0.58 megabyte.

8.2.3 *Digital Image Compression*

Because of the huge amount of data, images are normally stored in a compressed or compact format. Data compression is a very important field in digital signal and image processing. It allows for better and faster communication. Data compression algorithms can be categorized in two different ways, lossless and lossy methods. A lossless technique means that the restored data file is identical to the original. In comparison, a lossy technique introduces changes to the original image. However, the lossy techniques are much more effective at compression than the lossless methods. The higher the compression ratio, the more noise is added to the data. For example, the size of the grayscale Moon image (Fig. 8.1) in a TIFF (not compressed) file is about 190 kilo-bytes. If it is saved in a GIF format (a lossless data compression format), it will be about 160 kilo-bytes. If a lossy compression format is used, such as JPEG, the file size is reduced to approximately 12 kilo-bytes.

To estimate motion from a sequence of images, a non-compressed image format, such as Bitmap (BMP) or TIFF is required to obtain the relative deformation. More information about data compression and image formats can be found in Smith (1998).

8.3 Motion Estimation Methods

Motion estimation from a sequence of images, also referred to as optical flow estimation, is performed by treating the 2D image as a continuous mathematical function, $f(x,y)$, in which $f(x,y)$ equals to the light intensity at the position (x, y). Based on the mathematical calculation, estimation methods of two-dimensional translation were categorized into the Fourier method, matching, and the differential method by Huang and Tsai (1981). Anandan et al (1993) categorized them into fully parametric, quasi-parametric, and non-parametric model based on the characteristics of the mathematical models in representing the image intensity distribution. Following Huang and Tsai (1981), the basic motion estimation methods can be briefly explained in the following subsections.

8.3.1 The Fourier Method

The Fourier transform of an image function *f(x,y)* is *F(X,Y)*, which is given as follows,

$$F(X,Y) = \int_{-\infty}^{+\infty}\int_{-\infty}^{+\infty} f(x,y)e^{-i2\pi(Xx+Yy)}dxdy \tag{8.1}$$

where *(x, y)* denotes the spatial coordinates, *(X, Y)* the spatial frequency, and $i=\sqrt{(-1)}$. The Fourier transform of a shifted image, $g(x,y) = f(x+\Delta x, y+\Delta y)$, can also be calculated, where Δx and Δy are the relative movements in two directions.

$$G(X,Y) = \int_{-\infty}^{+\infty}\int_{-\infty}^{+\infty} g(x,y)e^{-i2\pi(Xx+Yy)}dxdy = F(X,Y)e^{i2\pi(X\Delta x+Y\Delta y)} \tag{8.2}$$

Comparison of Eqs. 8.1 and 8.2 shows that there is a phase shift between *F(X,Y)* and *G(X,Y)*.

$$\Delta\varphi(X,Y) = 2\pi(X\Delta x + Y\Delta y) \tag{8.3}$$

The relative displacement between the two images, (Δx,Δy), can be obtained by calculating the phase difference at each frequency (X, Y). The main advantage of Fourier method is that in the frequency domain translation, rotation, and scaling can be treated separately. However, caution is required in using the Fourier method. The phase angle may not be a mathematical singular. Further, the Fourier method is more applicable for situation with dark background, where the frequency function is very simple to process (Huang and Tsai 1981).

8.3.2 The Differential Method

The second method for motion estimation is the differential method, which relates time differentials to spatial differentials. In this method, the average gray intensities are assumed to be constant between images. This assumption is reasonable in many situations if the motion between the two images is small. Two images, $f(x,y)$ and $f(x+\Delta x, y+\Delta y)$, are taken at time t_0 and t_1, respectively. (Δx, Δy) is the displacement between two images in the x and y directions. The differential of the two images referring to time will be:

$$\frac{df}{dt} = \frac{f(x+\Delta x, y+\Delta y) - f(x,y)}{t_1 - t_0} \tag{8.4}$$

Because the movement between images is assumed to be small, the differential equation can be expanded by using Taylor series and ignoring the higher derivative terms as follows,

$$\frac{df}{dt} = \frac{\frac{\partial f}{\partial x}\Delta x + \frac{\partial f}{\partial y}\Delta y}{t_1 - t_0} = \frac{\partial f}{\partial x}u + \frac{\partial f}{\partial y}v \tag{8.5}$$

where u and v are the velocity components in the x and y direction. A relationship between the time differential and displacement through the spatial differentials is established in Eq. 8.5. By using the neighboring points to calculate the derivative at both directions, enough equations can be generated to solve for the unknowns (u and v).

8.3.3 The Matching Method

An accurate numerical differentiation as described in Section 8.3.2 may be impractical due to image noise, the small number of frames, or due to lack of the alignment between the images. Under these conditions, the matching method is more robust than the differential method (Barron et al 1992 and 1994). The most popular matching method is digital image cross-correlation (DIC).

The cross-correlation function, c, of two image functions, $f(x,y)$ and $g(x+\Delta x, y+\Delta y)$ and, is given as the follows:

$$c(\xi,\eta) = \int_{-\infty}^{+\infty}\int_{-\infty}^{+\infty} f(x,y)g(x+\xi, y+\eta)dxdy \tag{8.6}$$

The peak of function, c, is located at $(\xi,\eta)\,|\,c_{\max}$, which will coincide with (Δx, Δy). Locating the position of the peak indicates both the magnitude and direction of the displacement between two images.

Modified matching algorithms have also been used by Giachetti (2000) including sum of squared difference (SSD), sum of absolute differences (SAD), zero-meaned cross-correlation algorithm, and zero-meaned normalized cross-correlation algorithm (ZNCC).

Giachetti (2000) compared the performance of more than 10 different matching algorithms. It was found that different algorithms have better performance than others under different conditions. Generally speaking, the ZNCC algorithm performs the best among the many matching methods.

There is a close relationship between the matching method and the differential technique (Fleet 1992). Some researchers compared the performance of different algorithms under different situations (Barron et al 1992 and 1994, Little and Verri 1989). The matching methods produce a visually attractive flow field. They are also accurate enough for quantitative use. Differential methods are the best when the displacement in the images is small (≤2 pixels), while matching methods work well for moderate displacement. In engineering fields, the matching method is the most used movement estimation method from a sequence of images. Digital image cross-correlation (DIC) is the most commonly used matching method in engineering fields.

8.4 Digital Image Correlation

With the development of computers, CCD cameras, frame-grabbers, and laser light technology, digital image techniques have been improved tremendously and were applied in many engineering fields. Digital image processing makes the image processing much easier and faster than the time-consuming traditional film processing. The advantages of the digital image techniques are not only that the system is easier and faster to use. The sequence of images are downloaded and stored in a computer individually. The stored individual images allow correlating multiple separate images rather than only a couple of image used in traditional photography. More importantly, each image can easily be preprocessed before the analysis to produce better results (Sjödahl 1998).

Digital image correlation (DIC) uses the matching method mentioned in section 8.3.3. It has been widely used in many engineering fields, such as particle image velocimetry (PIV) in fluid dynamics measurements (Adrian 1991), surface displacement analysis (SDA) in solid mechanics (Chu et al 1985).

8.4.1 Discrete Cross-Correlation

The value of each pixel in the digital image represents a small region in the continuous image being digitized. The cross-correlation function shown in Eq. 8.6 is in the continuous form. In practice, all images have finite dimensions. For two digital images, $f(m,n)$ and $g(m,n)= f(m+\Delta m,\ n+\Delta n)$ and, with a dimension of M×N, the discrete cross-correlation function is given by:

$$c(l,k) = \sum_{m=1}^{M} \sum_{n=1}^{N} f(m,n) g(m+l, n+k) \tag{8.7}$$

where both the image matrices, f and g, have the same dimension of M by N.

8.4.2 Zero-Meaned Normalized Cross-Correlation

The standard cross-correlation function is shown in Eq. 8.7. When two images cross-correlate with each other, there are some regions out of the correlated area (Fig. 8.2). In the region outside the correlated area (patterned in the figure), one of the images has to be zero padded in this region in order to get the cross-correlation result. If the part outside the correlated area is padded by zeros, which is the default method used in most DIC calculation, the cross-correlation function will have a pyramidal shape with its maximum value at the origin. This will force the correlation peak position toward the origin and introduce a systematic error. Fig. 8.3 shows the cross-correlation function with zero padding. Two methods can be applied to reduce this systematic error. The first one is padding the sub-images with their mean value instead of zeros (Sjödahl 1998). The second is to subtract the mean value from the image, before the correlation. Thus the cross-correlation changes to cross-covariance, which gives a better prediction of location of the peak point (Fig. 8.4). The zero-meaned cross-correlation (ZCC) of two image functions can be normalized by their auto-correlation functions. The auto-correlation function is the cross-correlation function of an image with itself. The zero-meaned normalized cross-correlation (ZNCC) performs the best among the matching methods (Giachetti 2000), which is the algorithm used in this research. The discrete form of ZNCC is given as follows.

$$ZNCC(l,k) = \frac{\sum_{m=0}^{M-1}\sum_{n=0}^{N-1}[f(m,n)-\overline{f}][g(m+l,n+k)-\overline{g}]}{\sqrt{\sum_{m=0}^{M-1}\sum_{n=0}^{N-1}[f(m,n)-\overline{f}]^2\sum_{m=0}^{M-1}\sum_{n=0}^{N-1}[g(m,n)-\overline{g}]^2}} \tag{8.8}$$

where $\overline{f}$ and $\overline{g}$ are the average value of image functions, f and g.

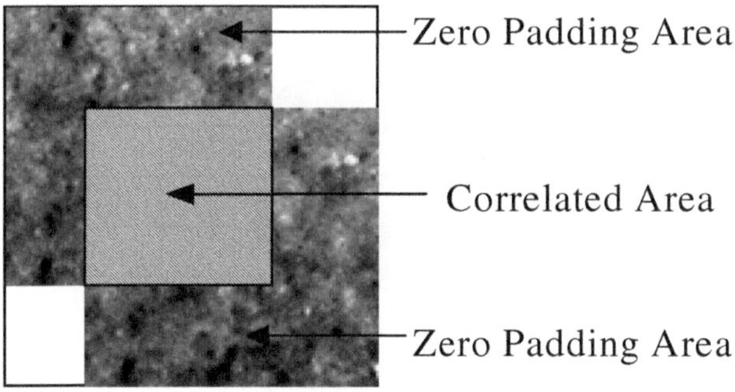

Fig. 8.2 Illustration of Correlated Area and Zero Padding

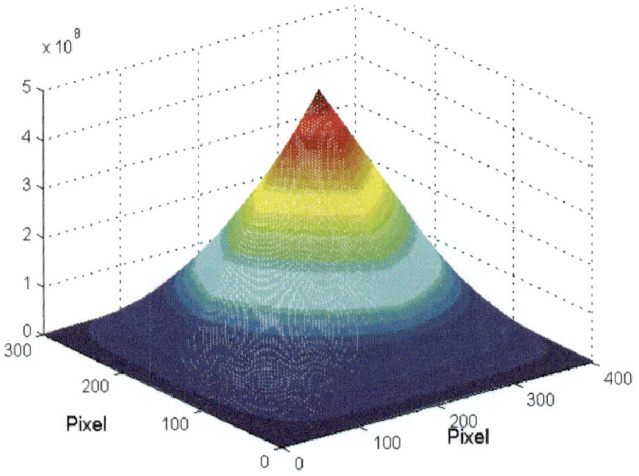

Fig. 8.3 The Cross-correlation of Two Images with Zero Padding

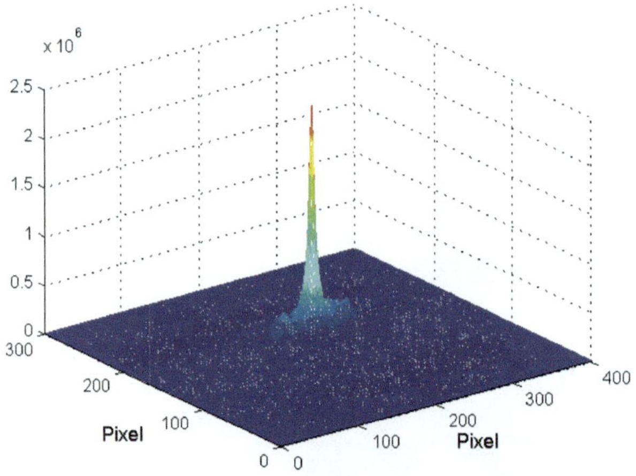

Fig. 8.4 The Cross-Correlation of Two Zero-Meaned Images

8.4.3 Execution of Cross-Correlation Using FFT

The fast Fourier transform (FFT) algorithm is normally performed to speed up cross-correlation calculations. One of the properties of the FFT is that the convolution of two continuous functions, $f\ (x,y)$ and $g\ (x,y)$, is the product of their frequency functions, $F(X,Y)$ and $G(X,Y)$. The convolution function, d, is shown below:

$$d(\xi,\eta) = f(x,y) \otimes g(x,y) = \int_{-\infty}^{+\infty}\int_{-\infty}^{+\infty} f(x,y)g(\xi - x,\eta - y)dxdy \qquad (8.9)$$

$$D(X,Y) = F(X,Y) \times G(X,Y) \qquad (8.10)$$

where *D(X,Y), F(X,Y)* and *G(X,Y)* are the Fourier transforms of *d(x,y), f(x,y),* and *g(x,y).*

Some modifications are required for the use of the FFT algorithm in the cross-correlation. For two images, *f(m,n)* and *g(m,n),* with a same dimension of *MxN*, the first step is to obtain *F(X,Y),* which is the Fourier transform of the image *f(m,n).* The second step is to obtain the conjugate of *G(X,Y)*, G*(X, Y), which can be achieved as the Fourier transform of rotated image *g*(m n)=g(M-m,N-n).* Next the cross correlation function of *f* and *g* is obtained as follows:

$$C_{fg}(\Delta x,\Delta y) = IFFT(F(X,Y) \times G^{*}(X,Y)) \qquad (8.11)$$

where IFFT stands for the inverse Fourier transform. Finally, the zero-meaned normalized cross-correlation function (ZNCC) is calculated as follows:

$$ZNCC = \frac{C_{fg}(\Delta x,\Delta y)}{\sqrt{C_{ff}(\Delta x,\Delta y)C_{gg}(\Delta x,\Delta y)}} \qquad (8.12)$$

where C_{ff} and C_{gg} are the auto-correlation functions of *f* and *g*.

8.4.4 How DIC Works

Locating the peak of cross-correlation function will give the displacement between two images. The procedure is illustrated next. Two images (Image 0 and Image 1 are shown in Fig 8.5. The peak of auto-correlation function of image 0 which is the cross-correlation of image 0 with itself is at the center of the region (the brightest point in Fig. 8.6a). The peak of the cross-correlation of image 0 and image 1 is a small distance away from the center (Fig. 8.6b).

Fig. 8.7 shows the horizontal displacement between the two images. The vector from the center of the region to the peaks of cross-correlation function gives the magnitude and direction of the relative movement between two images.

The displacement field throughout an image can be obtained by dividing the whole image into small interrogation windows (sub-images) and getting the corresponding movement in each interrogated region. Fig. 8.8 illustrates an example, where the whole image is divided into small interrogation windows with size 32×32 pixels. The correlated displacement filed is presented in form of arrows on the top of each interrogation windows.

(a) Image 0

(b) Image 1 (after displacement)

Fig. 8.5 Two Images used to Illustrate DIC (200×150 Pixels)

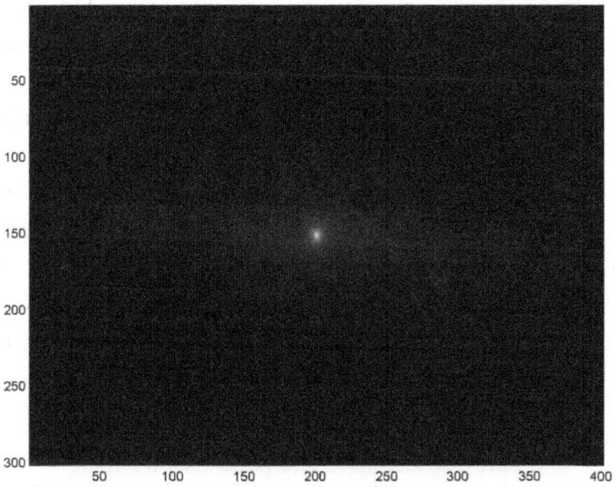

(a) Auto-Correlation

(b) Cross-Correlation Function

Fig. 8.6 Normalized Grayscale Images of Correlation Functions

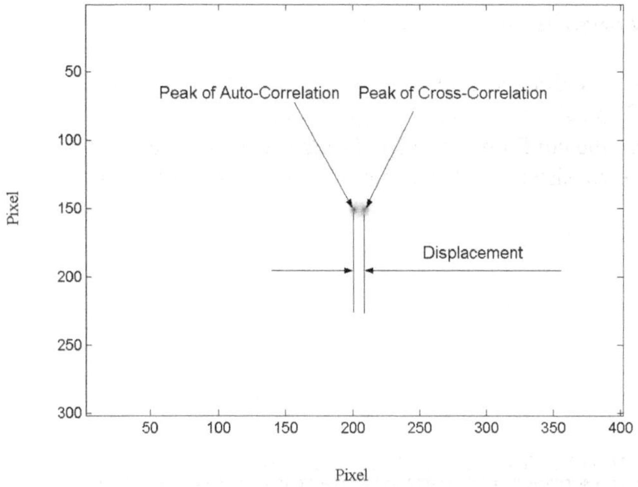

Fig. 8.7 The Displacement between Two Images

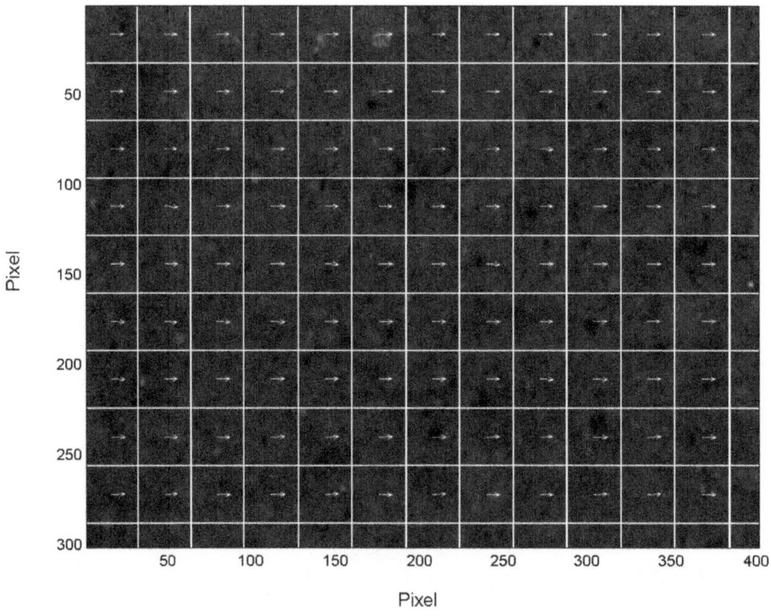

Fig. 8.8 The Whole Displacement Field Obtained by Correlating Each Small Region

8.4.5 *Sub-Pixel Resolution*

Locating the peak of cross-correlation function gives the displacement prediction with an accuracy of one integral pixel. A more accurate prediction can be given by approximating the cross-correlation peak function by a continuous function spanning the peak area. Several functions can be used including central-point calculation, bi-parabolic fitting, and splines. A Gaussian curve without baseline offset is the most commonly used function to interpolate the width, height and position of the peak. The Gaussian function gives a more accurate displacement than the discrete pixel location. A reliable displacement with an accuracy on the order of 0.1 pixel can then be achieved (Sjödahl and Benchert 1993).

8.5 DIC Error Analysis

8.5.1 *Conventional DIC*

The steps followed to obtain the displacement in each interrogation window by DIC can be summarized as follows:

- Obtain an array of interrogation windows from each image
- Calculate the FFT of the first interrogation window
- Calculate the FFT conjugate of the second interrogation window
- Calculate the cross-product of the FFT of the first window and the FFT conjugate of the second window
- Compute the IFFT of the above cross-product to obtain the cross-correlation function
- Locate the peak of cross-correlation function
- Obtain the movement between two interrogation windows
- Get a more accurate movement by sub-pixel resolution

It is not essential to apply FFT in digital image cross-correlation, but it speeds up the calculation. DIC calculation can be schematically shown in Fig. 8.9, where the interrogation windows between images are fixed without relative movement. A relative displacement is shown in Fig. 8.9b between the two images. Some of the information inside the original window area moved out of the window. The strength of the cross-correlation peak is proportional to the number of particle pairs between two interrogation windows. The number of particle pairs decreases as the relative movement increases.

The real images will not be as ideal as the image shown in Fig. 8.9 because of many factors. For example, there will be a particle pattern change between images, which becomes especially severe in high gradient displacement field. There will also be some particles moving out of the image plane, which causes image pair loss. The image noise during image capturing will also introduce some errors in the results.

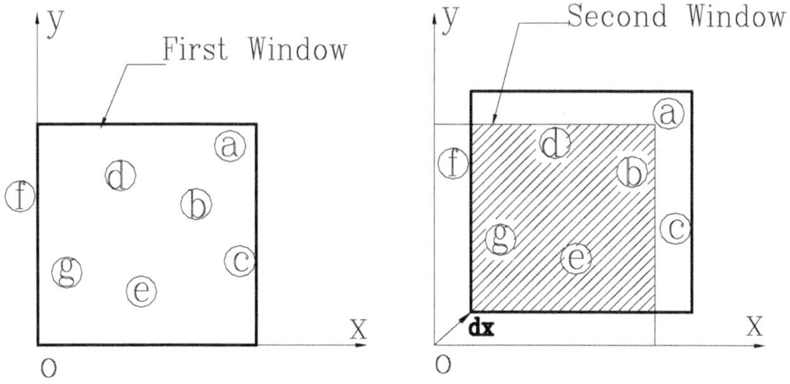

Fig. 8.9 orrelation between Two Images in DIC. (a) First Interrogation window –LHS (b) Second Window and Correlated Area – RHS.

8.5.2 Error Analysis

Errors occur primarily due to insufficient data, lack of texture (particles), or poor image quality. Errors may also occur from correlation anomalies resulting from unmatched particles due to in-plane motion or out-plane motion (Hart 2000, Adrian 1991). The random error, *e*, in the cross-correlation algorithm is estimated by Sjödahl (1998) as follows:

$$e = k\frac{\sigma^2}{B}\sqrt{\frac{1-v}{v}} \tag{8.13}$$

where σ is the average particle size in pixel, k is a constant that depends on the algorithm used, B is the size of interrogation window, and v is the mismatch correlation factor that is proportional to the correlated area ratio given by correlated area divided by the original area.

Eq. 8.13 indicates that increasing the interrogation window size B and increasing the correlated area between images can reduce the errors associated with the algorithm. There are two ways to increase the value of v. The first is to limit the relative movement between images (Huang and Tsai 1981). The second is to increase the information density, the texture contrast of the image.

8.5.3 Particle Density

In some cases, DIC is used for flow dynamic measurement. Micro particles are seeded inside the flow to trace the flow movement. The seeded particles are illuminated by pulse laser. Flow distribution can be obtained by cross-correlating the sequence of images. This technique is called particle image velocimetry (PIV) in fluid dynamics.

When the seeding density, i.e. the particle amount within a unit volume, increases, the information to predict the movement increases. Landreth and Adrian (1989) and Keane and Adrian (1990) demonstrated that obvious errors drop dramatically when the number of the particles inside the correlated region reaches more than 10. However, it is not always desirable to increase information (particle) density. It is also practically impossible in some cases to increase the particle density.

8.5.4 Interrogation Windows Size

Increasing the interrogated window size will increase the information inside the interrogated region. The correlated fraction increases with the same displacement because of the increased window size. However, this increment in window size will decrease the overall analysis resolution.

There are some limits to increase window size. The cross-correlation algorithm is based on the whole region's properties. The statistical properties of the interrogated regions are assumed consistent. When the interrogated window size increases, the particle pattern may change between images due to local deformation. Keane and Adrian (1993) demonstrated that significant errors due to asymmetry in the peak correlation are associated with this phenomenon.

8.6 Adaptive Cross-Correlation

Many researchers have studied the errors associated with the cross-correlation algorithm and advanced the algorithm in different ways to achieve better results. The methods used can be summarized into three categories. The first is to pre-process the image before the interrogation calculation (Hart 2000). The second is to advance the algorithm itself to account for the pattern change or lost image pairs (Tokumaru and Dimotakis 1995, Westerweel et al 1997, and Scarano and Riethmuller 1999). The last technique is to rule out some obvious errors from the result (Westerweel 1994). For example, Westerweel et al (1997) demonstrated that offset of interrogated window by one or more integral pixel in the displacement direction can reduce errors caused by unmatched particle pairs.

Scarano and Riethmuller (1999) used the increased window size to get global displacement pattern, then subdivided each region to smaller regions to refine the displacement field. Their approach takes full advantage of both window shifting and window sizing methods. Verification criteria are applied in each step to eliminate the spurious errors from the results. Their method is called Adaptive Cross-Correlation (ACC). It is implemented in Flow Manager software (Dantec 2001) which is intended for Particle Image Velocimetry (PIV). Flow Manager was adapted for this research to study soil deformations. In general, ACC differs from DIC in two issues, variable window size and window offset.

8.6.1 Variable Window Size

Increasing the interrogated window size will increase the information inside the interrogated window; thus increasing the peak value of the cross-correlation function and a more accurate prediction is achieved (Sadek 2002). DIC assumes that the statistical properties of the interrogated windows are constant (Giachetti 2000). This assumption holds for rigid body translations. However, geotechnical applications involve local deformations, which are diluted in large windows. If a large window size is used, local deformations in different directions may appear to have no deformation associated with a large standard deviation. Keane and Adrian (1993) demonstrated that this phenomenon causes asymmetry in the correlation function, which results in a prediction error. Therefore, in ACC, a large interrogation window size is first used to obtain the general direction of movement. Next, smaller windows are used to better define local displacements.

8.6.2 Window Offset

The position of two interrogated windows is fixed in conventional DIC, which causes the correlated area to decrease as the relative displacement increases. This effect is demonstrated in Fig. 8.9, where the correlated area is shaded. Typically, a high frame rate (sampling rate) is used to reduce the relative deformation between successive frames, thus reducing the prediction error. Westerweel et al (1997) demonstrated that offsetting windows during the analysis can reduce this type of error. The procedure involves shifting one of the images in the direction of the movement by a number of pixels that best approximates the deformation calculated in the previous iteration (Fig. 8.10). This allows for most of the information in the original window to be included in the second window.

There are two ways to shift the interrogation windows in ACC. The first is central difference (ACC-CD), where both windows have an offset of half of the required shift, but in opposite directions. The second is the forward difference (ACC-FD), where only the second image is offset by the required shift.

8.6.3 ACC Procedure

ACC is illustrated in Fig. 8.10. At the beginning, the image is divided into large interrogation windows say with a size 4×4 times the final window size. In this case, the analysis involves three steps. First, classical DIC is performed on these windows and the deformation is obtained for each window using Eq. 8.12. The second interrogation window is patterned in Fig. 8.10. The calculated movement from the first step, $dx0$, is shown as an arrow. The second step of the analysis involves dividing the original window into 4 smaller windows that are shifted from their original positions by the integer value of $dx0$ (ACC-FD is shown in Fig. 8.10). The calculated deformations for the 4 windows are shown as $dx1i$. This procedure is repeated for each of the smaller interrogation windows, until the final

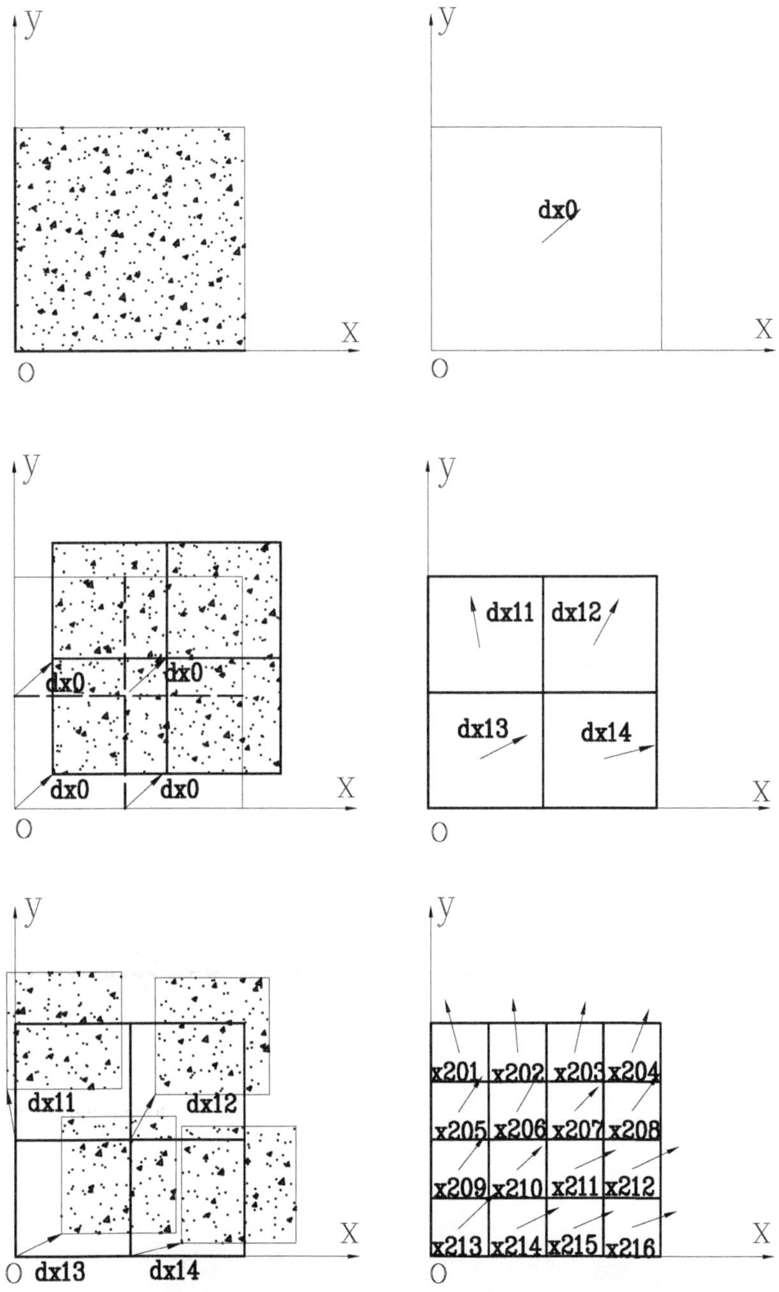

Fig. 8.10 Adaptive Cross-Correlation (ACC-FD shown)

window size is reached. Verification criteria are implemented in each step to root out most of the spurious errors from the results (Dantec 2001).

Near the peak of the cross-correlation function, a continuous Gaussian equation is used for approximating the cross-correlation function calculated at the pixel locations. Next, the position of the peak of the fitting function is obtained and a movement resolution on the order of 0.1 pixel can be calculated (Sjodahl and Benchert 1993).

A local median or a moving average can be used to reject spurious errors. The local median gives better results for gradual displacement, while the moving average is better suited for displacements with large gradients such as turbulent flow measurements (Westerweel 1994). Post-interrogation processing methods can correct for obvious errors, but cannot fundamentally reduce correlation errors caused by inadequate window size or frame rate.

8.7 Comparison between DIC and ACC

8.7.1 Verification of DIC and ACC Algorithms

An image of medium–coarse sand in a Plexiglas mold was acquired using a Nikon 990 digital camera (Fig. 8.11). The camera was oriented 20 cm away from the mold, with its optical axis perpendicular to the mold. The resolution of the image is 2046×1560 pixels, which corresponds to a scale of 0.077 mm per pixel. This image was used for verification of the DIC and ACC algorithms. The image was subjected to a digital shift, of a known horizontal and vertical displacements, u and v, respectively. The accuracy of the DIC and ACC algorithms in predicting displacement was tested by comparing original and shifted images.

Fig. 8.11 Typical Digital Image of Sand

The error in the analysis can be obtained by comparing the predicted horizontal or vertical displacement u or v with their known values, u_T and v_T. The final interrogation window was kept constant (32×32 pixels). The average values of the predicted displacements using DIC and ACC are shown in Table 8.1 for various digital shifts. For DIC, the maximum reliable displacement is approximately 8 pixels, which is about one fourth of the interrogation window size. This is consistent with a more comprehensive study of error in DIC measurements conducted by Sadek et al (2003). ACC is always more accurate than DIC. The maximum reliable displacement for conventional ACC employing both window sizing and window shifting was approximately 75% of the final interrogation window size. Next ACC was performed with window shifting only. Window shifting does not substantially increase the range of reliable predictions. It reduces the prediction errors and the standard deviation. For example, the standard deviations calculated using DIC, ACC-FD (no window sizing), and ACC-FD were 1.65, 0.55 and 0.03 pixels, respectively, for a movement of 8 pixels. The improved performance of ACC reflects not only the effects of window shifting and sizing but also the verification criteria used by flow manager to rule out the obvious errors.

Table 8.1 Comparison of DIC and ACC for Digital Translations

Applied u_T, pixels	Predicted u, pixels			Applied v_T, pixels	Predicted v, pixels		
	DIC 32×32 No Offset	ACC 32×32 6 Offsets	ACC 128×128, 64×64, 32×32 6 Offsets		DIC 32×32 No Offset	ACC 32×32 6 Offsets	ACC 128×128, 64×64, 32×32 6 Offsets
2	1.911	1.990	1.992	2	1.906	1.989	1.992
4	3.896	3.969	3.976	4	3.890	3.971	3.975
6	5.859	5.927	5.971	6	5.820	5.944	5.970
8	7.682	7.756	7.953	8	7.634	7.772	7.953
10	9.010	9.222	9.950	10	9.057	9.394	9.949
12	9.772	10.120	11.946	12	9.919	10.404	11.945
14	5.170	6.812	13.803	14	4.904	6.118	13.801
16	0.7007	2.286	15.292	16	0.759	2.321	15.269
24	-0.3780	0.678	22.653	24	-0.878	0.694	22.394
32	0.181	0.071	30.512	32	0.216	0.412	30.359

ACC-FD is used.

Predicted displacements are the average values of 961 interrogation windows.

Next, the image was cropped to 960x960 pixels and digitally rotated by 2° around its center. First, the rotation field was predicted using DIC by comparing the rotated image to the un-rotated one. The interrogation window size was 32x32 pixels. DIC was able to predict movements near the center, but failed near the edges of the image (Fig. 8.12). The rotation represents a movement of 17 pixels near the border of the image. This movement far exceeds the abilities of the DIC algorithm as described earlier. Next ACC was used to predict movement, starting with a 128x128 interrogation window size and ending with 32x32 window size. ACC was able to capture the magnitude and direction of displacement near the image edges even though these displacements represented approximately 50% of the final interrogation window size (Fig. 8.12).

The variable window size is credited with the improved performance. The average absolute error in u-u_T and v-v_T and the standard deviations, σ_u and σ_v, are used to quantify the results in Table 8.2. For example, the errors calculated when ACC was performed using a fixed 32x32 window and allowing for six window shifts was similar to DIC.

Next ACC was performed with 3 window sizes and without window shifting (i.e. 3 runs) and with window shifting while keeping the window size fixed for 6 iterations (Fig. 8.13). The analyses that involved window sizing without window shifting resulted in significantly fewer errors than the analyses that involved only window shifting, even though they involved half as many computer runs. The images of sand have less contrast than images used in PIV, which reduces the relative value of window shifting on the accuracy of the analysis.

Because the two images used in these verifications originated from the same image, their intensity is constant. The accuracy for real situation will be lower than the results shown above.

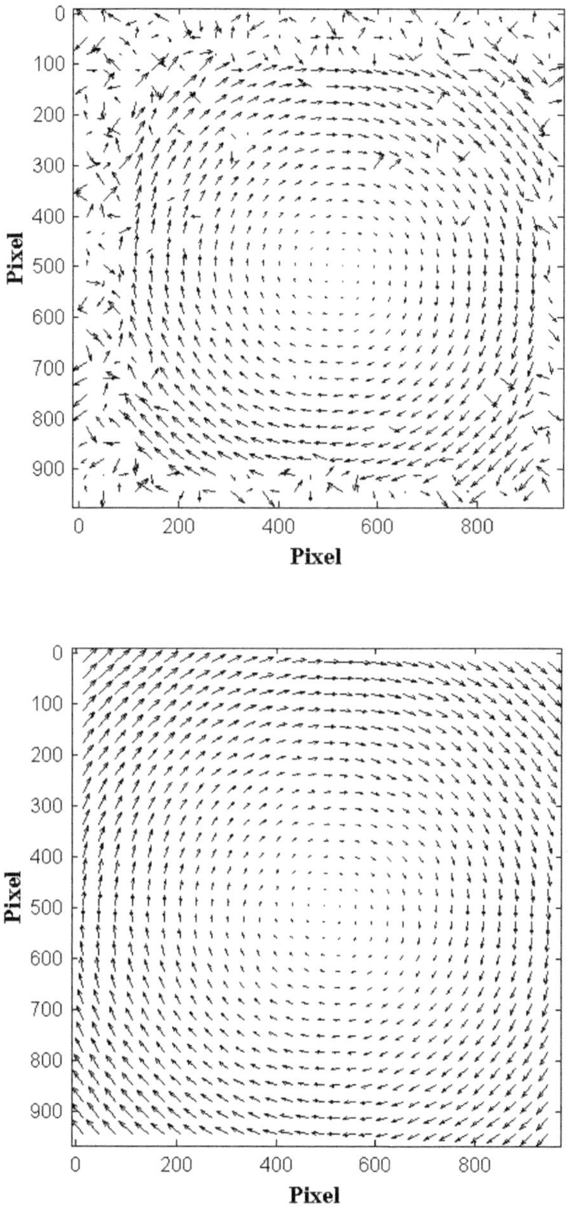

Fig. 8.12 Rotational Movement Calculated Using DIC on Top, (Window Size= 32x32. No Offset) and ACC at the bottom (Window Size= 128x128, 64x63, 32x32 Offset= 2 for each window size (Total 6))

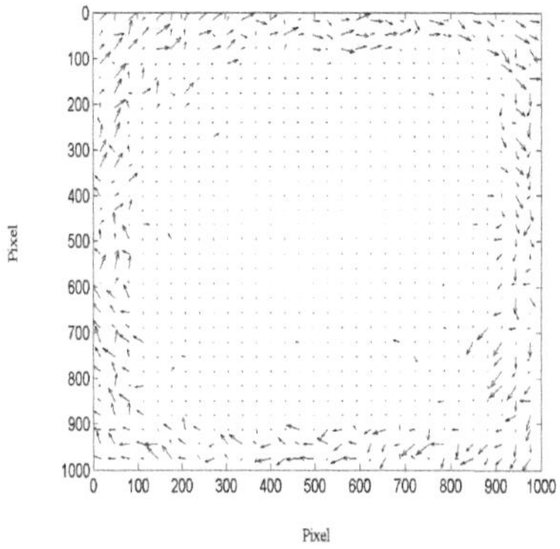

DIC
Window Size =32x32
Offset= None

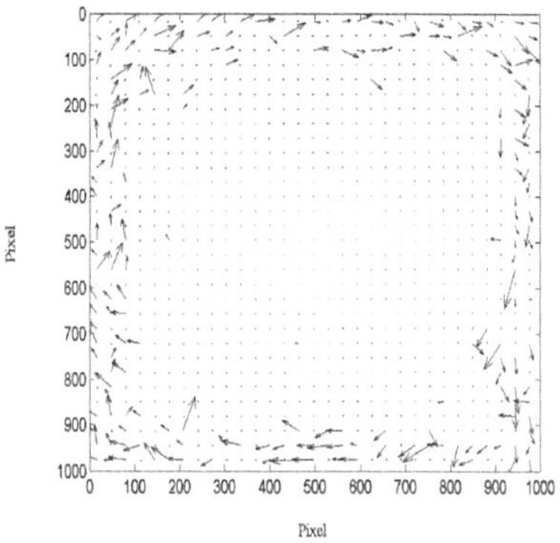

ACC
Window Size= 32x32
Offset= 6

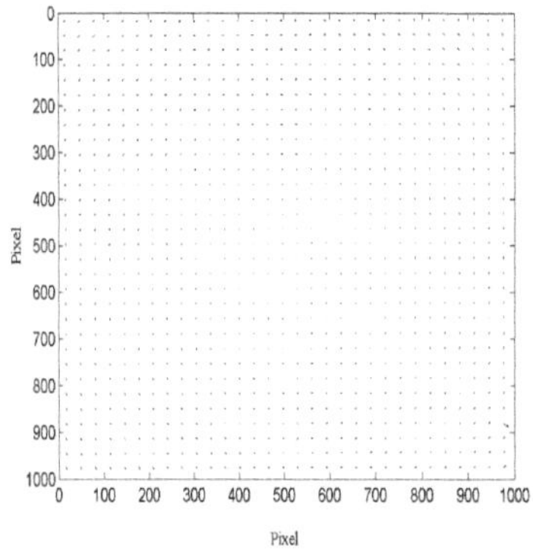

ACC
Window Size= 128x128, 64x63, 32x32
Offset= None

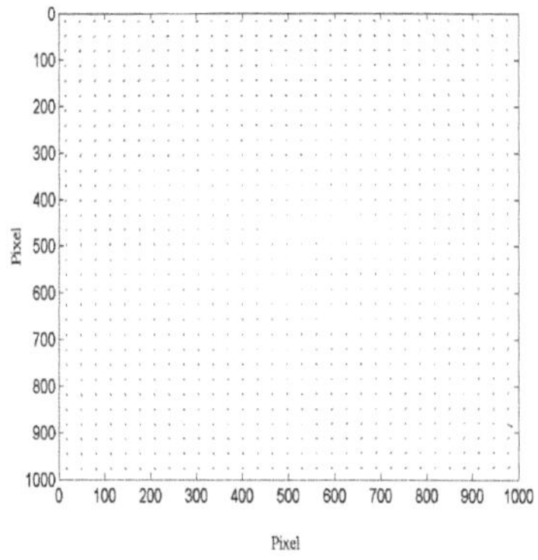

ACC
Window Size= 128x128, 64x63, 32x32
Offset= 2 for each window size (Total 6)

Fig. 8.13 Error Distribution from DIC and ACC in Rotational Movements

Table 8.2 Error Analysis in Prediction of Rotation using DIC and ACC

Algorithm	Window Size	Window Offset	U $\overline{\lvert u-u_T \rvert}$	U σ_u	V $\overline{\lvert v-v_T \rvert}$	V σ_v
DIC	32×32	0	1.986	4.768	1.922	4.228
ACC	32×32	6	1.967	4.728	1.909	4.215
ACC	128×128, 64×64, 32×32	0	0.100	0.172	0.106	0.224
ACC	128×128, 64×64, 32×32	6	0.088	0.141	0.087	0.165

ACC-FD is used.
Predicted displacements are the average values of 961 interrogation windows.

8.7.2 Performance of DIC and ACC Algorithms for Physical Movements

The intensity of images used for typical cross-correlation applications is influenced by ambient light, image noise, and refraction of light. Therefore, two images of the model captured consecutively were used to verify the performance of DIC and ACC algorithms under real test conditions. The model was placed on a precision linear stage and was moved by a fixed translation after the first image was captured. Next a second image was captured.

The spatial distribution of predicted displacement fields using DIC and ACC are presented in Fig. 8.14-15, along with the frequency histogram of predicted horizontal displacement. The median horizontal displacement was 0.354 mm (4.60 pixels) and 0.364 mm (4.72 pixels) for DIC and ACC respectively, which agreed well with the prescribed horizontal displacement of 0.36 mm (4.68 pixels).

ACC reduced obvious errors caused by the lack of texture in the model footing placed on top of the foundation sand. The standard deviation reduced from 1.2 pixels in DIC to 0.27 pixels in ACC (Table 8.3), where 1 pixel corresponds to 0.077 mm.

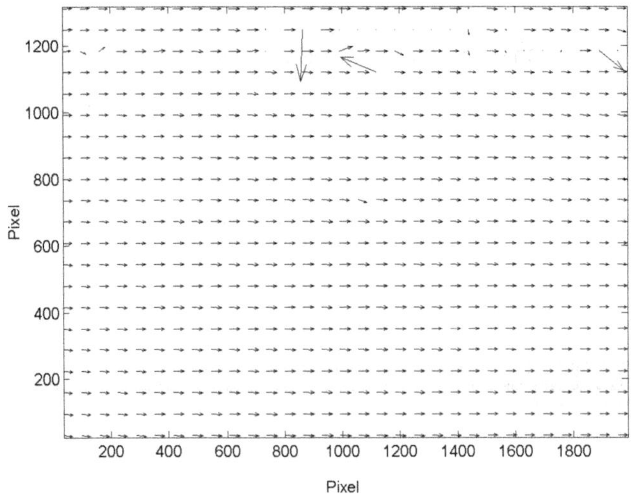

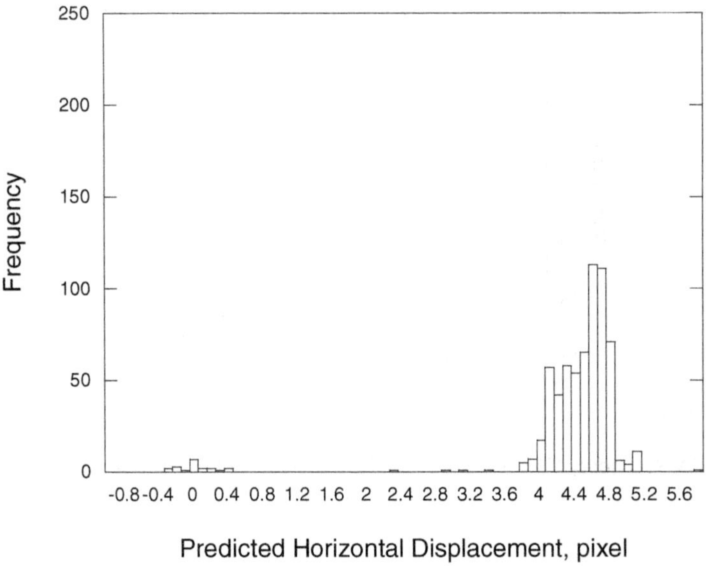

Fig. 8.14 Predicted Displacement Field using DIC (Top) and Histogram of Horizontal Displacement (RHS) using DIC. (Window Size 64x64 Pixel, Median: 4.60 pixels, Standard Deviation: 1.20 pixels).

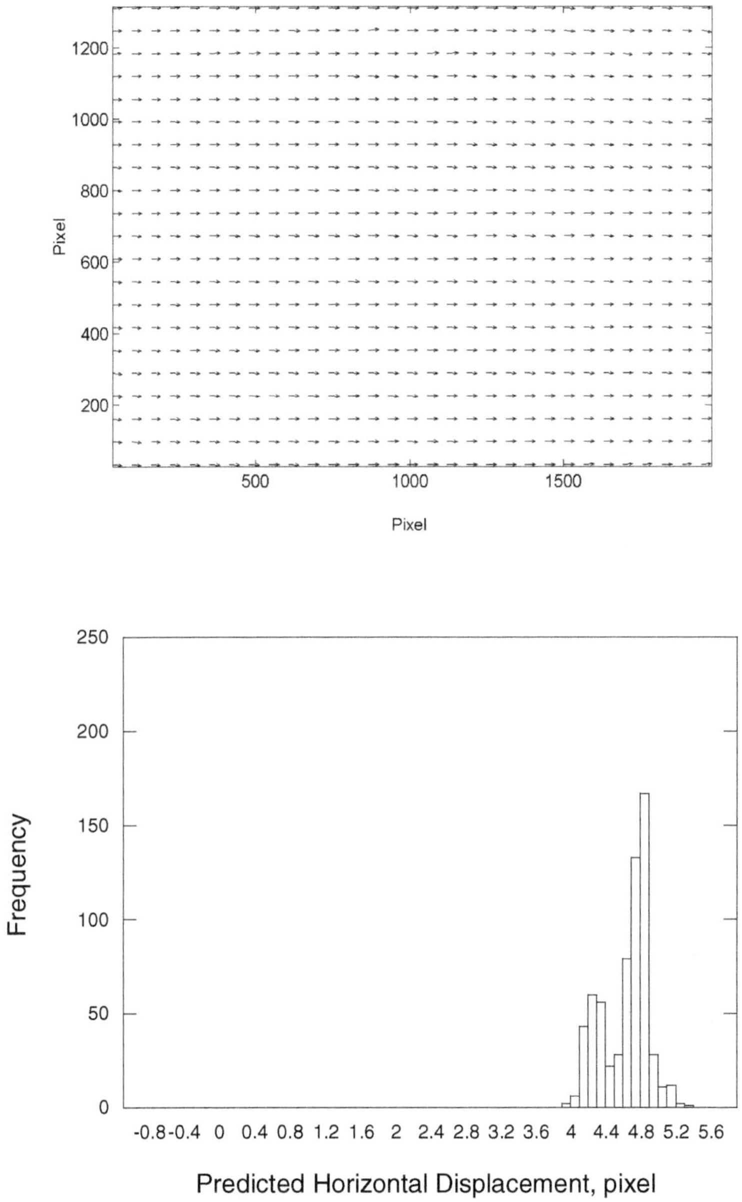

Fig. 8.15 Predicted Displacement Field using ACC (Top) and Histogram of Horizontal Displacement (Bottom) (Window Size: 256x256 to 64x64, Median: 4.72 pixels, Standard Deviation: 0.27 pixel).

Table 8.3 Comparison of DIC and ACC for Physical Translations

Algorithm	Window Size	Window Offset	Post-Interrogation Processing	u		v	
				$\overline{u-u_T}$	σ_u	$\overline{v-v_T}$	σ_v
DIC	64×64	0	None	4.383	1.199	-0.155	0.883
ACC	64×64	2	None	4.370	1.209	-0.166	0.878
ACC	128×128, 64×64	4	None	4.623	0.985	-0.165	0.666
ACC	256×256, 128×128, 64×64	6	None	4.623	0.384	-0.133	0.168
ACC	256×256, 128×128, 64×64	6	Local Median	4.651	0.268	-0.133	0.100
ACC	256×256, 128×128, 64×64	6	Moving Average	4.648	0.268	-0.134	0.097

Note: ACC-FD is used.
Predicted displacements are the average values of 744 interrogation windows.

8.8 Conclusions

In this chapter, the use of digital image cross-correlation for spatial deformation measurement is introduced and investigated. A non-intrusive continuous spatial deformation measurement is obtained by cross-correlating two images using DIC. DIC permitted measuring deformation with a resolution of 0.1 pixels, which corresponds to a displacement on the order of 0.008 mm using the developed optical setup. This accuracy can be improved by increasing the resolution of the camera or using a smaller focus area.

An advanced cross-correlation algorithm, adaptive cross-correlation (ACC), is investigated and compared with conventional DIC. Comparison demonstrated that ACC performs better in the analysis of soil images than DIC. ACC uses both variable window sizing and window shifting for a better result. Of the two, adjustable window sizing was found to be better suited for the analysis of local deformations. Adaptive cross-correlation was selected for measuring deformation in transparent synthetic soils non-intrusively.

References

Adrian, R.: Particle-imaging techniques for experimental fluid mechanics. Annual Review of Fluid Mechanics 23, 261–304 (1991)

Anandan, P., Bergen, J., Hanna, K., Hingorani, R.: Motion analysis and image sequence processing: Hierarchical model-based motion estimation. In: Sezan, M., Lagendijk, R. (eds.) Motion Analysis and Image Sequence Processing, pp. 1–22. Kluwer Academic Publishers, Boston (1993)

Barron, J., Fleet, D., Beauchemin, S., Burkitt, T.: Performance of optical flow techniques. In: Proc. IEEE Computer Vision and Pattern Recognition 1992, pp. 236–242 (1992)

Barron, J., Fleet, D., Beauchemin, S.: System and experiment: performance of optical flow techniques. Int. J. of Computer Vision 12(1), 43–77 (1994)

Bouchafa, S., Aubert, D., Bouzar, S.: Crowd motion estimation and motionless detection in subway corridors by image processing. In: IEEE Conference on Intelligent Transportation System, ITSC 1997, pp. 332–337 (1997)

Chu, T., Ranson, W., Sutton, M., Peters, W.: Application of digital image correlation techniques to experimental mechanics. Experimental Mechanics 25(3), 232–244 (1985)

Dantec Measurement Technology A/S, Flowmap® particle image velocimetry instrumentation-installation & user's guide, Tonsbakken 16-18, DK-2740 Skovlunde, Denmark (2000)

Fleet, D.: Measurement of Image Velocimetry. Kluwer Academic Publ., Norwell (1992)

Giachetti, A.: Matching techniques to compute image motion. IEEE Image Vision Computing 18, 247–260 (2000)

Hart, D.: PIV error correction. Experiments in Fluids 29(1), 12–22 (2000)

Hossain, I., Liu, J., You, J.: Tropical cyclone pattern recognition for intensity and forecasting analysis from satellite imagery. In: IEEE International Conference on Systems, Man, and Cybernetics, IEEE SMC 1999, vol. 6, pp. 851–856 (1999)

Huang, T., Tsai, R.: Image sequence analysis: motion estimation. In: Huang, T.S. (ed.) Image Sequence Analysis, pp. 1–18. Springer, Heidelberg (1981)

Keane, R., Adrian, R.: Optimization of particle image velocimeters. Part I: Double pulsed systems. Measurement Science and Technology 1, 1202–1220 (1990)

Keane, R., Adrian, R.: Theory of cross-correlation of PIV images. In: Nieuwstadt (ed.) Flow Visualization and Image Analysis, pp. 1–25. Kluwer Academic, Dordrecht (1993)

Landreth, C., Adrian, R.: Measurement and refinement of velocimetry data using high image density analysis in particle image velocimetry. In: Adrian, et al. (eds.) Applications of Laser Anemometry to Fluid Mechanics, pp. 484–497. Springer, Heidelberg (1989)

Leuck, H., Nagel, H.: Automatic differentiation facilitates of-integration into steering-angle-based road vehicle tracking. In: Proc. IEEE Computer Society Conference on Computer Vision and Pattern Recognition (CVPR 1999), vol. 2, pp. 360–365 (1999)

Little, J., Verri, A.: Analysis of differential and matching methods for optical flow. In: Proc. IEEE Visual Motion Workshop, pp. 173–180 (1989)

Loncaric, S., Majcenic, Z.: Optical flow algorithm for cardiac motion estimation. In: Proc. of the 22nd Annual EMBS Int. Conference, Chicago, IL, pp. 415–417 (2000)

Markandey, V., Reid, A., Wang, S.: Motion estimation for moving target detection. IEEE Transactions on Aerospace and Electronic Systems 32(3), 866–874 (1996)

Sadek, S.: Soil structure interaction in transparent synthetic soils using digital image correlation, Ph.D. Dissertation, Polytechnic University, New York, USA (2002)

Sadek, S., Iskander, M., Liu, J.: Accuracy of digital image correlation for measuring deformations in transparent media. ASCE J. of Computing in Civil Engineering 17(2), 88–96 (2003)

Scarano, F., Riethmuller, M.: Iterative multi-grid approach in PIV image processing with discrete window offset. Experiments in Fluids 26, 513–523 (1999)

Smith, S.: The scientist & engineer's guide to digital signal processing. California Technical Pub. (1998)

Sjödahl, M.: Some recent advances in electronic speckle photography. Optics and Lasers in Engineering 29, 125–144 (1998)

Sjödahl, M., Benckert, L.: Electronic speckle photography: analysis of an algorithm giving the displacement with sub-pixel accuracy. Applied Optics 32, 2278–2284 (1993)

Tokumaru, P., Dimotakis, P.: Image correlation velocimetry. Experiments in Fluids 19, 1–15 (1995)

Westerweel, J.: Efficient detection of spurious vectors in particle image Velocimetry data. Experiments in Fluids 16, 236–247 (1994)

Westerweel, J., Dabiri, D., Gharib, M.: The effect of a discrete window offset on the accuracy of cross-correlation analysis of digital PIV recording. Experiments in Fluids 23(1), 20–28 (1997)

Willersinn, D., Enkelmann, W.: Robust obstacle detection and tracking by motion analysis. In: IEEE Conf. on Intelligent Transportation System, ITSC 1997, pp. 717–722 (1997)

Chapter 9
Application of DIC for Measuring Deformations in Transparent Soils

Magued Iskander and Samer Sadek

Abstract. This chapter proposes a simple optical technique for measuring spatial deformations in transparent synthetic soil models using digital image correlation (DIC). The accuracy of the DIC technique was evaluated based on a scheme of predefined digital movements of synthetic soil images. Finally, a model consisting of a strip footing on a synthetic transparent soil is presented. The spatial deformations in the model are evaluated using the proposed DIC methodology and compared to the result of finite element analysis.

9.1 Introduction and Motivation

The use of markers and colored soil layers embedded in soil models is perhaps the oldest imaging technique reported for measuring spatial deformations in soil models. Movement of markers was tracked at the glass boundaries of plane strain models. Later on, lead markers were embedded in models and x-ray photography was used to track their motion (Robinsky and Morrison 1964, Butterfield and Andrawes 1971, Bourdeau 1993). However, lead has different physical properties than the surrounding soils which may affect the behavior of the model. Recently, digital imaging has allowed for computerized target tracking, which permits faster and more accurate analysis. For example, Raschke et al (1996), Gustafsson and Gustafsson (1996), and Guler et al (1999) studied micro deformation of granular soils by analyzing particle-level movement. Larger scale deformations can be tracked by monitoring a grid on soil samples (Alshibli and Sture 1999 and Saada et al 1999). Nevertheless, these advances do not permit monitoring deformation inside opaque soil models.

The fundamental premise of this research is that transparent synthetic surrogates with geotechnical properties similar to natural soils can be used in model tests to study three-dimensional deformation and flow problems using non-intrusive optical visualization techniques.

M. Iskander: Modeling with Transparent Soils, SSGG, pp. 165–179.
springerlink.com

The digital image correlation technique, discussed in Chapter 8, provides an ideal method for measuring spatial deformations, inside transparent soil models. The proposed methodology is outlined in this chapter.

9.2 Setup for DIC in Transparent Synthetic Soil Models

Transparent synthetic soil models are completely transparent, and transparent particles cannot be identified individually in the model. Models are sliced optically using laser light sheets (Fig. 9.1). The method presented in this chapter allows for measuring 2D deformations within any illuminated plane inside a transparent soil model. The "z-coordinate" of the measured plane can be readily identified as that where the laser penetrates the soil[1] (Fig 9.2).

The interaction between laser light and transparent soils produces a distinctive speckle pattern (Fig. 9.3). This speckle pattern does not mimic the soil grains or the voids. It is a unique signature of the interaction between the transparent soil matrix, impurities, entrapped air, and the laser. Small particle movement will result in change in the speckle distribution in the slice. If the deformation or particle flow is small, the contrast distribution resulting from the speckle effect will follow the particle movement.

Images captured before and after deformation are analyzed using the cross-correlation function discussed in Chapter 8. Comparing two images taken before and after movement indicates the spatial deformation throughout the slice. For large displacements, sequential images are required for accurate measurement.

A 3D model can be constructed by combining images from multiple slices in Matlab®. A true 3D model would require two perpendicular setups similar to the one shown in Fig. 9.1 in order to capture in and out of plane deformations.

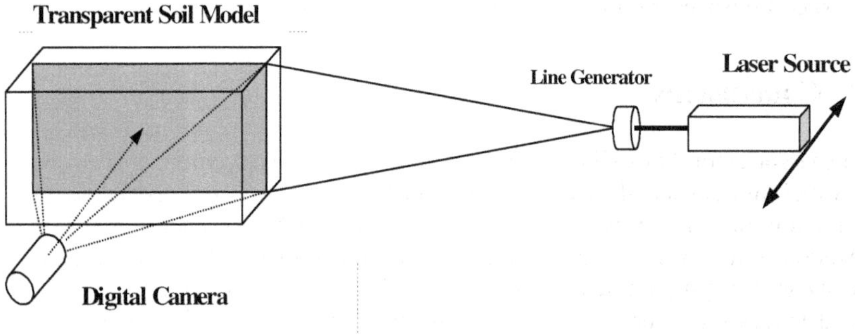

Fig. 9.1 Schematic Diagram Showing Setup for Slicing Transparent Synthetic Soils Optically

[1] Measurements where diffusion of the laser light sheet exceeds the particle size of the model soil are discounted.

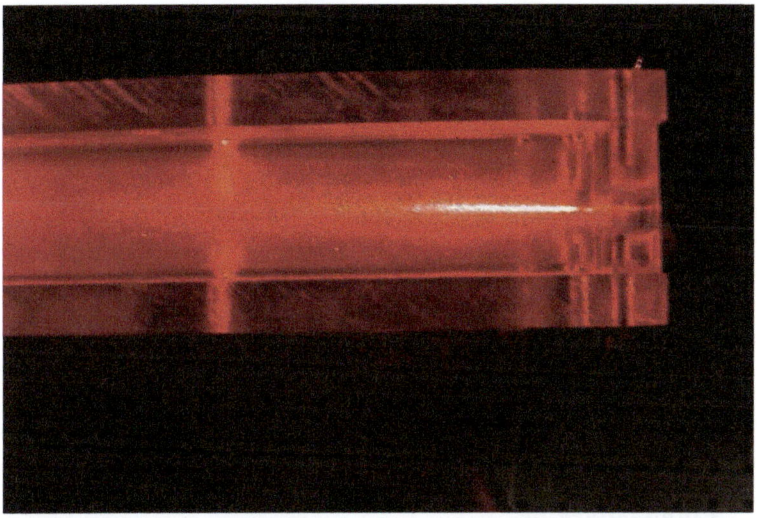

Fig. 9.2 Laser Light Sheet Penetrating into Transparent Soil Model (Top View)

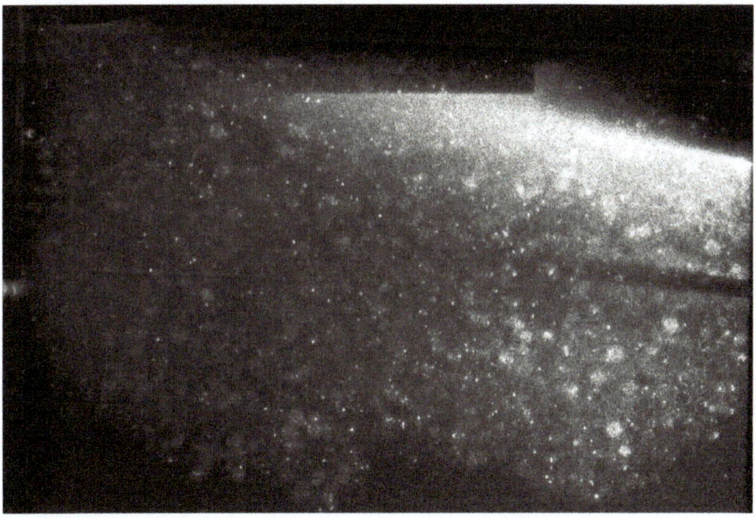

Fig. 9.3 Speckle Pattern Generated by the Interaction of the Laser Light Sheet and Transparent Soil Model

9.2.1 Transparent Soil Model

The model used to illustrate the methodology in this chapter is 5.1 cm (2 inches) thick, 30.5 cm (12 inches) wide, and 15.2 cm (6 inches) high. The model width is limited by the depth of penetration of the laser beam. The model thickness is limited by transparency degradation due to the imperfect matching of the refractive indices of the solid particles and the pore fluid. The model is made of Plexiglas, and was filled with fine silica gel (see properties of SG1 in Chapter 6).

9.2.2 Laser Beam

The depth of penetration of the laser beam in a transparent synthetic soil model depends on the laser power, and frequency of light (laser color). The laser beam used in this research is a Helium Neon Laser system made by Melles Griot, with output power of 35-75 mW, output beam diameter is 1.24 mm, and beam divergence of 0.68 mrad. The laser sheet penetrates 10 cm (4 inches) into a transparent synthetic soil model made of silica gel with an acceptable level of diffusion on the order of 1.5 mm (Fig. 9.2). This was considered acceptable since the grain size of the transparent material used to make the soil model ranged from 0.5 to 1.5 mm (Sadek 2002).

9.2.3 Line Generator

A line generator lens is used for generating the laser sheet. The line generator is a cylindrical lens that expands the laser beam in one direction only. The lens was obtained from Edmond Industrial Optics with a lens holder (lens stock number L43-473 and holder stock number J54-622).

9.2.4 Digital Camera

Any digital camera can be used as long as it can save in a lossless format. Different cameras are used for different demonstrations throughout this book. The camera used in this chapter is a Nikon Coolpix 990 with a resolution of 2048x1360 pixels with a remote cord for camera control. A remote cord and camera stand were used to illiminate camera movement during image capture and to ensure that images are taken from the same exact location. Measurement resolution can be increased by decreasing the area of the captured image, for example, the maximum zoom out for the Nikon 990, has a resolution of 9 pixels/mm for a focus distance of 250 mm. The resolution can be increased to 27 pixels/mm, at the same focal distance, by using the maximum optical zoom.

Though still cameras can give high resolutions, their main disadvantage is that downloading the images requires stopping the experiment and connecting the camera to a PC. Comparatively priced industrial CCD cameras offer a lower resolution but more convenience[2]. A system consisting of an industrial CCD camera and frame grabber is presented in Chapter 10.

Zoom is applied digitally in many digital cameras. The speckle information is lost if a digital zoom is used and consequently, particle movement information is lost. Therefore a digital zoom can not be used. Also, images have to be stored in a lossless format such as tiff, to prevent loss of image texture during compression.

---

[2] Modern still cameras that communicate wirelessly with a PC may offer the best of both worlds.

9.2.5 Correlation Algorithm

DIC has been explained in Chapter 8. Commercial software, which uses DIC for deformational measurement or particle image velocimetry (PIV), is available. The algorithm used in this chapter is an open source code originally written by Liberzon et al (2000) using Matlab,® [3] and modified by Sadek et al (2003) to adequately analyze the problem.

The algorithm applies a cross-correlation function on two interrogated images. The peak of the correlation function is identified for each interrogated window of the image. The displacement of the interrogated window is assumed to be at the location of the peak of the cross correlation function.

The algorithm has five extra features for better results. First, the cross correlation function is calculated on a Fast Fourier Transform (FFT) of the image for faster analysis. Second, the peak value of the cross correlation function is compared to the noise value. The noise value is defined as the larger of either the second highest peak or the mean value of the cross correlation function. Next, a signal-to-noise-ratio is calculated for evaluating the validity of the result. The user can identify a limit for accepting the results such that high noise values can be rejected. Third, all peak values located at the borders are rejected since accuracy of cross correlation function is lost at the borders of the image. Fourth, the peak is identified by first curve fitting the peak region and then the peak is identified mathematically, thus allowing sub-pixel displacement estimation. Finally, if the vector of a certain window is rejected, the algorithm interpolates the vector displacement from neighboring vectors. This fifth feature is optional and the output contains the original results and the interpolated ones.

9.3 Calibration of DIC for Deformation Measurement in Transparent Soils

Westerweel et al (1997) showed that the accuracy of correlation based techniques depends on the size of the particle moving and the magnitude of the movement. Thus, applying imaging correlation in transparent synthetic soil models requires a comprehensive study for quantifying system limitations, optimum use, and anticipated errors.

9.3.1 Calibration Methodology

Two identical images were captured for a laser sheet crossing a transparent synthetic soil model. The images were then cropped to 512x512 at different locations such that the "tested image" is shifted a distance "S" pixels from the original image as shown in Fig. 9.4. "S" was varied to be 1, 2, 4, 8, 10, 15, 20, 30, 40, 50, 60,

---

[3] The original software URIPIV has been succeeded by OpenPIV an open source software available at http://www.openpiv.net/ (last accessed October 2009).

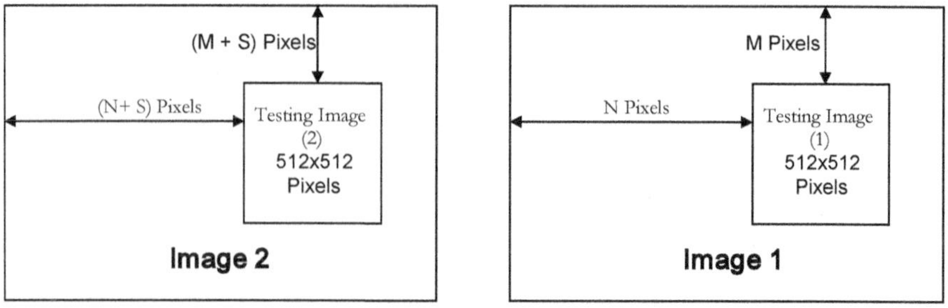

Fig. 9.4 Location of Cropped Images used for Running the Accuracy Analysis

and 80 pixels. For each image shift "S," analysis was performed using interrogation windows of sizes 16x16, 32x32, 64x64, 128x128, and 256x256 pixels.

9.3.2 Calibration Results

Fig. 9.5 shows representative results, for movements of 15, 30 and 40 pixels. The window size used in these figures is 64x64 pixels with spacing equals to 64 pixels (no overlap). The zoom level corresponding to this analysis was 1mm = 9 pixels. Analysis of the results shows that the correlation system can predict accurate results up to a threshold movement. Any movement beyond this threshold results in a high probability of having misleading results. For example use of a 64x64 window predicted movements up to 15 pixels with a high accuracy (Fig. 9.5a). For a 30-pixel movement, 57% of the windows calculated acceptable results (Fig. 9.5b), and for an image shift of 40 pixels, only 16 % of the windows calculated acceptable results. The results (Fig. 9.6) also indicate that the calculated movements using the correlation system have a constant error on the order of 0.1 pixel.

The same error analysis scheme was repeated (Fig. 9.7) for the same model using another zoom level of the camera (1 mm = 27 pixels). The main observation in the second set of analyses is that the accuracy is lost at smaller displacements. For example, the 64x64 pixel window showed good accuracy up to 15 pixels displacement for the low-resolution case (1mm = 9 pixels), while this analysis has accurate result up to displacements equal to 10 pixels only.

In summary, a small window size is required for high-resolution analysis in order to increase the number of interrogation windows. On the other hand, the previous analyses indicate that the probability of having large errors resulting from DIC becomes larger as the window size becomes smaller, thus accurate analysis requires large window sizes. Very small windows do not detect the difference in deformation because windows become similar and correlation is lost. This discrepancy can be resolved by using adaptive cross correlation as shown in Chapter 8 and 10.

Table 9.1 and 9.2 show the maximum calculated displacement for each window size used in the analysis at different accuracy levels. The performance of the analysis is very accurate at small displacements. The calculated root mean square

(RMS) error for all interrogated window sizes indicates that a small RMS error on the order of the order of 0.1 pixel occurs up to a threshold displacement. Beyond the threshold displacement, the RMS error jumps to very high values (Fig 9.6 and 9.7).

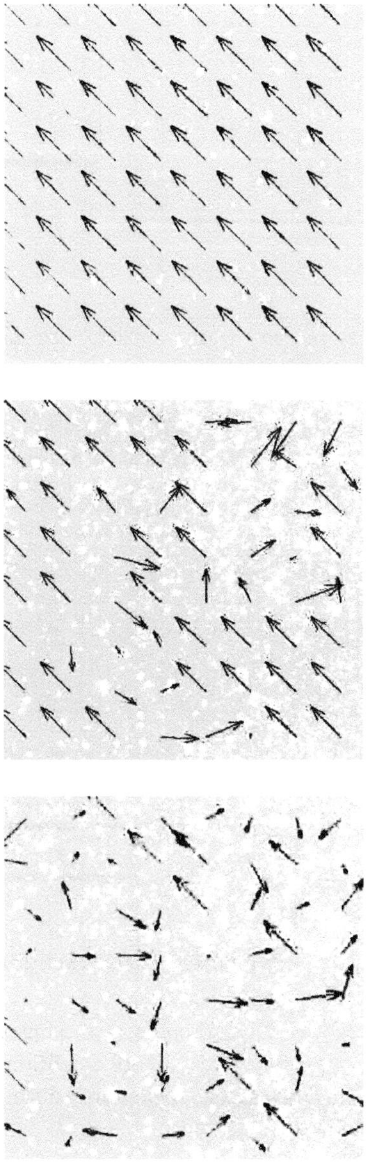

(a)

(b)

(c)

Fig. 9.5 Analysis Results using 64x64 Pixel Interrogation Window for (a)15 Pixels-Movement, b) 30 Pixels-Movement, and, c) 40 Pixels Movement

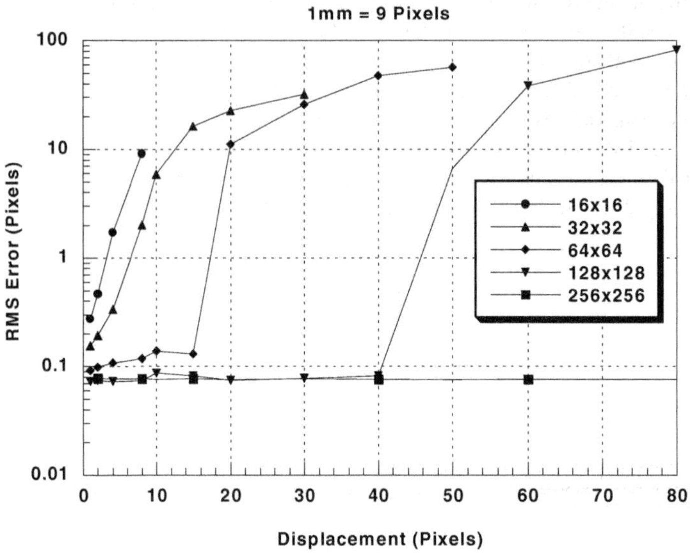

Fig. 9.6 Root Mean Square Error for All Calculated Displacements (1mm = 9 Pixels)

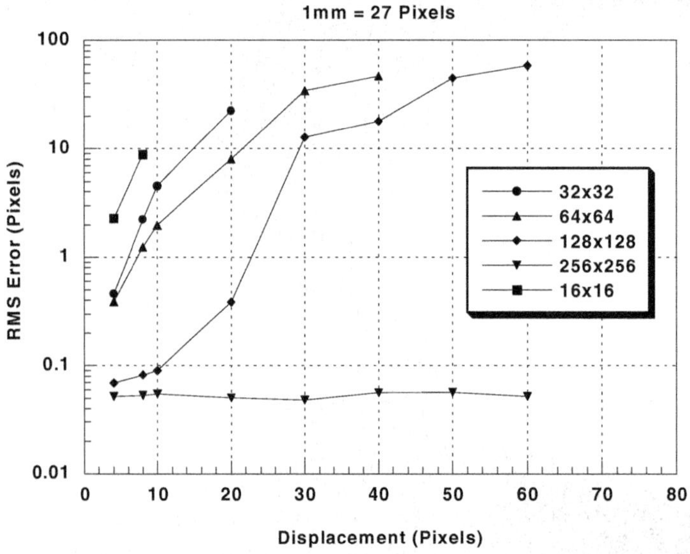

Fig. 9.7 Root Mean Square Error for All Calculated Displacements (1mm = 27 Pixels)

Table 9.1 Maximum Calculated Displacement for Different Window Sizes (1mm = 9 Pixels)

Window Size	80 % Probability Error is less than 20 %	80 % Probability Error is less than 10 %	95 % Probability Error is less than 5 %
16x16	4 Pixels	---	---
32x32	10 Pixels	8 Pixels	4 Pixels
64x64	20 Pixels	20 Pixels	15 Pixels
128x128	50 Pixels	50 Pixels	50 Pixels
256x256	(More than 80)**	(More than 80)**	(More than 80)**

** Out of analysis range.

Table 9.2 Maximum Calculated Displacement for Different Window Sizes (1 mm = 27 Pixels)

Window Size	80 % Probability Error is less than 20 %	80 % Probability Error is less than 10 %	95 % Probability Error is less than 5 %
32x32	8 Pixels	8 Pixels	4 Pixels
64x64	20 Pixels	10 Pixels	10 Pixels
128x128	40 Pixels	30 Pixels	20 Pixels
256x256	(More than 60)**	(More than 60)**	(More than 60)**

9.4 Other Errors

9.4.1 Reflection

Light reflection from model walls, and surrounding objects is a major source of error. A dark room is preferred but black sheets surrounding the setup can sufficiently improve the quality of the analysis. Failure to eliminate reflections from ambient light will render the experiment useless. The structural members studied in the model may also reflect laser light inside the soil model. These reflections result in loss of accuracy in some location in the soil model, especially near foundation elements. Finishing model foundations in rough black surfaces can help minimize these reflections.

9.4.2 Alignment Error

The optical axis of the camera must be set perpendicular to the image plane. The error in movement estimation increases if the camera is misaligned.

9.4.3 Focus Errors

Because images are taken for the laser sheet inside the transparent synthetic soil model; particle movement in front of the image slice may influence the analysis. Cameras with small depth of field will perform better.

9.4.4 Out-of-Plane and Rotational Movement Errors

One of the main assumptions in digital image correlation is that the contrast will follow the particle movement. Out-of-plane movement may result in losing the identity of the image part under consideration. Out-of-plane movements can be easily ignored in the case of rigid surface displacement. In the case of transparent synthetic soils, movement of particles out of plane cannot be prevented. In addition, rotation of the particles sometimes results in changing the speckle pattern. These errors can be reduced by acquiring more images and minimizing the movement step per image.

9.5 Application of DIC in Modeling Soil Structure Interaction

A setup consists of a transparent soil model, laser source, sheet generator lens, and a digital camera; all mounted on an optical table. The model dimensions are 305x152x50 mm (12x6x2 inch). The model is made of Plexiglas, and was filled with fine silica gel (particle size 0.5-1.5 mm). A rigid footing with a footprint measuring 50x50 mm (2x2 in) was mounted on the surface of the silica gel. The footing was pushed into the model using a system that insures that loading is axial with no torque. The total displacement recorded by the LVDT was 2.41 mm (0.095 in).

9.5.1 DIC Analysis Results

Two images for analysis were captured, before and after displacement. The DIC algorithm previously explained in Section 9.2.5 was used to compare the two images and obtain the displacement field. The output of the algorithm is shown in Fig. 9.8, superimposed on the speckle pattern generated by the interaction of laser light and transparent silica gel prior to deformation. The arrows indicate the displacement of each interrogation window within the image. Arrows are plotted at the centers of the window they represent. Analysis was performed using a window size of 256x256 pixels (28.4mm x 28.4mm). The spacing between windows was 64 pixels (7.1mm), i.e. the overlap between adjacent windows was 192 pixels (21 mm).

The calculated movements are consistent with classical bearing capacity theory with a wedge failure clearly visible beneath the footing. The lack of symmetry in calculated movement is due to the diffusion of laser light and loss of power as the laser penetrates the soil material. This error can be minimized using a stronger laser.

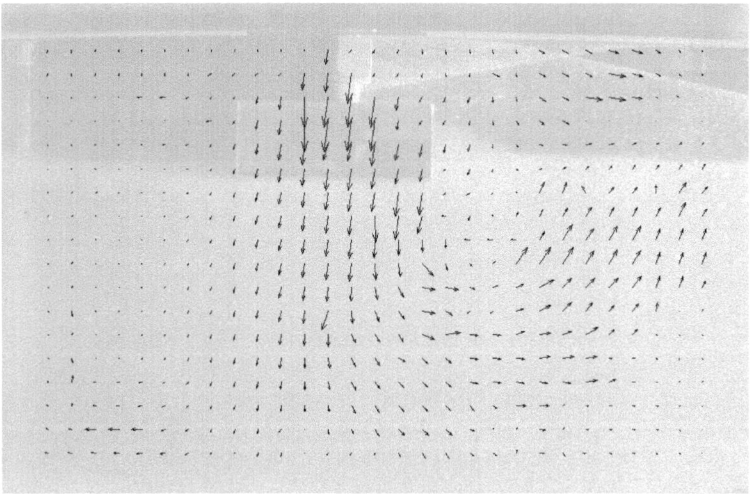

Fig. 9.8 DIC Analysis Results. Arrows Indicate Spatial Displacements in a Transparent Soil Slice

9.5.2 Comparison with FEM

The finite element program Plaxis was used to compare with the DIC analysis results. Plaxis can only simulate plane strain or axisymmetric problems using 4 built in constitutive material models developed for various soils (Brinkgreve and Vermeer 1998). The hardening model was used for numerical simulation of the transparent soil. The input parameters were based on drained triaxial tests performed on loose silica gel available in Chapter 6. First, a triaxial test performed with a confining pressure (σ_3) of 140 kPa (20 psi) was used to obtain the input values for Plaxis shown in Table 9.3. Next, a test performed with a confining pressure of 280 kPa (40 psi) was used to verify the input values (Fig. 9.9). The simulated test results are in a good agreement with the measured triaxial test results.

A rigid footing was simulated on the top of a soil model having the same dimensions of the model shown in Fig. 9.8. A prescribed settlement of 2.41mm was applied to the top of the footing.

The results from Plaxis are presented along with measured DIC results for two horizontal cross sections. The first is located 36 mm below the initial soil surface (Fig. 9.9), and the second is 64 mm below the initial soil surface (Fig. 9.10). The cross sections show that the displacement trend is the same in both analyses. The values of displacement are somewhat different, which might be due to averaging of the movement calculation in DIC, since a window of 256x256 pixels (28x 28 mm) was used in the analysis.

The FEM simulation proves that DIC can capture the actual movement in the soil mass. However, more work is needed to improve the analysis results quantitatively by using high-resolution cameras and smaller area of focus. Additionally, improvements in experimental setup by using a higher power laser, purer silica, and better alignment of the camera setup are expected to increase the versatility of the system.

Table 9.3 Assumed Parameters for Plaxis Hardening Soil Model

Symbol	Name	Assumed Value
φ	Angle of internal friction	33°
ψ	Angle of dilatancy	0°
C	Cohesion	0 lb/in^2
E_{50}^{ref}	Secant stiffness from drained triaxial test	3300 psi
E_{oed}^{ref}	Tangent stiffness for primary oedometer loading	3000 psi
M	Power for stress-level dependency of stiffness	0.51
E_{ur}^{ref}	Unloading/reloading stiffness	10000 psi
υ_{ur}	Poisson's ratio for unloading and reloading	0.45
p^{ref}	Reference stress for stiffness	20 psi
K_0^{nc}	K_0-value for normally consolidated	1-sinφ
$\sigma_{tension}$	Tensile strength	0 psi
$c_{increment}$	Increase in cohesion with depth	0 psi/in

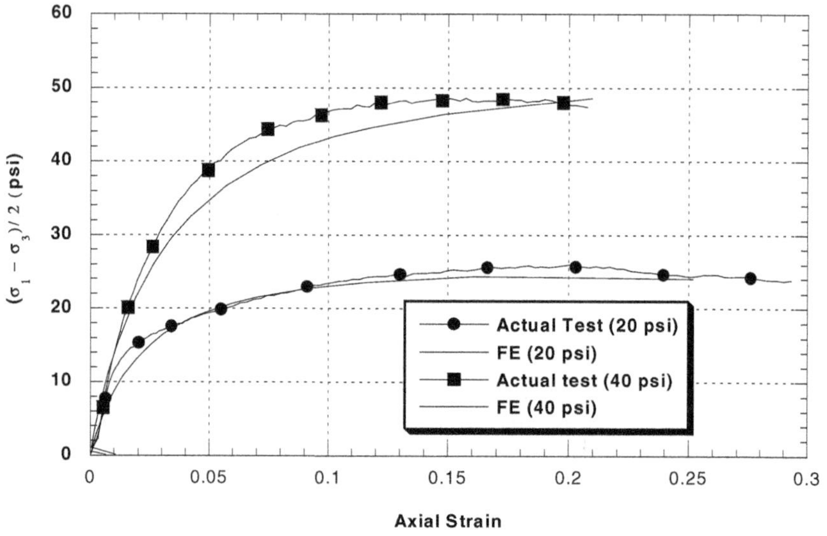

Fig. 9.9 Simulation of Stress-Strain Curves for 2 Triaxial Tests at σ_3 = 140 and 280 kPa (20 and 40 psi)

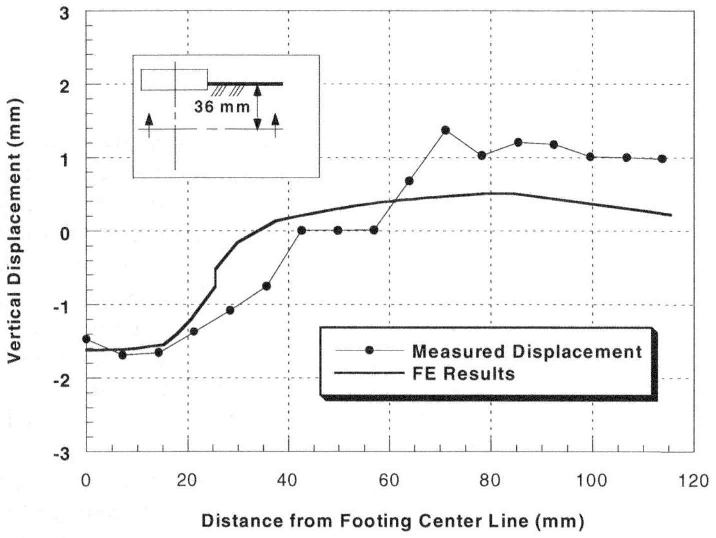

Fig. 9.10 Vertical Displacement Distribution at a Horizontal Cross Section 36mm Below the Initial Soil Surface

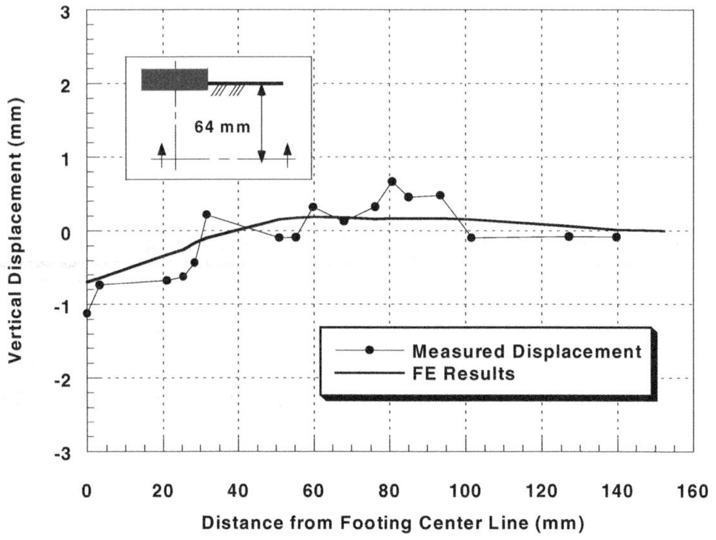

Fig. 9.11 Vertical Displacement Distribution at a Horizontal Cross Section 64 mm Below the Initial Soil Surface

9.6 System Limitations

Presence of impurities in transparent soils combined with the limited power of the available laser resulted in attenuation of the laser light sheet for distance greater than 10 cm. A stronger laser should greatly increase the size of the analysis area.

Out-of-plane movement cannot be measured using this system. Additionally, the presence of large out-of-plane movement may result in de-correlation.

9.7 Conclusions

DIC was shown to be a reliable technique for measuring deformation in transparent media. The accuracy of the technique depends on the displacement magnitude, and the window size used in the analysis. In general, the technique provides accurate results for small displacements when using large interrogation window sizes. It was shown that for every window size there is a maximum reliable displacement that can be calculated. Finite element analysis of a strip footing on a transparent synthetic soil model showed good agreement with the DIC results. DIC can provide a valuable input in analyzing complicated models by visualizing modes of failure and ranges of influence.

References

Alshibli, K., Sture, S.: Sand shear band thickness measurement by digital imaging techniques. J. of Computing in Civil Engineering 13(2), 103–109 (1999)

Bourdeau, P.: Radiographic visualization in experimental soil mechanics. In: Proc. Digital Image Processing: Techniques and Applications in Civil Engineering, ASCE, pp. 125–134 (1993)

Brinkgreve, R., Vermeer, P.: Plaxis Finite Element Code for Soil and Rock Analysis Plaxis Manual Version 7, Balkema, Rotterdam/Brookfield (1998)

Butterfield, R., Andrawes, K.: The visualization of planer displacement field. In: Parry, R. (ed.) Proc. Roscoe Memorial Symposium: Stress Strain Behavior of Soils, Cambridge University, pp. 467–475 (1971)

Guler, M., Edil, T., Bosscher, P.: Measurement of particle movement in granular soils using image analysis. J. of Computing in Civil Engineering, ASCE 13(2), 116–122 (1999)

Gustafsson, L., Gustafsson, P.: Studying mixed granular flows by image analysis. In: Proc. 11th Conference ASCE Engineering Mechanics Division, pp. 614–617 (1996)

Liberzon, A., Gurka, R., Shavit, U.: URAPIV^{TM} Particle Image Velocimetry Research Group (2000), `http://www.openpiv.net/` (last accessed October 2009)

Raschke, S., Hryciw, R., Donohoe, G.: Microdeformations in sands by digital image processing and analysis. Transportation Research Record 1548, 31–37 (1996)

Robinsky, E., Morrison, C.: Sand displacement and compaction around model friction piles. Canadian Geotechnical J. 1(2), 81–91 (1964)

Saada, A., Liang, L., Figueroa, J., Cope, T.: Bifurcation shear band propagation in sands. Geotechnique 49(3), 367–385 (1999)

Sadek, S.: Soil Structure Interaction in Transparent Synthetic Soils Using Digital Image Correlation, Ph.D. Dissertation, Polytechnic University, Brooklyn, NY (2002)

Sadek, S., Iskander, M., Liu, J.: Accuracy of digital image correlation for measuring deformations in transparent media. ASCE J. of Computing in Civil Engineering 17(2), 88–96 (2003)

Westerweel, J., Dabiri, D., Gharib, M.: The effect of a discrete window offset on the accuracy of cross-correlation analysis of digital PIV recording. Experiments in Fluids 23(1), 20–28 (1997)

Chapter 10
Validation of Measured 2D Deformations

Jinyuan Liu and Magued Iskander

Abstract. The modeling capacity of transparent soils was studied by comparing 2D soil deformations under a model footing on silica gel and on sand. General agreement between deformation in silica gel and sand was found in almost every aspect. Non-intrusive measurements of deformation inside a soil model were also realized by cross-correlating speckle images generated by interaction of laser light and transparent synthetic soils. These results presented in this chapter validate the premise that synthetic transparent soils are a suitable surrogate for natural soils in model studies.

10.1 Introduction

The development of transparent materials, which can be customized to meet model test requirements in terms of strength and deformation, is the necessary step to model natural soils in deformation and strength studies. Two compatible families of transparent materials have been developed by matching refractive indices of silica gels or powders with pore fluid. The first family, discussed in Chapter 5, is made of amorphous silica powder for modeling clays (Iskander et al 2002a, Liu et al 2002). The second family, discussed in Chapter 6, is made of amorphous silica gel for modeling sands (Iskander et al 2002b).

The discrepancy between the theoretical bearing capacity derived by Terzaghi (1943) or that modified by Meyerhof (1951) and the measured performance of footings is well known among practicing engineers. Few studies were performed on soil deformations under shallow foundations (Eggstad 1963 and Marangos 1994). In this chapter, soil deformations under a model footing are used to illustrate the potential application of transparent synthetic soils in geotechnical engineering research.

In this chapter, comparison of 2D deformation in sand and transparent synthetic soil was performed. The comparison study was performed under both dry and saturated conditions. Comparison was done for surface deformation of sand and silica gel. In the dry condition, silica gel is opaque and a light bulb was used to illuminate the surface. Under the saturated condition, a laser light sheet, generated

M. Iskander: Modeling with Transparent Soils, SSGG, pp. 181–225.
springerlink.com

by passing laser beam through a special line generator lens, was used to illuminate specific cross-sections inside synthetic soil models, as discussed in Chapter 9. A light bulb was still used for illuminating the saturated sand surface. Two additional cases, loose sand over dense sand and sand over soft clay, were modeled. Finally, 2D deformations inside transparent soils were studied using both silica gel and silica powder. All test groups are summarized in Table 10.1.

Table 10.1 Summary of Test Program (Liu 2003)

Group No.	Light	Material	Saturation Condition	Density	Cross-Section	Purpose
Group 1	Light bulb	Sand, Fine silica gel	Dry	Loose	Surface	Comparison
Group 2	Light bulb	Sand, Fine silica gel	Dry	Dense	Surface	Comparison
Group 3	Light bulb and laser	Sand, Fine silica gel	Saturated	Dense	Surface	Comparison
Group 4	Laser	Coarse, fine silica gel	Saturated	Dense	Surface	Modeling loose over dense Sand
	Laser	Fine silica gel, silica powder	Saturated	Dense	Surface	Modeling sand over soft clay
Group 5	Laser	Fine silica gel	Saturated	---	Inside	Modeling sand
	Laser	Silica powder	Saturated	---	Inside	Modeling clay

10.2 Test Program

10.2.1 Proposed Method

Digital image cross-correlation (DIC) has been investigated in Chapter 8 for use in measuring deformation from a sequence of images. In this chapter, DIC is used to obtain soil deformations under a model footing. The surface deformation of both silica gel and sand are obtained using DIC and compared to evaluate the modeling capacity of transparent soils. The deformations inside a transparent synthetic soil model are studied by slicing the model using a laser light sheet, as discussed in Chapter 9. The speckle image from the interaction between laser light and transparent synthetic soil is analyzed to obtain the deformation inside the model.

10.2.2 Set-Up

The test set-up, shown in Fig. 10.1, consist of a Cohu 2622 black & white CCD camera, a laser light source, a line generator lens, a loading frame, a test table, and a PC for image processing. The camera has a resolution of 640×480 pixels. It is controlled by the PC through a Matrox Meteor 2/4 frame grabber. A macro-zoom lens with a variable focus length from 18-108 mm was mounted on the CCD camera, which was set 20 cm away from the model with its optical axis perpendicular to the model. The control software was customized under Visual C^{++} environment using the Matrox Mil-Lite software commands. The laser was a 35mW *Melles Griot* He-Ne with a wavelength of 632.8 nm. Additional details of the test setup are available in Chapter 9.

A Plexiglas mode with dimensions of 50×150×300 mm was used in the tests presented in this chapter. A model footing with a plan dimension of 50 mm by 25 mm was used. The footing simulated a continuous footing with a width of 25mm. Load was applied vertically through a precise screw mechanism. A load cell and an LVDT were connected to the load-frame to measure load and deformation during the test.

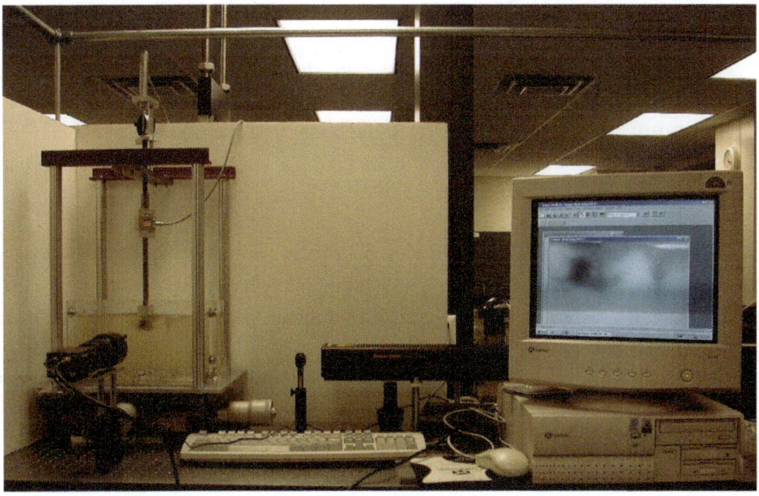

Fig. 10.1 Test Set-up

10.2.3 Test Program

The fine silica gel discussed in Chapter 6 was used. Sand with a matching grain size distribution was used in order to obtain comparable test results (Fig. 10.2). Two density conditions were studied, one was loose; and the other was dense.

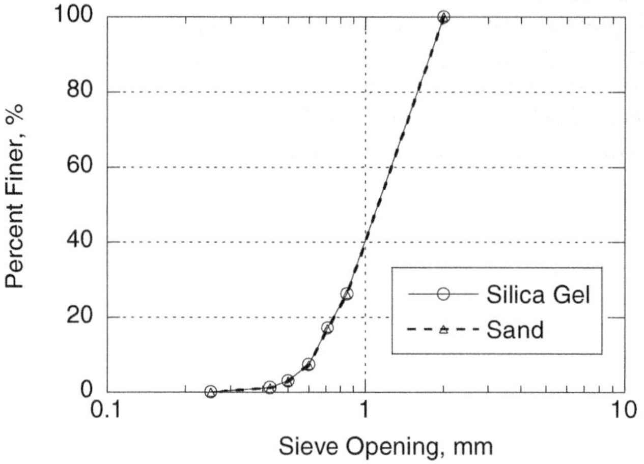

Fig. 10.2 Grain Size Distribution Curves for Sand and Silica Gel

For the tests with sand, the maximum density determined according to ASTM standard D698 was γ_{max}=16.29 kN/m^3. The minimum density determined according to ASTM Standard D4254 was γ_{min}=14.93 kN/m^3. In the model tests, the density of loose sand was γ=15.19 kN/m^3 corresponding to a relative density of 20.8%. The density of dense sand in the model test was γ=15.87 kN/m^3 corresponding to a relative density of 71.3%.

The minimum density of silica gel determined in accordance to ASTM Standard D4254 was γ_{min}= 6.89 kN/m^3. Because of the porous structure of silica gel, ASTM Standard D698 was not applicable in silica gel. The maximum density of silica gel obtained using a vibrator compaction was γ_{max}=7.51 kN/m^3. The maximum density of sand compacted using a similar vibrator procedure was 16.67 kN/m^3, which was slightly higher than 16.29 kN/m^3 determined from ASTM D698. The density of loose silica gel in the model test was γ=7.00 kN/m^3 corresponding to a relative density of 19 %. The density of dense silica gel was γ=7.34 kN/m^3 corresponding to a relative density of 74 %. The relative densities of sand and silica gel in both dense and loose conditions were comparable to each other.

Direct shear tests were performed on the sand according to ASTM Standard D3080. The friction angles derived from dry direct shear tests were approximate 38° for sand and 33° for fine silica gel (See Chapter 6).

10.3 Comparison under Dry Loose Condition

10.3.1 Failure Mode

The load settlement p-z curves of the footing on dry loose silica gel and dry loose sand are shown in Fig. 10.3. Load and settlement were measured from a load cell and an LVDT connected to the loading frame system. Images captured during the loading process were labeled on the load-settlement curve with numbers.

The failures in both sand and silica gel appear more like punching and local shear failures instead of general shear failure based on the research by Vesic (1973). The loads measured in sand were higher than in silica gel because of the density of sand was heavier than silica gel.

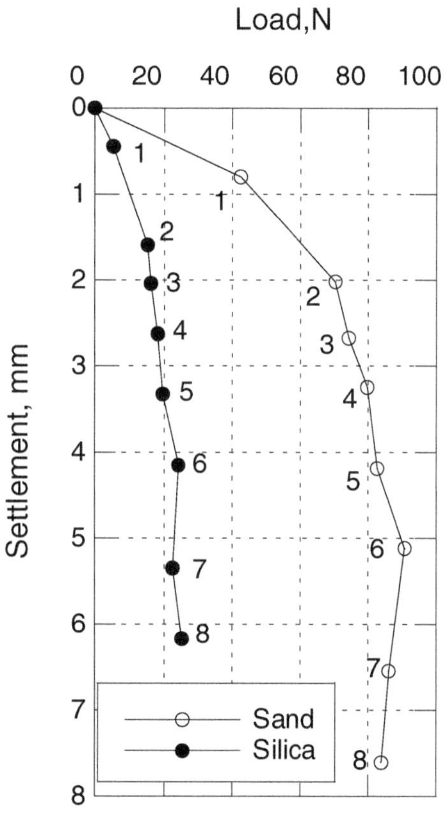

Fig. 10.3 p-z Curves for Footing on Dry Loose Sand and Silica Gel

10.3.2 Displacement Field

The measured displacement fields are presented in Fig. 10.4 for sand and Fig. 10.5 for silica gel with a typical image as the background. Because of the difficulty in identifying the corresponding points in both silica gel and sand, the comparison study was focused mainly on the pre-failure and post-failure. The pre-failure was the deformation between point 2 and 3 in Fig. 10.3. The post-failure was the deformation between the points of 6 and 7 in the Fig. 10.3.

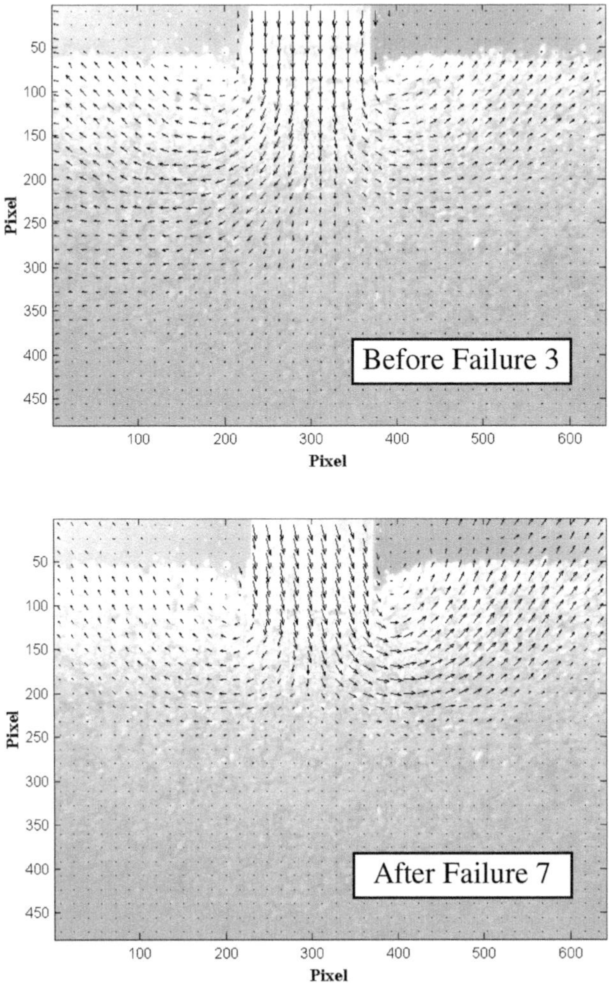

Fig. 10.4 Displacement Field in Dry Loose Sand (ACC Window Size= 128×128, 64×64, 32x32, 16×16, Offset=8)

The developed displacement field in loose silica gel is similar to that of loose sand. In both cases, large zones are mobilized by the movement of the footing. As the loading increases, the influence region shrinks to a limited area after the failure. The influence region is smaller in silica gel than that in sand, because silica gel has a smaller friction angle than sand.

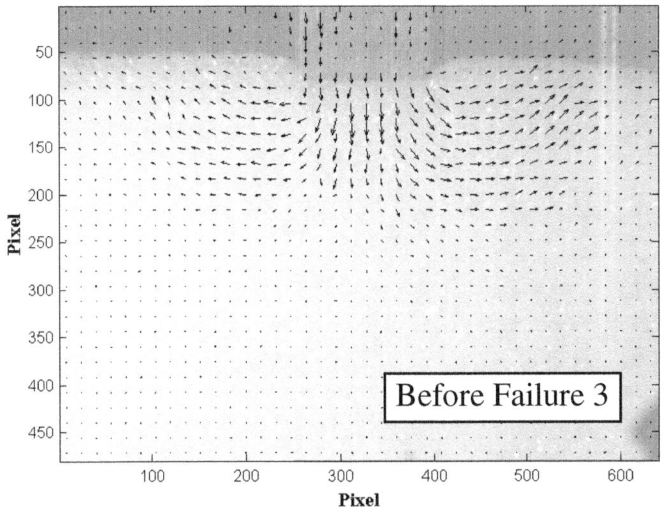

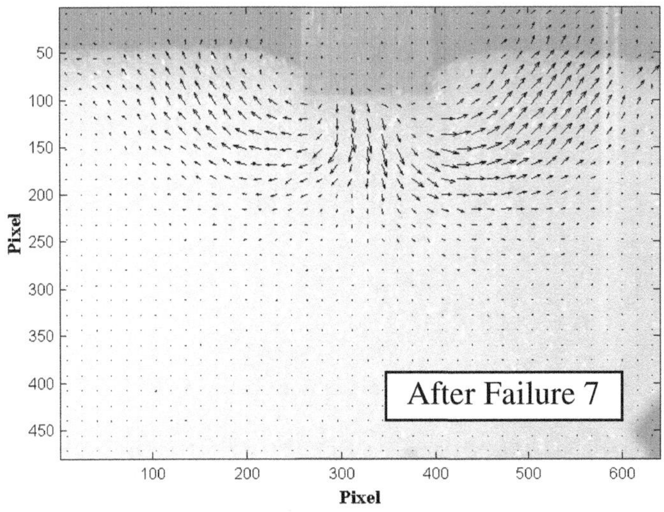

Fig. 10.5 Displacement Field in Dry Loose Silica Gel (ACC Window Size= 128×128, 64×64, 32x32, 16×16, Offset=8)

10.3.3 Maximum Horizontal Displacement

The maximum horizontal displacement was normalized by the deformation of the footing. It was plotted against the distance from the footing center. Comparison between dry loose sand and silica gel is shown in Fig. 10.6. The magnitude of maximum normalized horizontal displacement in silica gel is larger than that in

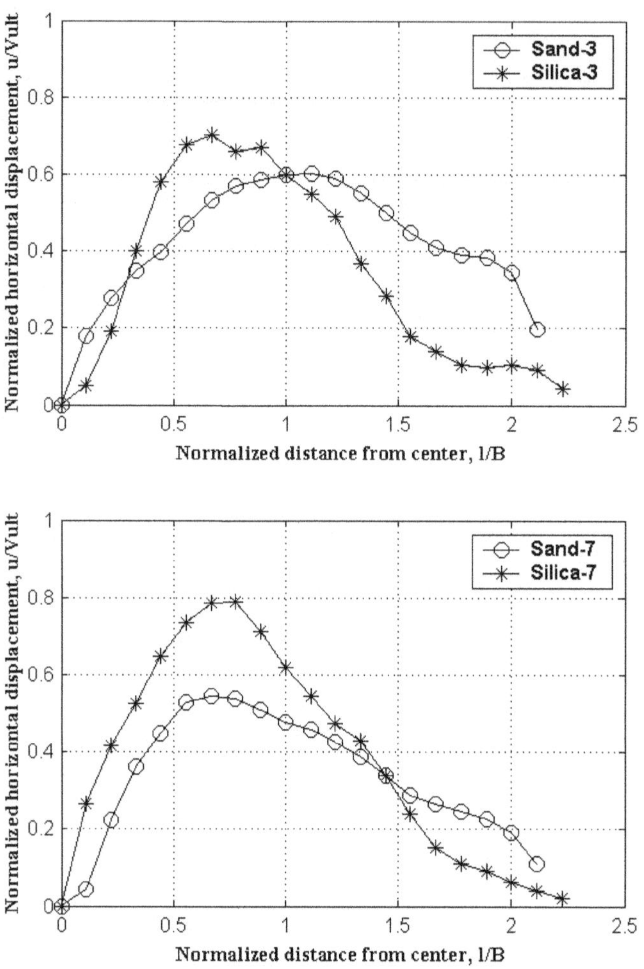

Fig. 10.6 Maximum Horizontal Displacement of Sand and Silica Gel

sand. The maximum normalized horizontal displacement in silica gel reduced with distance at a rate higher than that in sand. It becomes negligible at the distance around 1.75B from footing edge. In both cases, the maximum horizontal displacement occurs within 0.5B from the footing edge.

The depth of the maximum horizontal displacement is shown in Fig. 10.7, where the depth from the footing bottom was normalized by the footing width, B. The depth of maximum horizontal displacement in sand is deeper than the depth in silica gel. The depth in both sand and silica gel tends to reduce as the loading increases. The errors in identifying the corresponding points are believed to contribute to the difference of the magnitude.

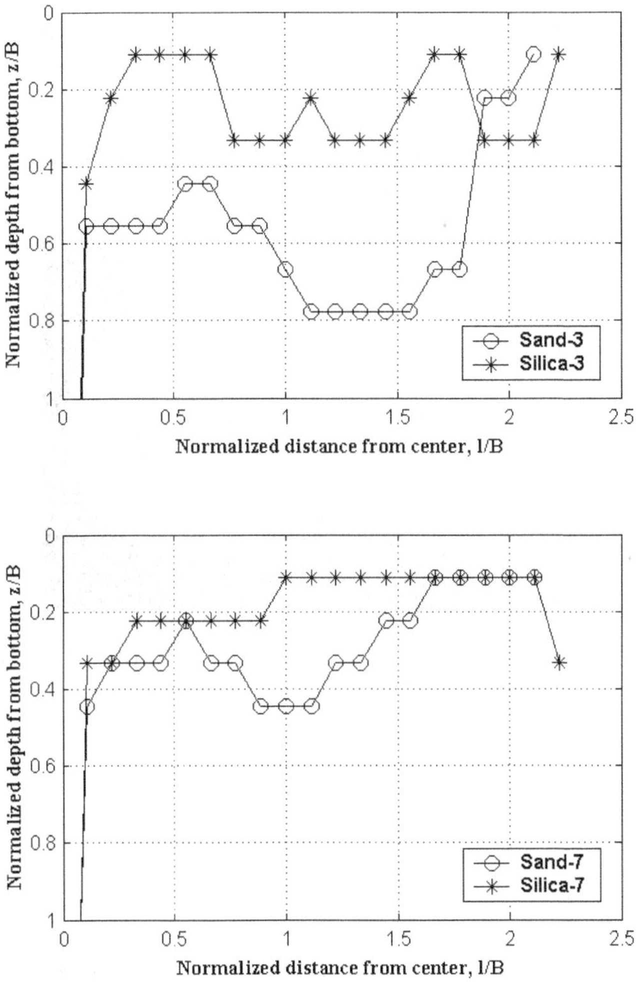

Fig. 10.7 Position of Maximum Horizontal Displacement

10.3.4 Settlement Distribution

Settlement along the cross-section across the footing centerline is shown in Fig. 10.8. The maximum settlement in silica gel appears some depth below the footing because of DIC errors around footing as evident in Fig. 10.4. The settlements in both cases were normalized for comparison. Except for the errors just beneath the footing, the distribution of settlement with depth is similar to each other.

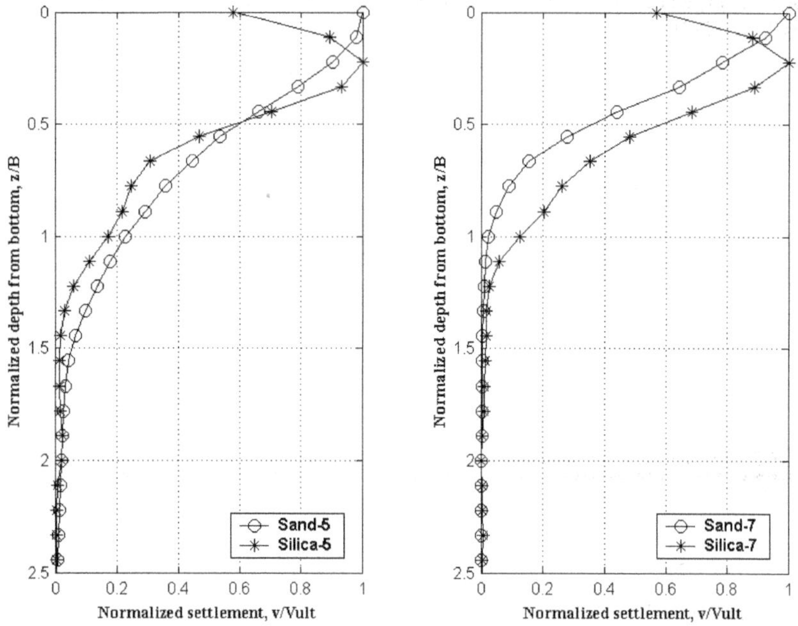

Fig. 10.8 Comparison of Settlement along Footing Center Line

Settlements along one horizontal cross-section 0.5B from footing bottom are presented in Fig. 10.9, where the settlements were normalized by the settlement of the footing. The settlement in silica gel is slightly larger than that in sand. Both the distribution of settlement along the horizontal plane and the influence area are similar to each other.

10.3.5 Strain Field

Strains can be calculated using the predicted displacement field per Roscoe et al (1963) and Yamamoto and Kusuda (2001). The displacement in the center of the region is assumed to be the same as that calculated based on region's movement. This assumption can be proved to be reasonable if the movement is relatively small. Green's strains are used regardless of the magnitude of the displacement, as follows.

$$\varepsilon_x = \frac{\partial u}{\partial x} + \frac{1}{2}\left[\left(\frac{\partial u}{\partial x}\right)^2 + \left(\frac{\partial v}{\partial x}\right)^2\right] \tag{10.1}$$

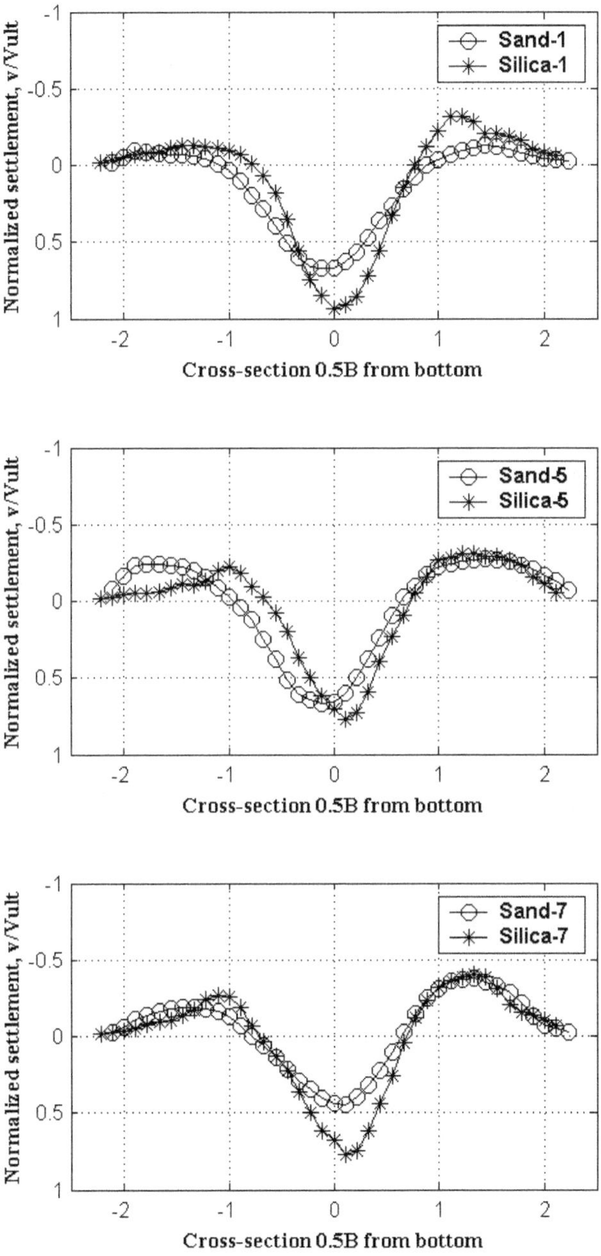

Fig. 10.9 Settlement of a Cross-Section 0.5B Deep in Loose Sand and Silica Gel

$$\varepsilon_y = \frac{\partial v}{\partial y} + \frac{1}{2}\left[\left(\frac{\partial u}{\partial y}\right)^2 + \left(\frac{\partial v}{\partial y}\right)^2\right] \tag{10.2}$$

$$\gamma_{xy} = \frac{\partial u}{\partial y} + \frac{\partial v}{\partial x} + \left[\frac{\partial u}{\partial x}\frac{\partial u}{\partial y} + \frac{\partial v}{\partial x}\frac{\partial v}{\partial y}\right] \tag{10.3}$$

where u and v are the displacement components in the x and y directions. ε_x and ε_y are the axial strains in the x and y directions. γ_{xy} is the shear strain. The maximum shear strain, γ_{max}, can be calculated using the strains above.

$$\gamma_{\max} = \sqrt{(\varepsilon_x - \varepsilon_y)^2 + \gamma_{xy}{}^2} \tag{10.4}$$

The volume strain, ε_{volume} , can be computed as follows,

$$\varepsilon_{volume} = \varepsilon_x + \varepsilon_y \tag{10.5}$$

The maximum shear strain fields are shown in Fig. 10.10 for sand and Fig. 10.11 for silica gel. The failure plane was studied by locating the peak of the maximum shear strain field, which was assumed to coincide with failure plane.

At the beginning stage, the shear strain bulb appeared in the sand. The maximum shear strain was concentrated only under the footing. No continuous failure envelope was found at this stage. As the load approached the ultimate bearing capacity of the footing, the shear strain at both sides of the footing was reduced. The peak of the maximum shear strain developed and converged from both sides along one failure plane.

Based on Terzaghi (1943) and Vesic (1973), the failure plane under a footing can be divided into three different parts, triangular zone just beneath the footing, two radial shear zones, and two Rankine passive zones.

The triangular zone immediately under the footing is gradually reduced as the loading increases. The angle at the footing edge was approximately 49° at stage 3 and was reduced to approximately 33° at stage 7 in sand. This is different from the assumed constant angle in the assumptions made by Terzaghi (1943), Meyerhof (1951), and Vesic (1973). The influence zone shrinks as the loading increases, which is consistent with that found in displacement field. Similar development was found in silica gel (Fig. 10.11).

10.3.6 Volume Strain

In order to illustrate clearly the different compression and expansion regions, the volume strain fields of Eq.10.5 are shown in Fig. 10.12 for sand and Fig. 10.13 for

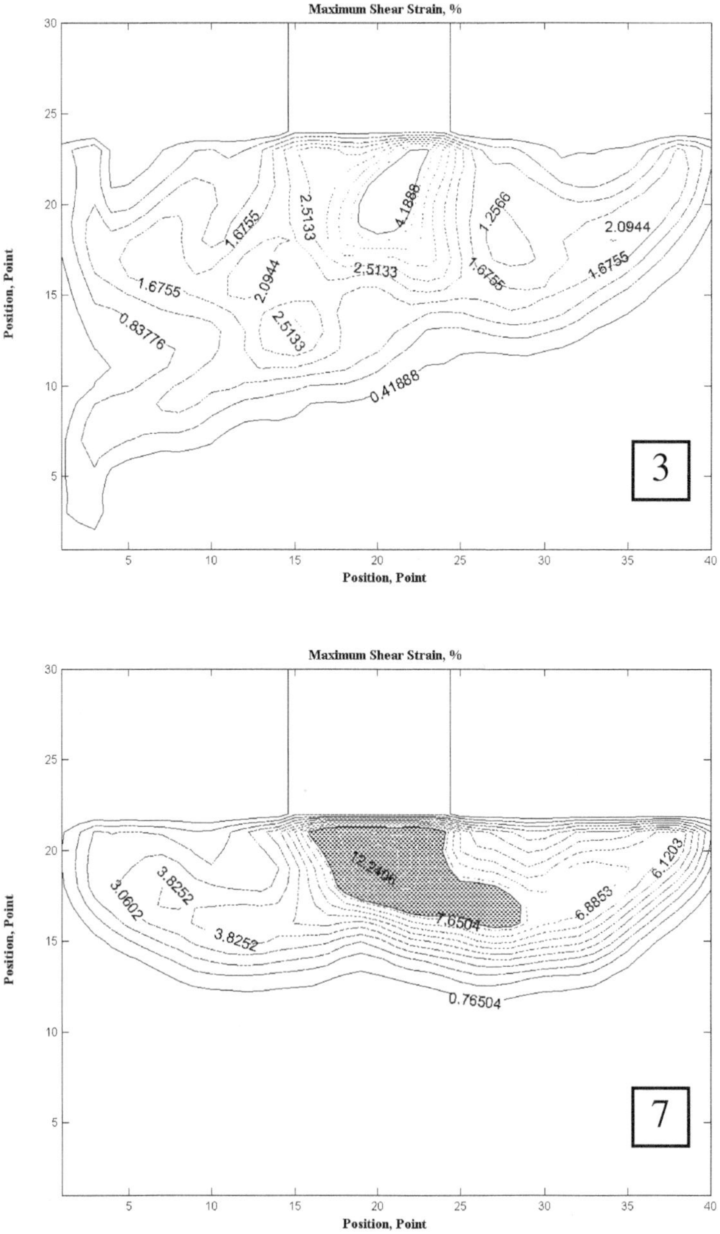

Fig. 10.10 Maximum Shear Strain in Dry Loose Sand

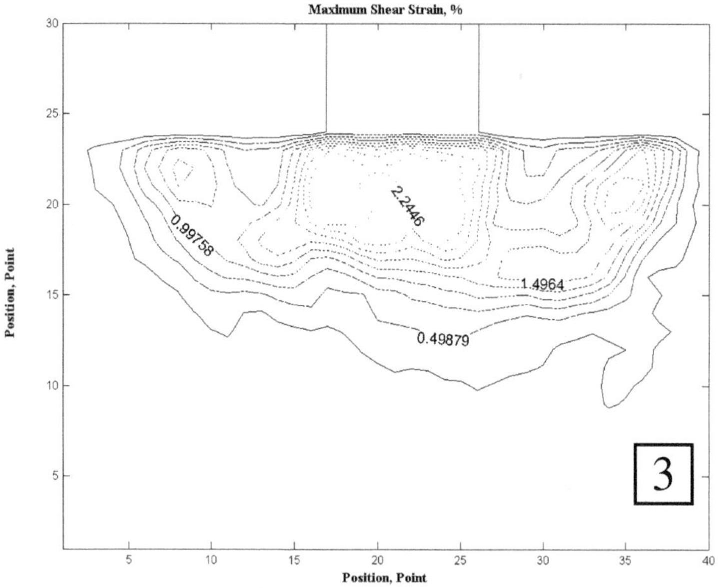

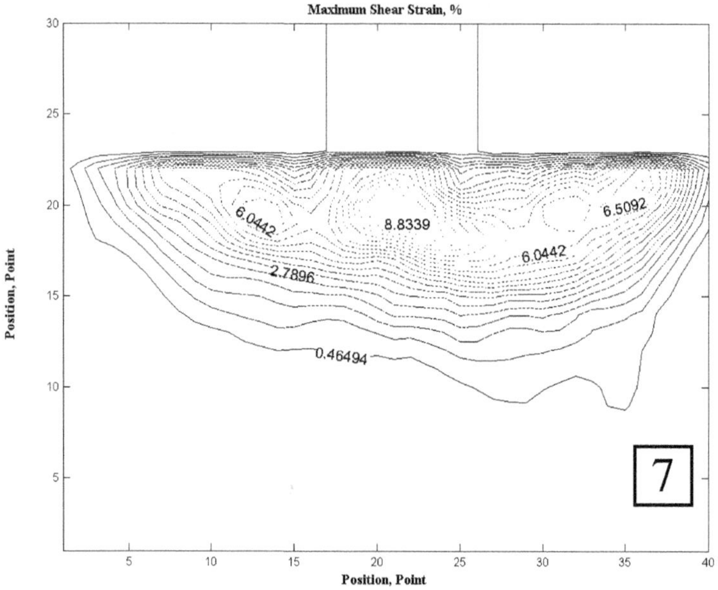

Fig. 10.11 Maximum Shear Strain in Dry Loose Silica Gel

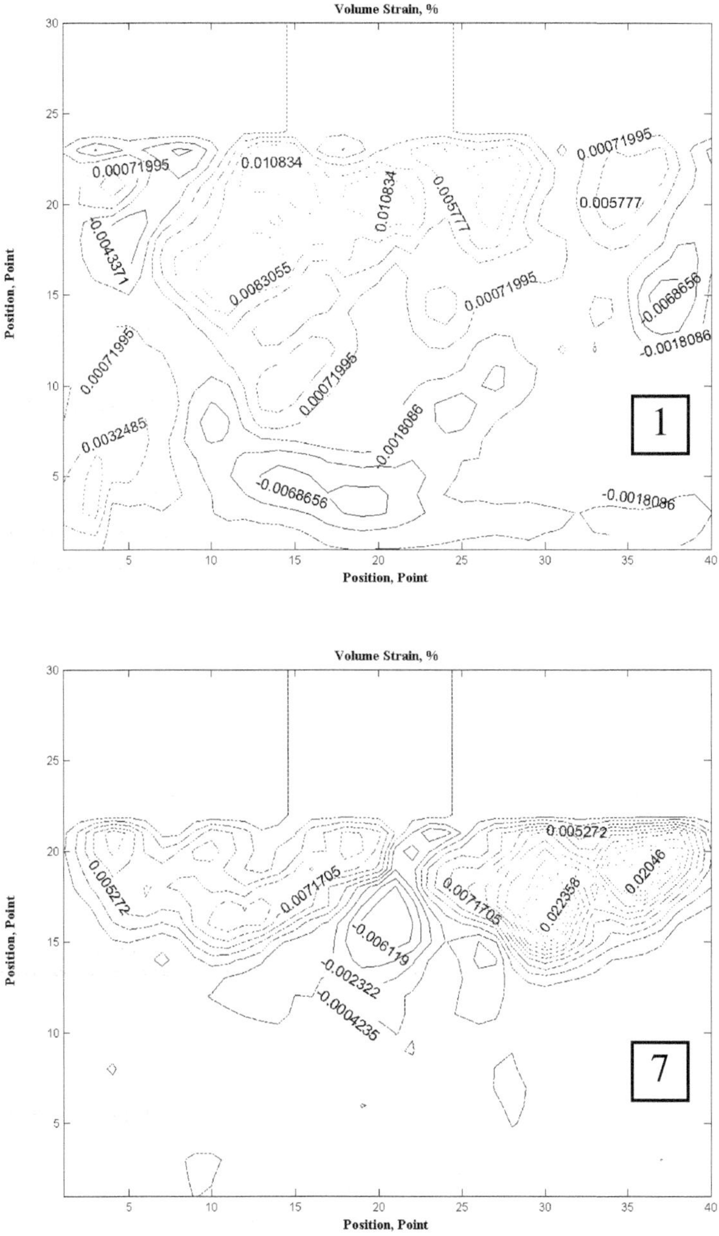

Fig. 10.12 Volume Strain in Dry Loose Sand

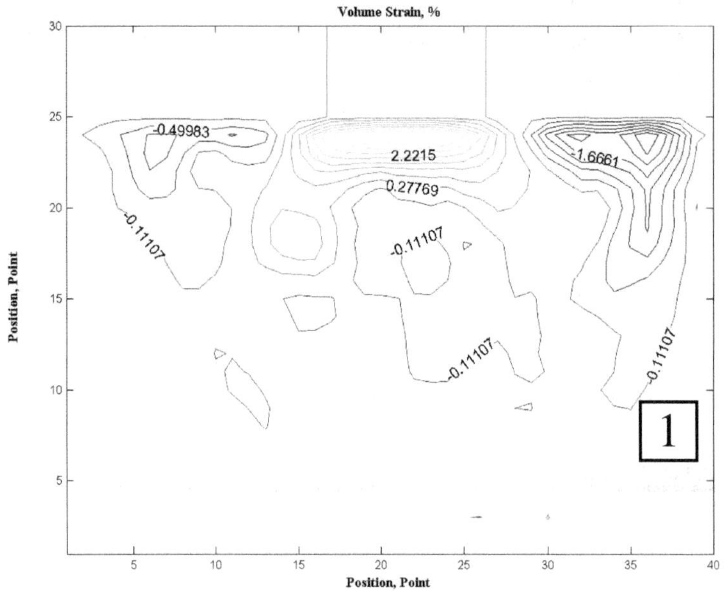

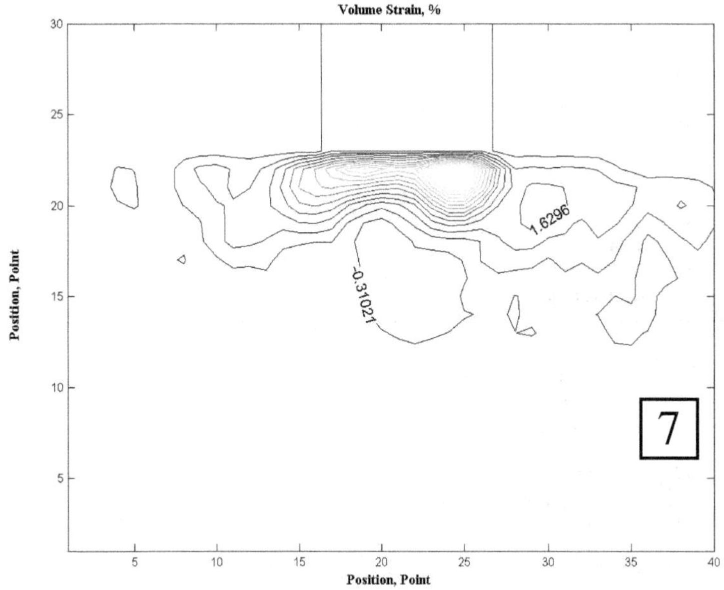

Fig. 10.13 Volume Strain in Dry Loose Silica Gel

silica gel, where the positive value shows volume expansion and negative value means volume compression. There is expansion immediately under the model footing in both silica gel and sand. The behaviors are more similar to each other after the loading exceeded the ultimate capacity. In silica gel, the zone of expansion is more localized around the footing than in sand.

As the loading increases, the expansion region shrinks to the zone immediately beneath the footing and the rest of the area tends to be in compression. The development of volume strain in silica gel is similar to that of natural sand. The expansion in silica gel proved that transparent synthetic soil can be used in modeling sand, including dilatancy. The difference in sample densities is believed to contribute to any discrepancies.

10.3.7 Vertical Strain

Comparison of the vertical strain along cross section passing though the footing centerline is shown in Fig. 10.14. The magnitudes are similar but not identical because of the different magnitude of footing settlement. The maximum strain in both cases occur at a depth of 0.25 B from the footing bottom, which is smaller than 0.75B given by Eggstad (1963) on homogeneous sand under a circular footing with a diameter of B.

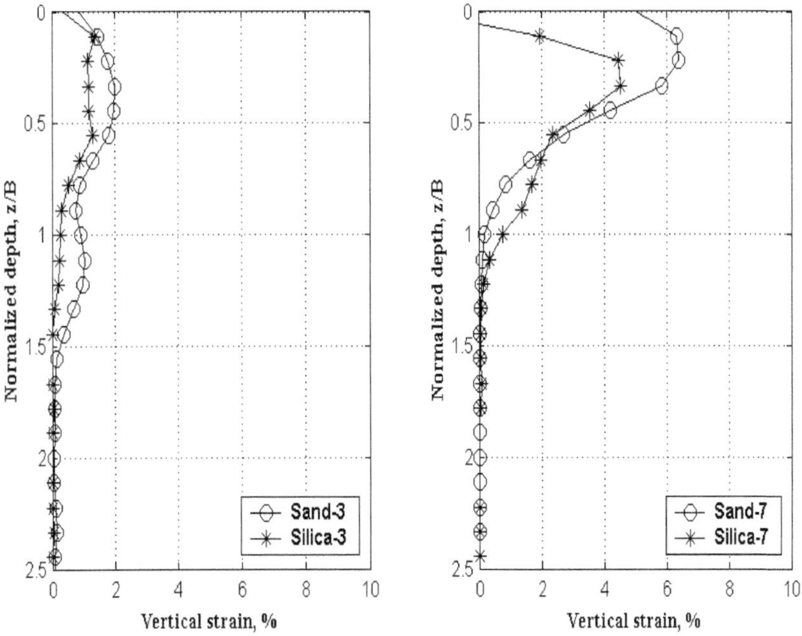

Fig. 10.14 Vertical Strain along the Footing Center Line in Dry Loose Condition

The depth of influence zone decreases from approximate 1.5B to 1B at the final failure stage. Generally speaking, the distribution of vertical strain along the centerline resembles each other.

10.4 Comparison under Dry Dense Condition

The p-z curves of footing on both dry dense silica gel and sand are shown in Fig. 10.15. The images corresponding to various loading stages were marked on the load settlement curve. A punching or local shear failure is evident for both sand and silica gel. The difference on the magnitude of load is caused by the differences in material properties.

The predicted displacement fields are shown in Fig. 10.16 for dry dense sand and Fig. 10.17 for dry dense silica gel. The displacement fields are similar to each other. At the beginning, the movement of the footing causes a large influence zone, which shrinks as the loading increases. At the failure stage, the depth of influence is reduced to approximately 1 B from the footing bottom. An asymmetrical deformation pattern also appeared in the tests in dry dense sand. In the region around the model footing, there were errors in the deformation prediction because of light reflection from the footing. The errors in the displacement field around the footing caused errors in the following strain analyses.

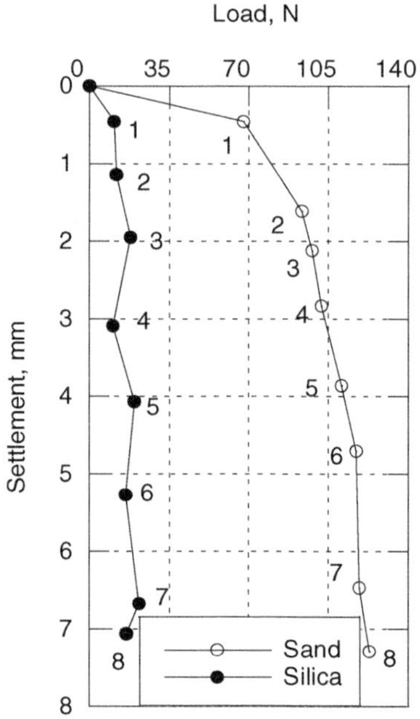

Fig. 10.15 p-z Curves of Footing on Dry Dense Sand and Silica Gel

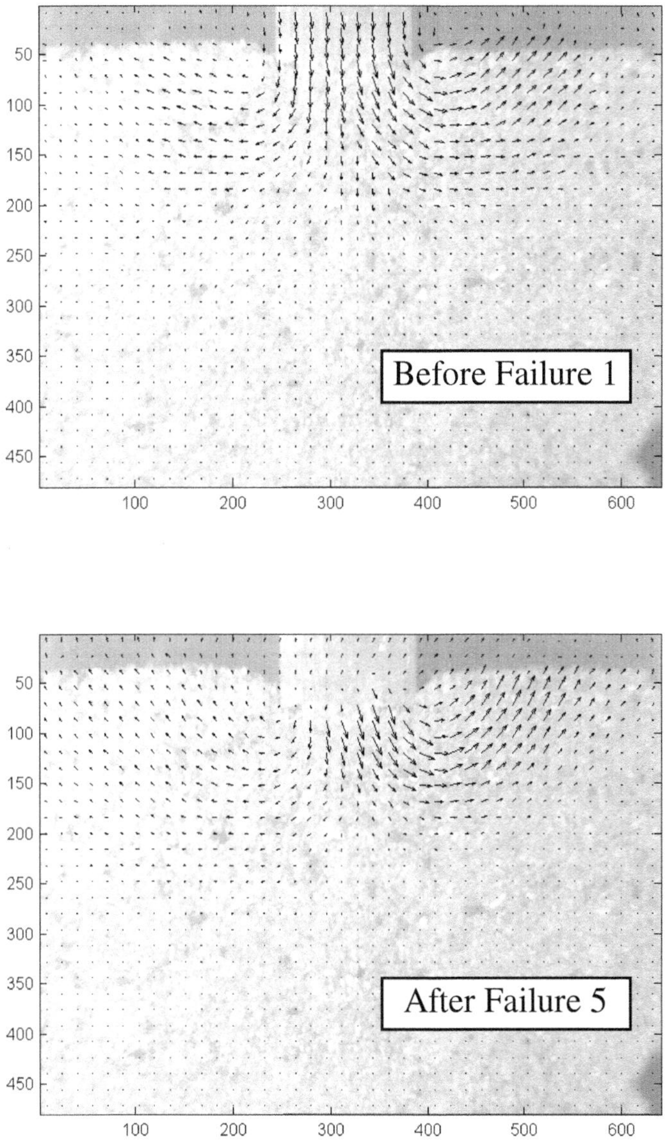

Fig. 10.16 Predicted Soil Displacement under Footing in Dry Dense Sand (ACC Window Size= 128×128, 64×64, 32x32, 16×16, Offset=8)

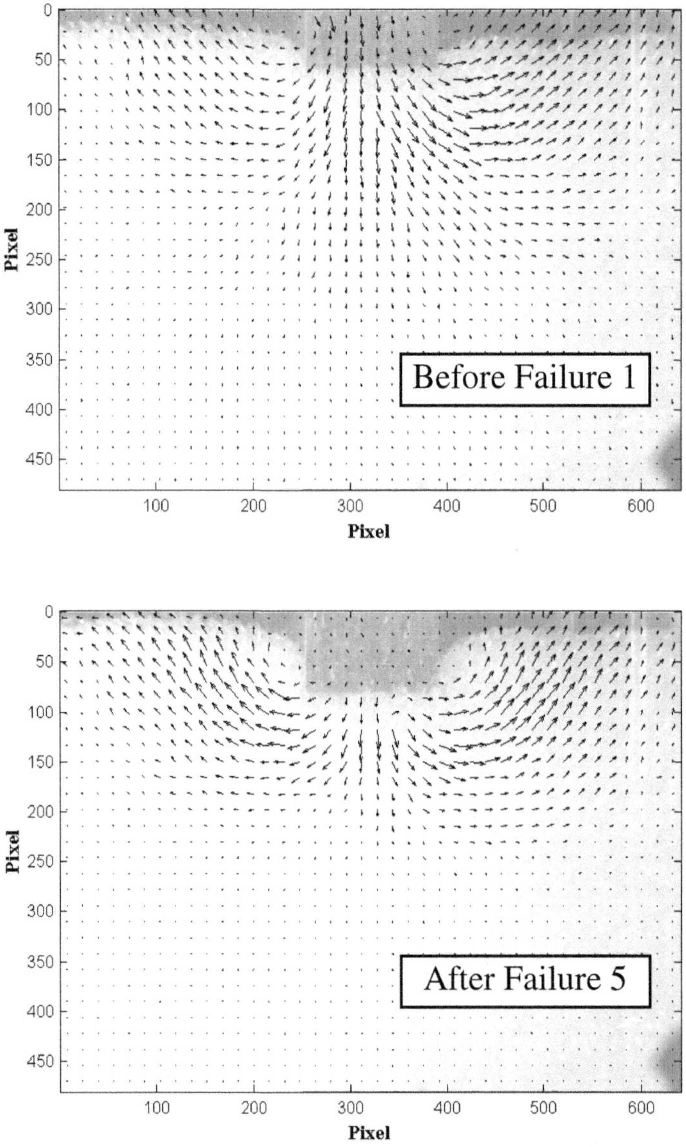

Fig. 10.17 Predicted Soil Displacement under Footing in Dry Dense Silica (ACC Window Size= 128×128, 64×64, 32x32, 16×16, Offset=8)

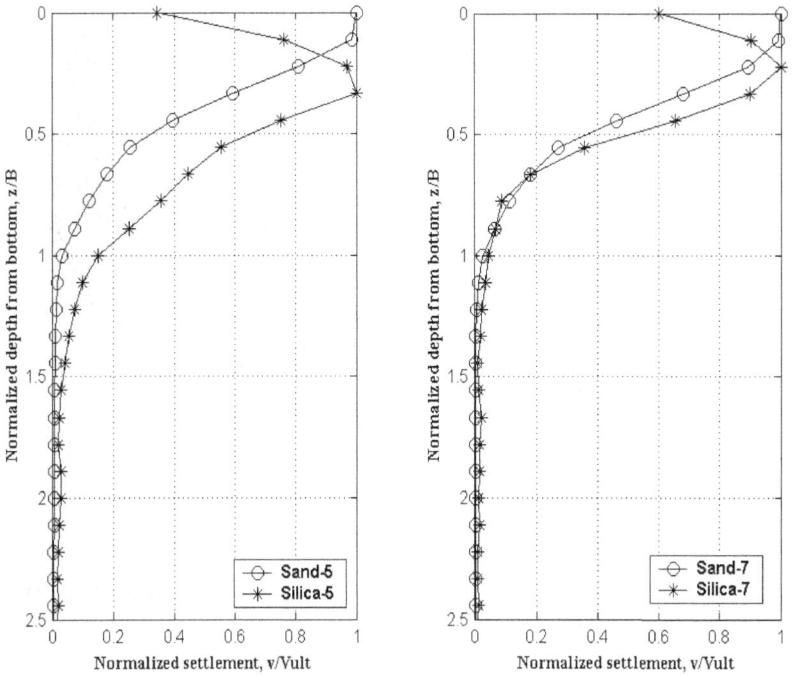

Fig. 10.18 Comparison of Settlement along Footing Center Line

10.4.1 Settlement Comparison

Comparison of settlement along the footing centerline is shown in Fig. 10.18. The distribution of normalized settlement with depth in silica gel is similar to that in sand, though the magnitudes of both cases are not comparable partially due to the differences in material properties. Another reason was the prediction errors in the region around model footing, which caused the maximum normalized settlement to appear at a depth 0.2 B from the footing instead of at the footing bottom.

Comparison of normalized settlement along the cross-section 0.5B from the footing bottom is shown in Fig. 10.19. The distribution of normalized settlement along the horizontal distance from footing center was similar to each other. The magnitudes in both cases are not identical because of the errors in the footing settlement prediction in silica gel. Differences in loading stage and density of the materials are also believed to contribute partially to the discrepancy.

10.4.2 Maximum Horizontal Displacement

Comparison of the maximum horizontal displacement is presented in Fig. 10.20, where the horizontal displacement was normalized by footing settlement from

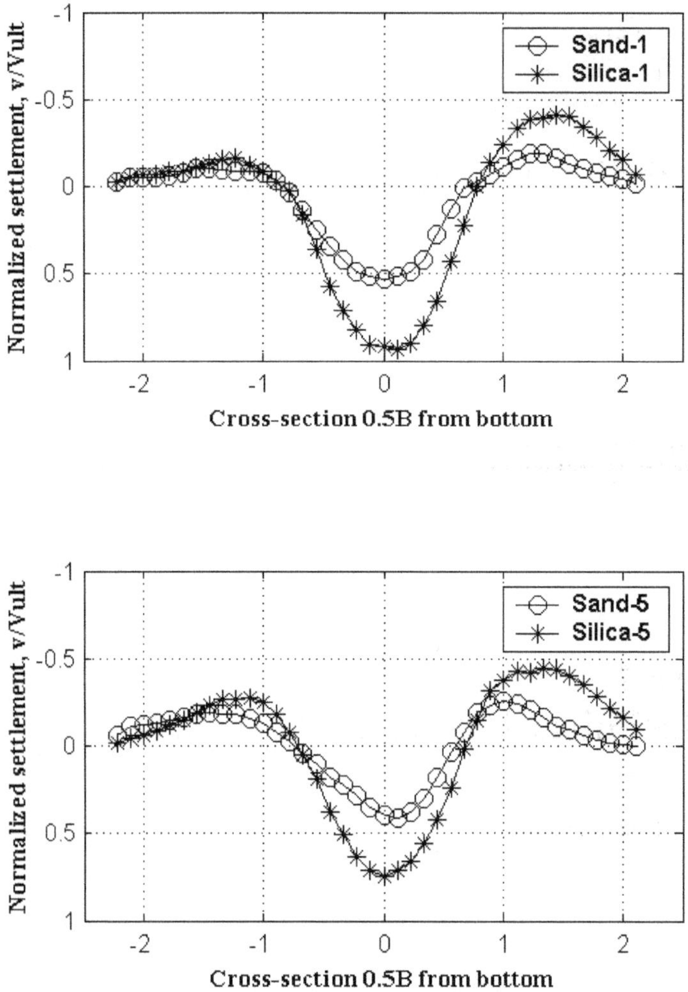

Fig. 10.19 Comparison of Settlement on 0.5B from Footing Bottom

LVDT data. The magnitudes of these two tests were comparable. The magnitude of maximum horizontal displacement reduces as the load increases. As the loading increases, the maximum horizontal displacement approaches the footing edge (B/2).

The depth of the maximum horizontal displacement in dry dense sand is deeper than that in dry dense silica gel (Fig. 10.21), similar to the dry loose condition. After failure, the depth of the maximum horizontal displacement in both sand and silica gel coincides after the distance 1 B from footing edge.

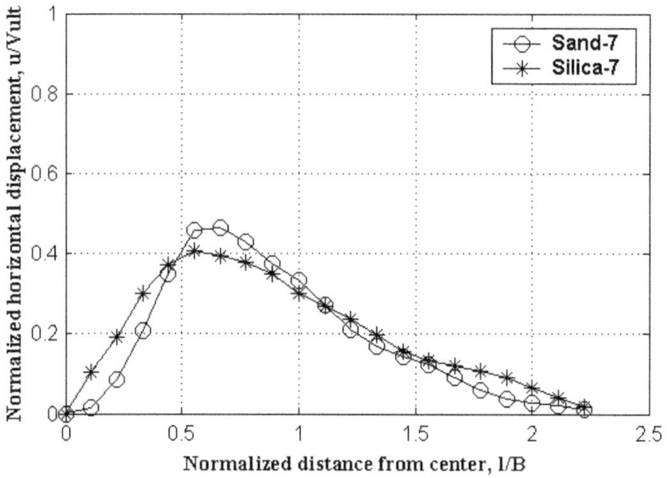

Fig. 10.20 Comparison of the Maximum Horizontal Displacement

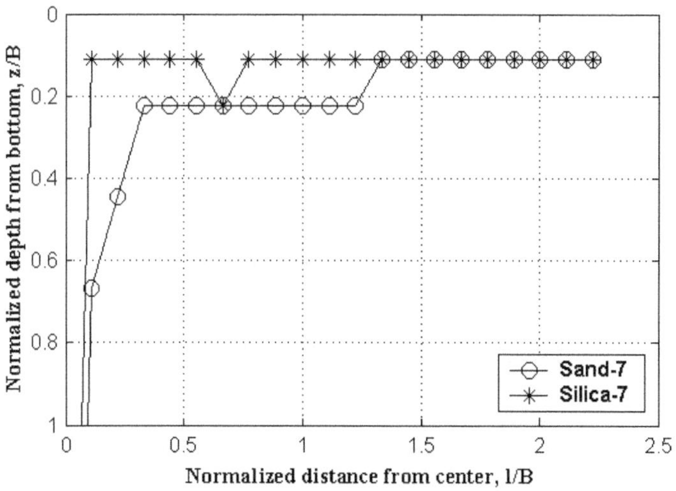

Fig. 10.21 Depth of Maximum Horizontal Displacement

10.5 Comparison under Saturated Dense Condition

10.5.1 2D Deformation Measurement in Transparent Soil

The pore fluid used to saturate the silica gel consisted of a blend of mineral oil and a paraffinic solvent. The pore fluid used to saturate sand was water. Under saturated condition, silica gel becomes transparent and light bulb cannot illuminate the surface of silica gel, though it was still used to illuminate the surface of saturated sand. A 35 mw He-Ne laser with a wavelength of 632.8nm was used to illuminate the specific cross-section inside transparent soil model. A laser light sheet was generated by passing the laser beam through a special line generator lens. Optically slicing the transparent synthetic soil model with the laser sheet produced the distinctive speckle pattern, shown in Fig. 10.22, where the top was through coarse silica gel and the bottom part was through fine silica gel.

The speckle pattern created by a coherent light beam, such as a laser, scattered from a rough surface or from particles in a liquid, has a well-defined spatial structure (Goodman 1975). The speckle properties typically depend on both the roughness and reflectance of the surface. Laser speckle has been used extensively in the study of surface deformation of solids (Erf 1978, Sutton et al 1983). It has also been used to study many engineering problems, including heat and mass transfer problems (Kihm 1997), in-plane movement of a diffusing object (Schnell et al 1998), and the detection of fatigue damage (Kato 2000). The speckle pattern from interaction between laser light and transparent synthetic soils has been used in studying "soil" deformation under a model footing in Chapter 9, where the conventional digital image cross-correlation (DIC) was used in the analyses. In this chapter, Adaptive Cross Correlation (ACC) discussed in Chapter 8 is employed.

Fig. 10.22 Typical Image of Speckles In Silica Gels

Under the saturated condition, three studies were performed. First, comparison of the surface deformation in sand and silica gel was performed to study the modeling ability of transparent synthetic soil (Group 3 in Table 10.1). Second, transparent soils were used to simulate the stratum situation in the field, where surface deformations were used for analyses (Group 4). Finally, 2D deformations inside a transparent soil were studied, where both silica gel and silica powder were used to model sands and clays (Group 5).

10.5.2 Model Preparation

For the preparation of silica gel models, silica gel was first immersed in the pore fluid. Second, vacuum was applied to de-air the mix until the mixture turned to transparent. Time for vacuum depends on the depth of silica gel and the power of the vacuum pump. Approximately 24 hours was required to fully saturate the silica gel. Third, silica gel was packed into the Plexiglas mold to form a transparent silica gel sample. The mold was partially filled with pore fluid at the beginning and de-aired silica gel was slowly poured into the box. At the same time, the silica gel sample was stirred to release gas bubbles entrapped during pouring. The box was vibrated while being packed and silica gel was placed at 30 mm lifts. The amplitude of the vibration was 0.1 mm and duration for vibration was 2 minutes for each lift, same as the sand.

The sand shown in Fig. 10.1 was used again. For the saturated sand model preparation, first, a dry dense sand sample was prepared by using the vibrator with a lift of 3 cm. Second, de-aired water was infiltrated through one end of the model gradually to saturate the model. The results from the model preparation shown above were not successful, because presence of water in the sand reduced the texture of the image and caused substantial errors in the displacement field. Therefore, sand was colored using an oil-soluble red dye before sample preparation. Sand was intentionally colored non-uniformly to give a better contrast. An oil-soluble dye was selected to prevent the separation of dye with the particles in the water.

The load settlement p-z curves of the model footing on saturated dense silica gel and sand are shown in Fig. 10.23. The images captured at the corresponding time were labeled in the figure. The failures in both sand and silica gel appear more like general shear failure instead of punching and local shear failures observed in the tests with dry condition.

10.5.3 Displacement Field

Typical predicted displacement fields are presented in Fig. 10.24 for fine silica gel and Fig. 10.25 for sand. The displacement fields are similar to each other. At the beginning, the movement of the footing causes a larger influence zone, which shrinks as the loading increases. At the failure stage, the depth of influence is reduced to approximate 1 B from the footing bottom.

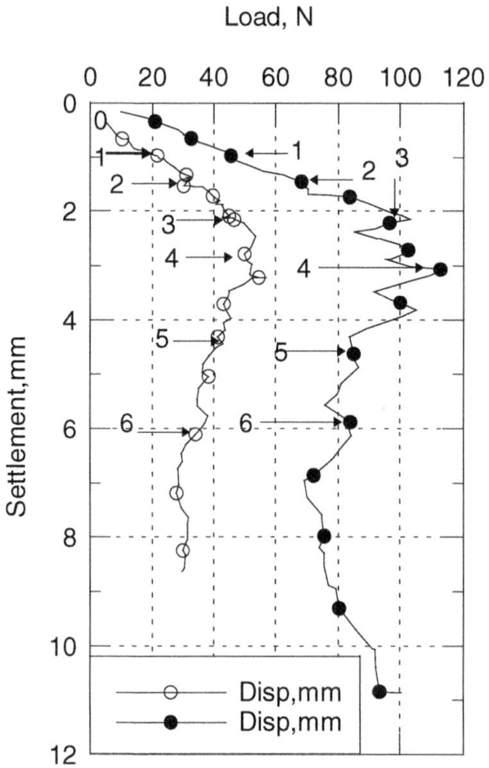

Fig. 10.23 p-z Curves for Footing on Saturated Dense Sand (Solid Legend) and Silica Gel (Hollow Legend)

10.5.4 Vertical Displacement

Settlements along the cross-section across the footing centerline before failure (between point 1 and 2 in Fig. 10.23) and after failure (between point 4 and 5 in Fig. 10.23) are on the left and right side of Fig. 10.26, where settlements in both cases were normalized by footing settlement. The depth of influence before failure in silica gel was deeper than that in sand. After failure the settlement along the centerline agreed very well.

Settlements along a horizontal cross-section 0.5B from footing bottom are presented in Fig. 10.27, where the distance from footing center and settlement were normalized by the footing width, B, and footing settlement, v_{ULT}, respectively. Silica gel in saturated dense condition agrees very well with sand in both the magnitude and distribution along the distance from footing center.

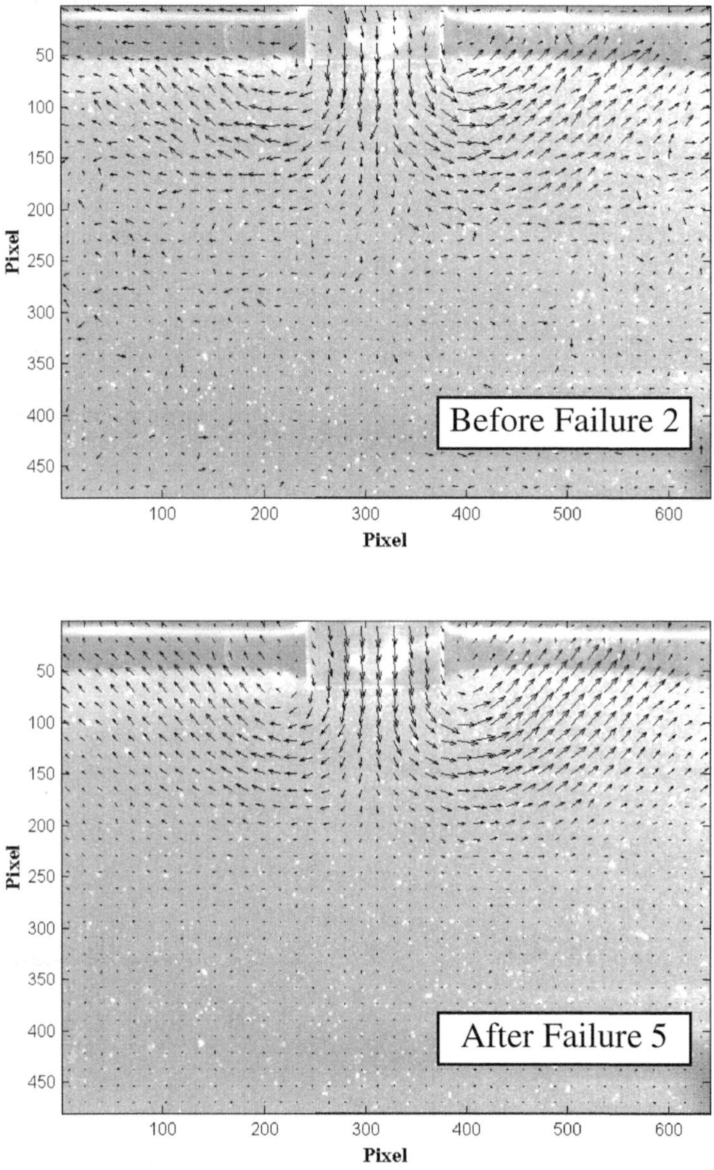

Fig. 10.24 Displacement Field in Saturated Dense Silica Gel. (ACC Window Size= 128×128, 64×64, 32x32, 16×16, Offset=8).

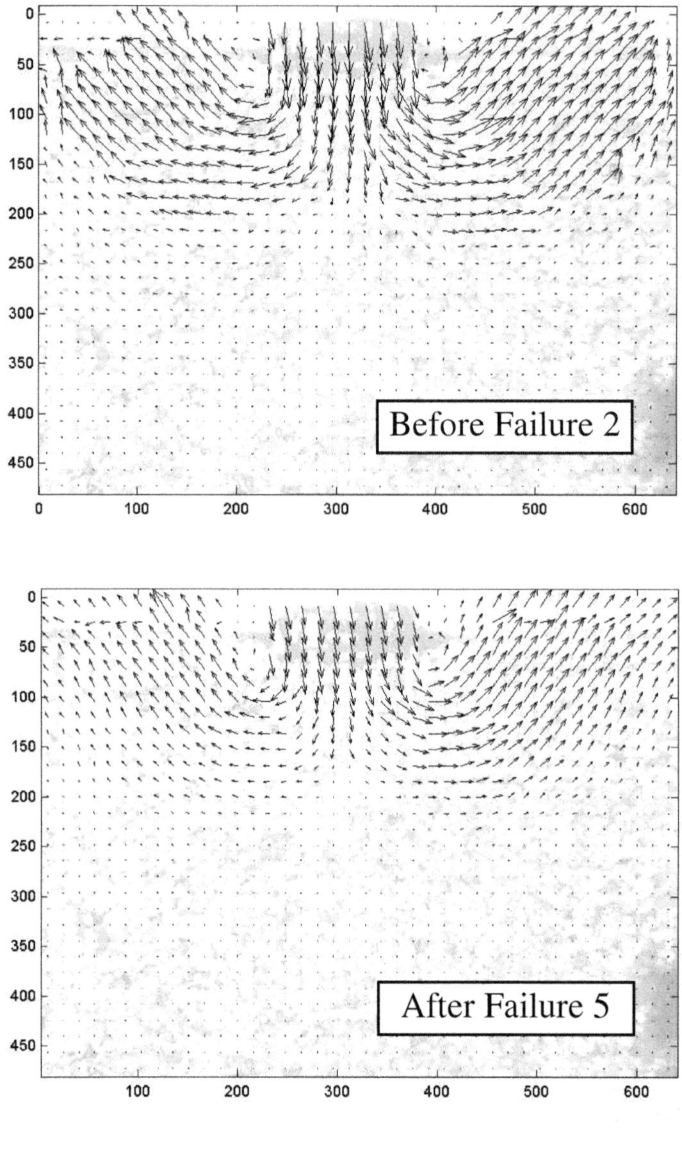

Fig. 10.25 Displacement Field in Saturated Dense Sand (ACC Window Size= 128×128, 64×64, 32x32, 16×16, Offset=8)

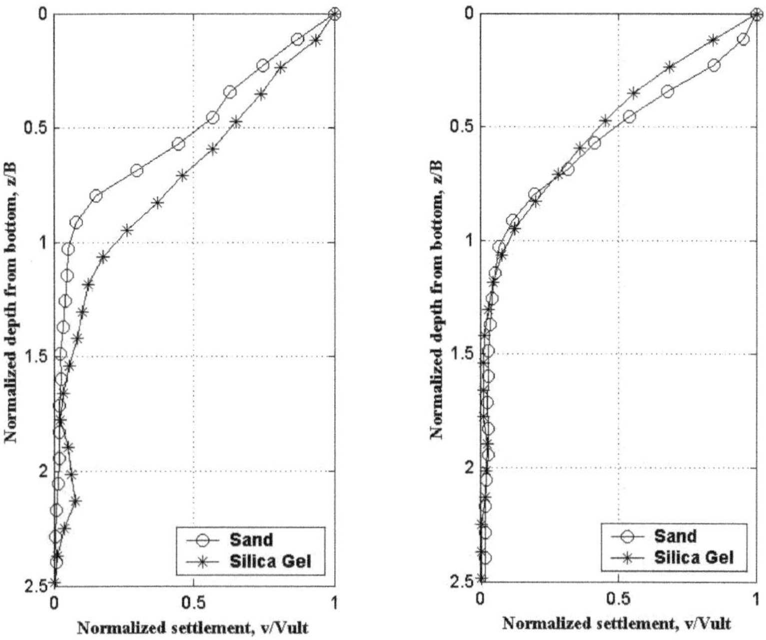

Fig. 10.26 Comparison of Settlement along Footing Center Line Before (LHS) and After (RHS) Failure

10.5.5 Maximum Horizontal Displacement

The normalized maximum horizontal displacements are plotted against the distance from the footing center (Fig. 10.28), where the distance and the horizontal displacement were normalized by footing width, B, and footing settlement, v_{ULT}, respectively. The positions of maximum horizontal displacement are similar to each other in both before failure and after failure, which are near the footing edge. The magnitude and distribution of maximum horizontal displacement are comparable at the point before failure. As loading increases, the magnitude in sand is reduced and there are discrepancies in the magnitude between sand and silica gel, which is caused by the differences in the properties of the materials.

The position of the maximum horizontal displacement is shown in Fig. 10.29, where both the distance and the depth from the footing were normalized by the footing width, B. The depth of maximum horizontal displacement in silica gel is slightly deeper than that in sand. Both tend to reduce as the loading increases. It becomes the same at the distance more than 1 B from the footing edge after failure.

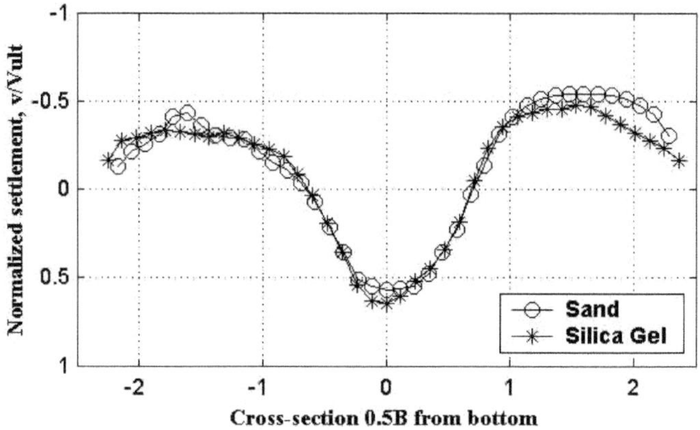

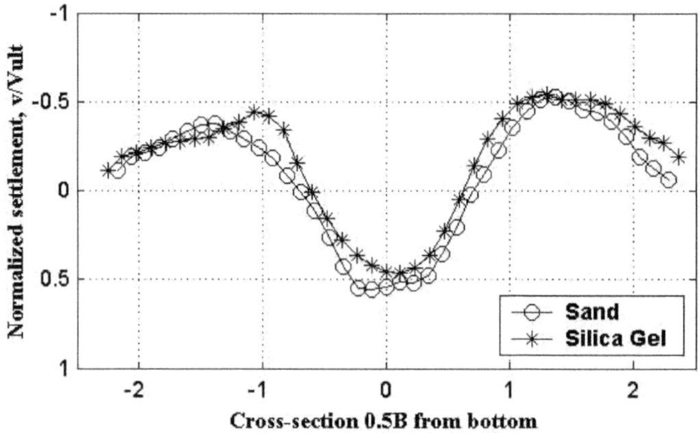

Fig. 10.27 Settlement along Cross-Section 0.5B from Footing Bottom. Before Failure (top) and After Failure (Bottom).

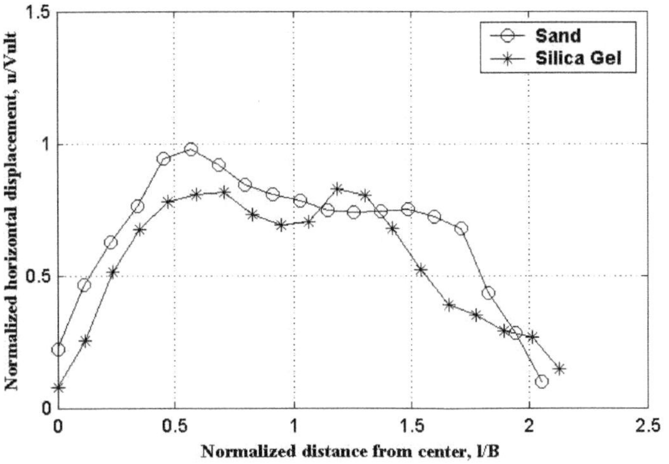

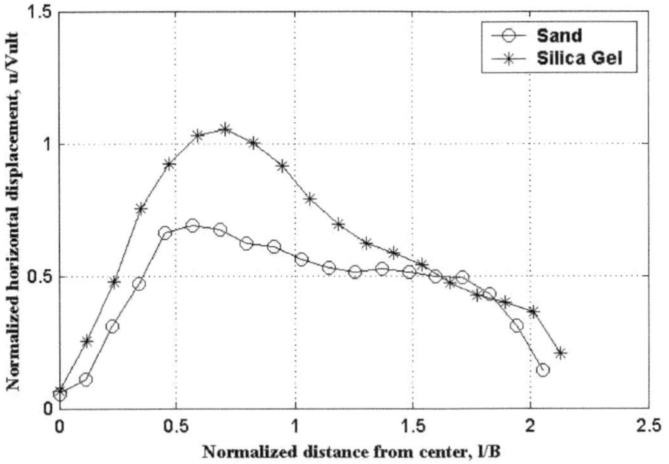

Fig. 10.28 Comparison of Maximum Horizontal Displacement. Before Failure (Top) and After Failure (Bottom).

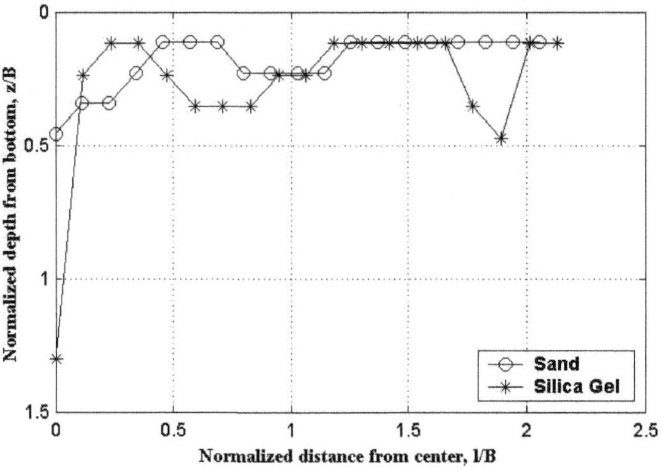

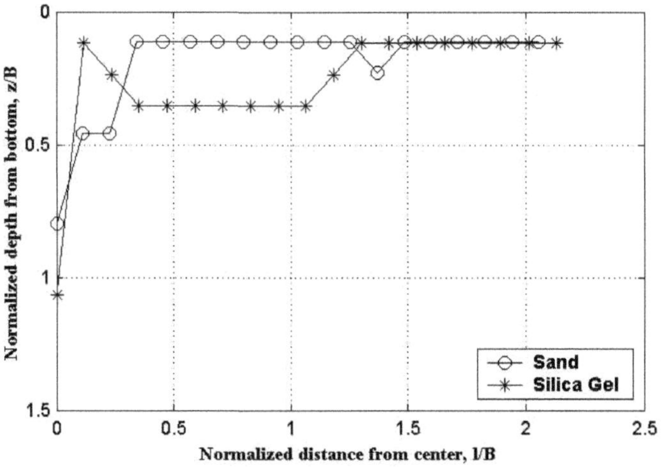

Fig. 10.29 Position of Maximum Horizontal Displacement. Before Failure (Top) and After Failure (Bottom).

10.5.6 Maximum Shear Strain

Strains were calculated using the predicted displacement field. The maximum shear strain fields are shown in Fig. 10.30 for silica gel and Fig. 10.31 for sand both before failure and after failure, where the maximum shear regions were shaded. The maximum shear strain appears at the same region at both sides of footing. The failure envelope in sand is deeper than that in silica gel. The differences in sample density and material properties are believed to contribute to this difference.

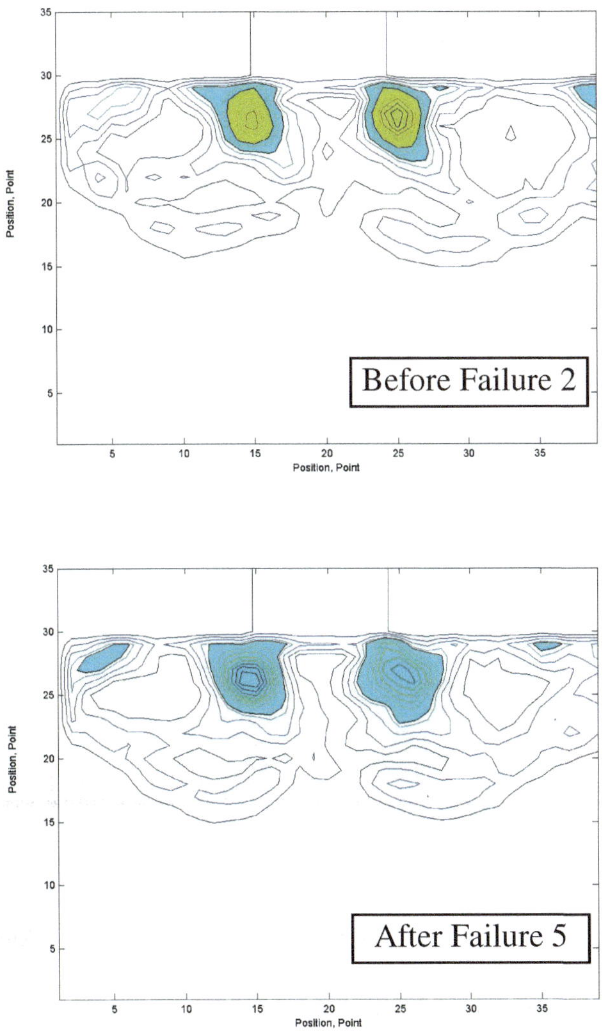

Fig. 10.30 Failure Envelope in Saturated Dense Sand

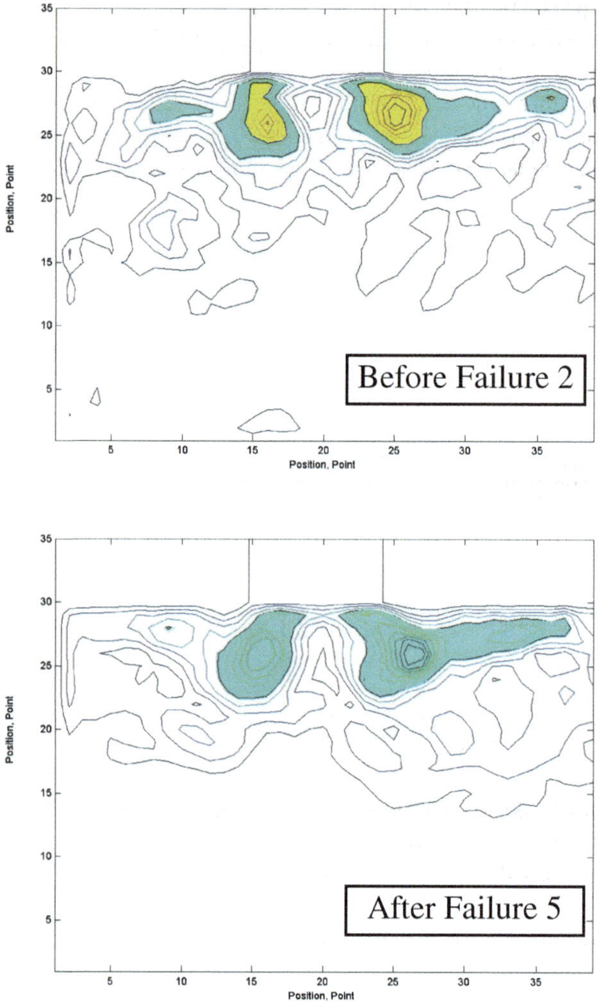

Fig. 10.31 Failure Envelope in Saturated Dense Silica Gel

10.6 Modeling Stratified Soils Using Transparent Surrogates

Two families of transparent synthetic soils were developed to model sands and clays, respectively. One advantage of these materials is that they have the same refractive index, which permits modeling the stratum problems. Two examples of footing under different configurations are presented below.

10.6.1 Modeling Loose "Sand" Over Dense "Sand"

Loose "sand" over dense "sand" was simulated by overlying coarse silica gel, round with an average diameter of 2.5mm, over fine silica gel, angular with an average diameter of 1 mm. The properties of both materials are available in Chapter 6. The depth of coarse silica gel approximately equals to the footing width, 2.5cm.

Typical displacement fields are presented in Fig. 10.32 for before failure (Top) and after failure (Bottom) conditions, with the speckle image as the background.

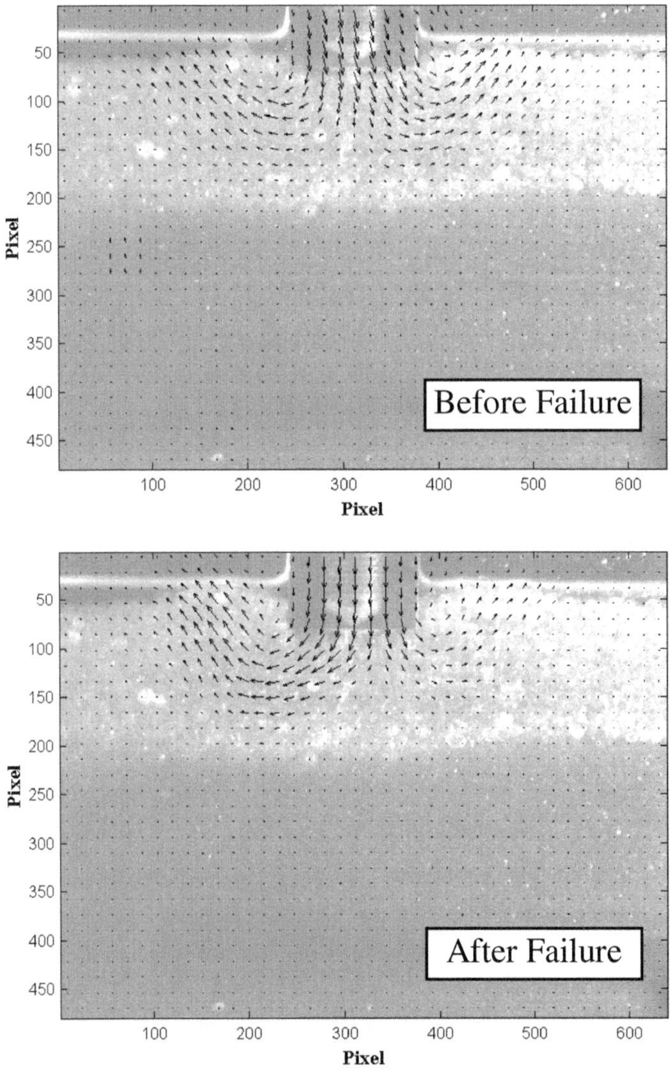

Fig. 10.32 Deformation in 1B Deep Coarse Silica Gel over Fine Silica Gel (ACC Window Size= 128×128, 64×64, 32x32, 16×16, Offset=8)

The speckles generated in the top coarse silica gel layer were brighter and coarser than those in the lower fine silica gel. Coarse silica gel exhibits a lower friction than fine silica gel. Accordingly, the displacement occurred mostly in the top coarse silica gel layer. A symmetrical displacement pattern appeared at the beginning, which became asymmetrical at the failure stage. The influence region of the maximum shear strain is smaller than that in the uniform fine silica gel (Fig. 10.33). The maximum shear strain occurred primarily within top loose layer.

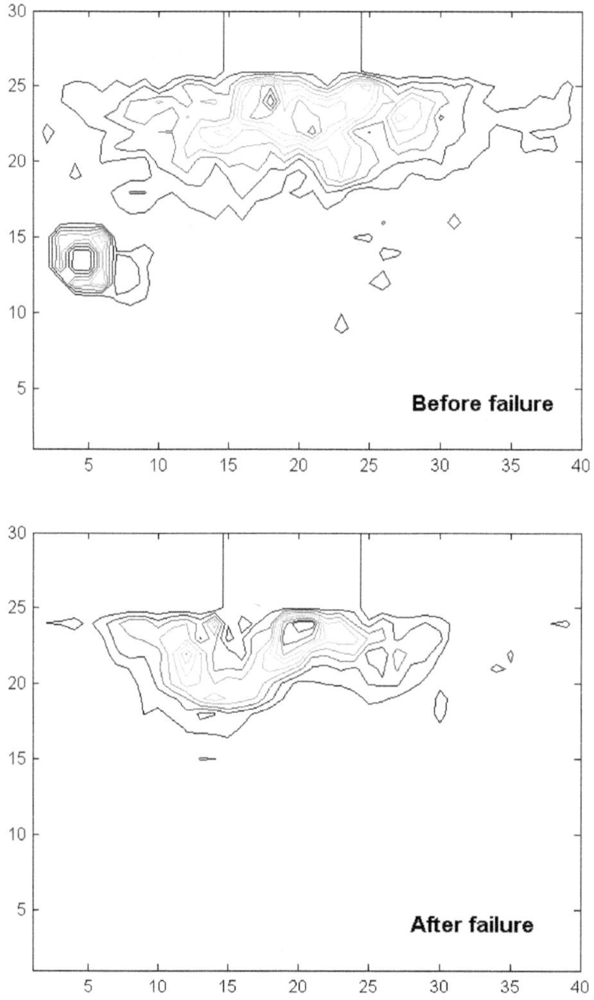

Fig. 10.33 Maximum Shear Strain in 1B Deep Coarse Silica Gel over Fine Silica Gel

10.6.2 Modeling "Sand" over Soft "Clay"

Shallow foundations are sometimes located in a sand stratum of limited thickness on top of a deep bed of clay. The ultimate bearing capacity of footing on sand layer overlying clay was investigated by Meyerhof (1974). Model tests along with field test data from former researchers were presented and discussed in his paper.

A footing on stiff sand overlying soft clay was simulated using fine silica gel over consolidated silica powder. The preparation of bottom silica powder layer followed a special procedure as follows. A de-aired suspension of amorphous silica powder (Flo-Gard SP) with mineral oil blend was poured inside the mold slowly. The solution was stirred occasionally to release entrapped air bubbles. The solution was consolidated in the Plexiglas mold using a precise plate. There were holes in the plate to let the pore fluid dissipate from the sample. Loading was applied and increased gradually to prevent the plate from suddenly sinking into the sample. 10 days were required to consolidate the bottom silica powder layer. Next, the plate was removed and de-aired silica gel was poured onto the top of silica powder to the required depth.

After sample preparation, the laser light sheet was used to slice the sample immediately behind surface of the Plexiglas mold. The load settlement, p-z, curve of a model footing on this sample showed a punching shear or local shear failure mode, which is consistent with the simulated problem, strong sand over soft clay. The predicted displacement field with the speckle image as the background is shown in Fig. 10.34, where the top image is the displacement field before failure and the image is after sample failure.

The displacement field is similar to that presented by Meyerhof (1974). The displacement in the top silica gel layer was dominated by the downward displacement near the footing. The outward displacements at both sides of footing were not as prominent as that in the uniform silica gel test, especially after failure. The displacement punched through the top layer and spread to the bottom layer. The failure mechanism is evident in the maximum shear strain fields, shown in Fig. 10.35. The failure mode of footing on stiff "sand" over soft "clay" can be described as follows. Punching shear failure occurred in top stiff "sand" layer. The failure envelope developed downward from both sides of footing to the surface of bottom soft clay layer. Beyond that the failure envelope similar to that under a model footing in uniform "clay" layer was developed in the bottom layer.

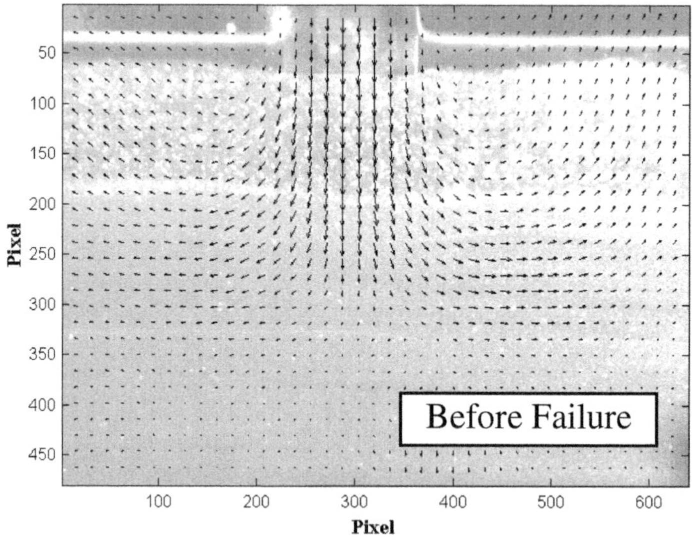

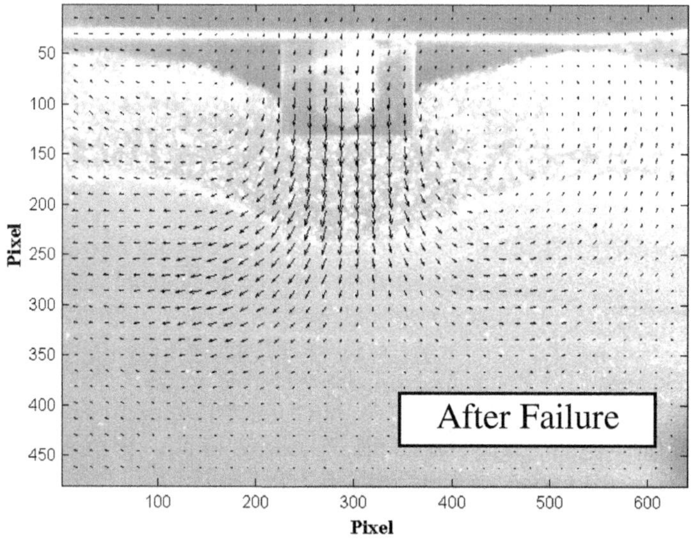

Fig. 10.34 Surface Displacement of Silica Gel over Silica Powder (ACC Window Size= 128×128, 64×64, 32x32, 16×16, Offset=8)

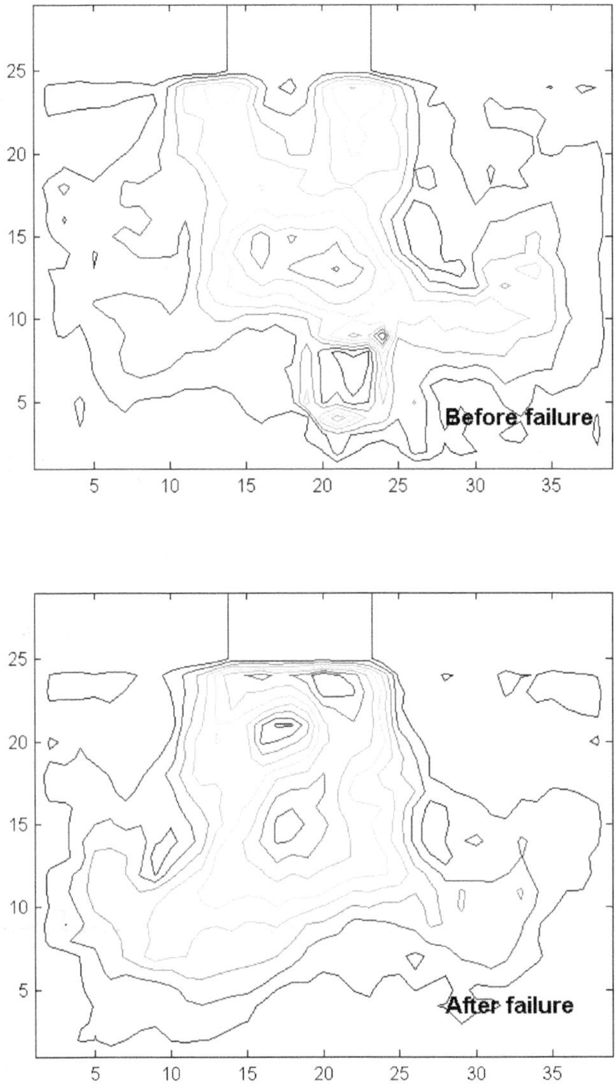

Fig. 10.35 Maximum Shear Strain on Surface of Silica Gel over Silica Powder

10.7 Deformations Inside Transparent Synthetic Soil Models

10.7.1 Internal Deformations in Silica Gel Representing Sand

Deformation measurements inside the model were obtained using the procedure discussed in Chapter 9. Typical displacement fields on the cross-section half inch inside silica gel model are shown in Fig. 10.36. The deformation patterns are similar to those obtained in surface measurement. The maximum shear strain fields (Fig. 10.37) are also similar to those obtained from the sample surface. Optical system for inside deformation measurement in transparent soil is proved to be effective by the similarities in both the displacement field and the strain field, where the simulated problem was a typical plane strain problem.

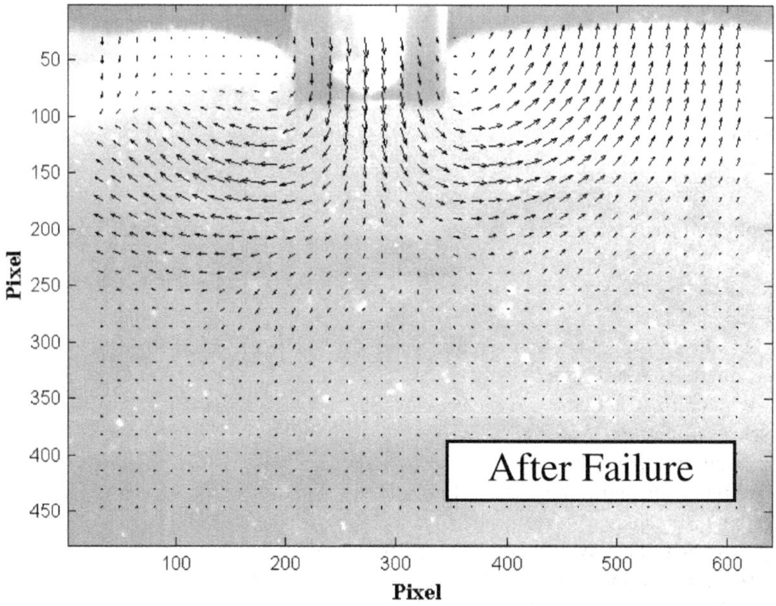

Fig. 10.36 Displacement Field Half-Inch inside Silica Gel (ACC Window Size= 128×128, 64×64, 32x32, 16×16, Offset=8)

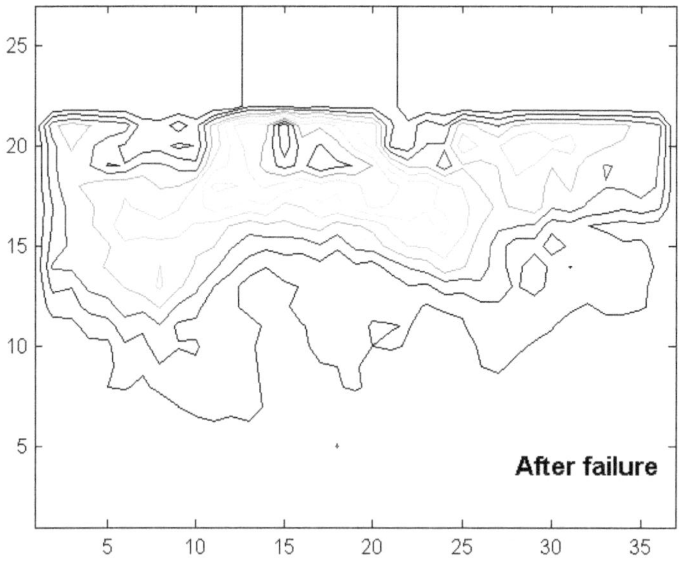

Fig. 10.37 Maximum Shear Strain Half-Inch Inside Silica Gel

10.7.2 Internal Deformations in Amorphous Silica Powder Representing Clay

Two typical deformation fields of a cross-section half-inch inside transparent silica powder are shown in Fig. 10.38 with the image as the background. In silica powder, the adaptive cross-correlation algorithm used for displacement field calculation was slightly different from the analyses used before. The window sizing varied from 256×256 to final window size of 32×32 instead of 128×128 to the final window size of 16×16. The window spacing was only half of the window size, which means that there was a 50% window overlapping between two neighboring interrogation windows. In the previous analyses, there was no window overlapping. The reason for these changes was that the texture in the image from silica powder was not as distinctive as that in silica gel, which caused errors in the displacement field. The final window size was increased from 16×16 to 32×32 to reduce the errors in prediction by increasing the peak of cross-correlation function. The window overlapping was intended to give a smoother displacement field.

The displacement fields and strain fields (Fig. 10.39) were similar to those in silica gel. The failure envelopes went deeper than that in silica gel, which is probably caused by the differences in the materials.

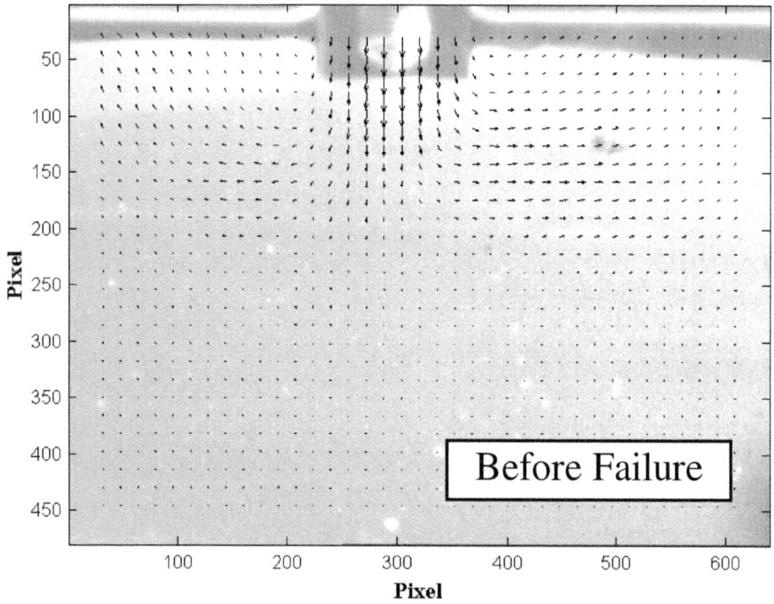

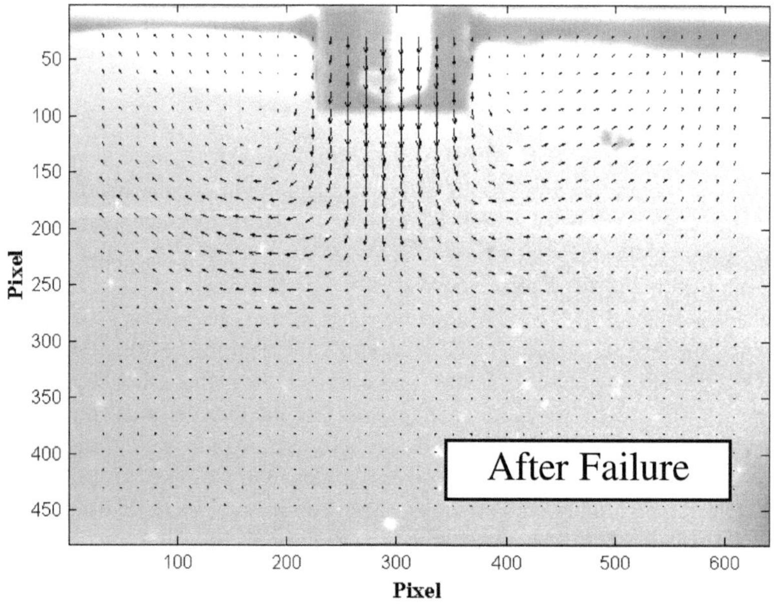

Fig. 10.38 Displacement Field Half-Inch Inside Silica Powder (ACC Window Size=256 ×256, 128×128, 64×64, 32x32, 50%Overlapping,Offset=8)

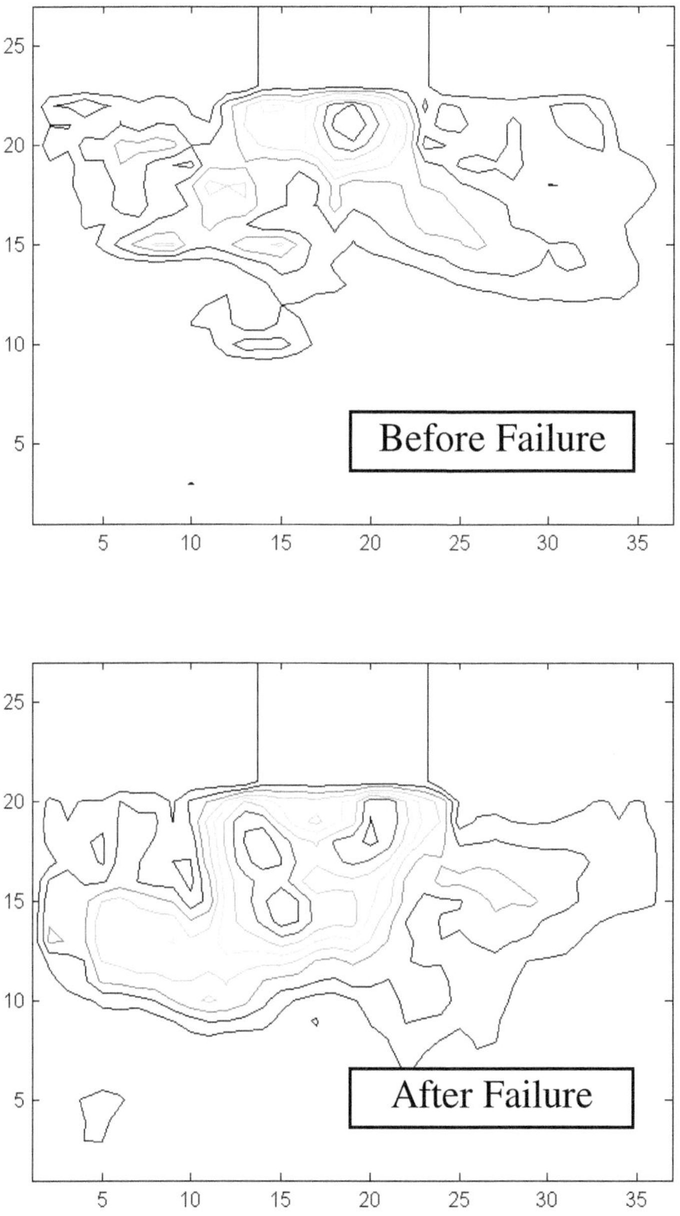

Fig. 10.39 Maximum Shear Strain Half-Inch Inside Silica Powder

10.8 Conclusions and Recommendations

Transparent synthetic soils not only exhibit macro geotechnical properties similar to natural soils, but also can be used in studying soil problems. Because of a lighter unit weight and a porous structure, silica gel has a lower density than sand. The difficulties in obtaining the samples with a same density and identifying the corresponding points in the load settlement curves are believed to contribute to the discrepancies in the results. Further research is required to standardize the sample preparation and quantify the sample properties. A study of similitude is also required to help in interpreting and extrapolating the model test results to natural soil behavior in both laboratory and the field scales.

References

ASTM D698, Test method for laboratory compaction characteristics of soil using standard effort (12,400 ft-lbf/ft3(600 kN-m/m3)), 1993 Annual Book of ASTM Standards, vol. 04.08, pp. 164–172 (1993)

ASTM D3080, Standard Test Method for Direct Shear Test of Soils under Consolidated Drained Conditions, 1998 Annual Book of ASTM Standards, vol. 04.08, 6 p. (1998)

ASTM D4254. Standard test method for minimum index density and unit weight of soils and calculation of relative density, 1993 Annual Book of ASTM Standards, vol. 04.08, pp. 674–681 (1993)

Erf, R.: Speckle Metrology. Academic Press, New York (1978)

Eggstad, A.: Deformation measurements below a footing on the surface of dry sand. In: Proc. of the Eur. Conf. on Soil Mech. and Found. Eng., Wiesbaden, Germany, vol. 1, pp. 233–239 (1963)

Goodman, J.: Statistical properties of laser speckle patterns. In: Dainty (ed.) Laser Speckle and Related Phenomena, pp. 9–75. Springer, Heidelberg (1975)

Kato, A.: Detection of fatigue damage in steel using laser speckle. Optics and Laser in Engineering 33, 323–333 (2000)

Kihm, K.: Laser speckle photography technique applied for heat and mass transfer problems. Advances in Heat Transfer 30, 255–311 (1997)

Iskander, M., Liu, J., Sadek, S.: Transparent amorphous silica to model clay. ASCE J. Geotech. & Geoenv. Eng. 128(3), 262–273 (2002a)

Iskander, M., Sadek, S., Liu, J.: Optical measurement of deformation using transparent silica gel to model sand. Int. J. of Physical Modeling in Geotechnics 2(4), 27–40 (2002b)

Liu, J., Iskander, M., Sadek, S.: Consolidation and permeability of transparent amorphous silica. ASTM Geotechnical Testing J. 26(4), 390–401 (2003)

Marangos, C.: Vertical and horizontal deformations for foundations on sand: an experimental study. In: Proc. of the Conf. on Vert.& Horiz. Deformations of Foundations and Embankments, Part 1, pp. 179–189 (1994)

Meyerhof, G.: The ultimate bearing capacity of foundations. Geotechnique 2, 301–332 (1951)

Meyerhof, G.: Ultimate bearing capacity of footings on sand layer overlying clay. Canadian Geotechnical J. 11(2), 224–229 (1974)

Roscoe, K., Arthur, J., James, R.: The determination of strains in soils by an X-ray method. Civil Engineering and Public Works Review 58, 873–876, 1009–1012 (1963)

Schnell, U., Piot, J., Dändliker: Detection of movement with laser speckle patterns: statistical properties. J. of the Optical Society of America A 15(1), 207–216 (1998)

Sutton, M., Wolters, W., Peters, W., Ranson, W., Mcneill, S.: Determining of displacements using an improved digital correlation method. Image Vision Computing 1(3), 133–139 (1983)

Terzaghi, K.: Theoretical Soil Mechanics. Wiley, New York (1943)

Yamamoto, K., Kusuda, K.: Failure mechanism and bearing capacities of reinforced foundations. Geotextiles and Geomembranes 19(3), 127–162 (2001)

Chapter 11
3D Deformation Measurement

Jinyuan Liu and Magued Iskander

Abstract. In this chapter, three-dimensional deformations inside a transparent synthetic soil model were measured non-intrusively by cross-correlating the speckle images. The speckles were generated by interaction between laser light and transparent synthetic soils. A soil profile consisting of "sand" overlying soft "clay" was simulated. The deformation and strain field under the model footing were similar to those published. Error analysis of the mechanism and system showed that the proposed method is suitable for measuring spatial deformations non-intrusively.

11.1 Introduction

Visualization of three-dimensional (3D) deformations or strain fields can significantly improve the interpretation and understanding of many geotechnical engineering problems and help in solving difficult challenges.

Transparent synthetic soil has been shown to exhibit macro geotechnical properties similar to those of natural soils in chapters 5-7. Two-dimensional (2D) non-intrusive measurement of deformation inside transparent soil model is illustrated in chapter 9, where the laser speckles generated by the interaction of laser light and transparent silicas are used to obtain the displacement field optically. Comparison of soil deformation below a model footing in transparent and natural soils in chapter 10 also demonstrates that transparent synthetic soil can be used in modeling natural soils under both dry and wet conditions.

In this chapter, the 3D deformation field inside transparent synthetic are investigated non-intrusively using the optical method developed in chapter 9. A simple model test is used to illustrate the methodology.

11.2 3D Measurements

3D visualization techniques are extremely popular in the biomedical field. Imaging techniques including x-ray radiography, computerized tomography (CT), magnetic resonance imaging (MRI) or nuclear magnetic resonance (NMR) are

M. Iskander: Modeling with Transparent Soils, SSGG, pp. 227–258.
springerlink.com

routinely used to probe the interiors of the human body in biomedical applications. Most of these techniques involve "slicing" the body part of interest to obtain the image of the sliced cross-section. 3D view of the body is rendered on the computer. Normally there are two ways to analyze the data, one is to create a 3D view of suspected areas and examine thin slice of those areas. Another is to let the computer sort through data, identify suspected pathologies using pattern matching algorithms, pull out just those areas of potential interest, and reconstruct them in 3D for human evaluation. Computers can also measure the volume and dimensions of pathologic tissues to plan surgery or judge the effectiveness of therapy. It can also calculate the diameter of blood vessels at specific location to help physicians locate and open clogged arteries (Spiesberger and Tasto 1981, Udupa and Herman 2002).

3D visualization technique provides an improved understanding of the spatial variability of information that is not possible with 2D graphics. 3D visualization techniques are also being successfully applied in many other engineering fields, for example,

- In geoscience and geology, Liner (1999) used the public domain software of National Institute of Health to interpret 2D seismic geophysical data to construct a 3D image. Repin and Pavlis (1997) applied 3D volume visualization techniques for displaying and analyzing both body waves and surface waves and 3D particle motion.
- In the groundwater and pollutant transport field, 3D visualization techniques were used to characterize the extent and distribution of contaminants over time within the shallow subsurface based on high-resolution shallow time-lapse 3D reflection seismic data (McKenna et al. 2001).
- In the hydraulic engineering field, 3D flow fields around a 90° open-channel junction were visualized using data from acoustic Doppler velocity meters located at well-defined positions. The result was used to verify 3D computational fluid dynamics codes (Weber et al. 2001).

11.2.1 Methodology

In geotechnical engineering, visualization of internal soil information has been aided by radiography for many years. A picture of movement that occurred at failure in clay under a loaded plate was first attempted by Bergfelt (1956) using X-ray radiography. Roscoe et al (1963) adapted the X-ray method to study the strain fields inside soils. Similar techniques were applied later by James and Bransby (1970) and Heinz (1971) in the study of passive earth pressure and the density distribution of sand under strip footings. Microscopic behavior of soil under shear was also aided by radiography. For example, shear bands in granular materials were studied by Nemat-Nasser and Okada (2001) using radiography.

3D visualization of stress and deformation fields inside models has been investigated mostly using the optical slicing technique. Konagai et al. (1992) used crushed glass and a pore fluid with a matching refractive index to study deformation of an embankment under dynamic loading non-intrusively. The embankment model was optically sliced by laser light sheet and deformation on each sliced cross-section was illustrated qualitatively. The tests, however, was limited by the poor transparency of the model due to the entrapped air in the tiny cracks in the particles. Dupre and Lagarde (1997) and Zenina et al. (1999) studied 3D stress field of a photoelastic model using the optical slicing methodology. 3D flow fields within an opaque material were developed using a combination of digital speckle photography and flash X-rays by Grantham et al (2000).

The optical slicing technique can be illustrated using Fig. 11.1. MRI images of different cross-sections are shown on the top, and then data of the images are imported to a computer to visualize in 3D view of objects on the bottom, where a few slices present a clearer spatial variation of the human head.

11.2.2 Test Setup

The procedure outlined in chapter 9 is followed. The setup, shown in Fig. 11.2, consists of a CCD camera, a laser light source, a loading frame, a precision stage, and a PC for image processing.

A light source was used to illuminate the cross-section of interest. An intense laser-light sheet was generated by passing a laser beam through a line generator lens. The laser was 35mW Melles Griot He-Ne with a wavelength of 632.8nm.

A Cohu 2622 black & white CCD camera with a resolution of 640×480 pixels was set approximately 20cm away from the laser sheet with its optical axis perpendicular to the laser sheet plane. A macro-zoom lens with a focus length variable from 18 to 108 mm was used to adjust the size of captured area. The images were captured and imported to a PC through a Matrox Meteor 2/4 frame grabber.

A Plexiglas mold with a dimension of 50×150×300 mm was used in the model tests. A model footing with a plan dimension of 50×25 mm was used to simulate a continuous footing with a width of 25 mm. Load was applied vertically through a mechanism that consists of a screw linear stage and a load cell.

A precise linear stage was used to move the transparent soil mold perpendicularly to the light sheet, which allows for slicing the model. The stage moves 0.635 mm per roll. Each roll was divided into 72 divisions giving a resolution of 0.009 mm per division.

The test setup allows for measuring 2D deformations and strains within an illuminated plane inside a transparent soil model. Multiple planes can be combined in a computer model to construct a psudu-3D model. Out of plane deformations are not captured using the setup shown in Fig 11.2. However, out of plane deformations can be captured using a perpendicular identical setup consisting of 2 lasers and two cameras.

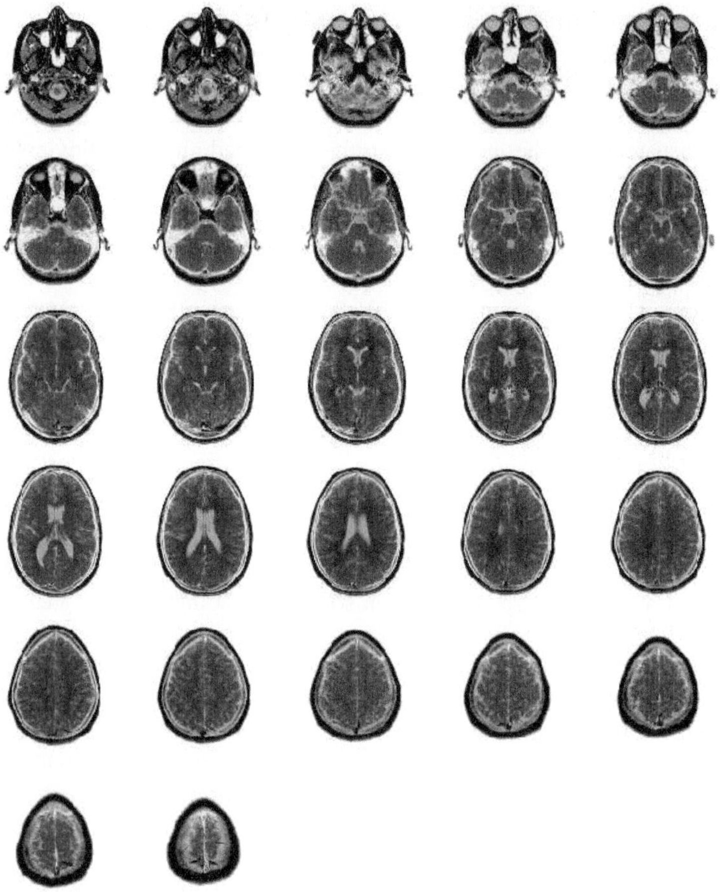

Images of Slices

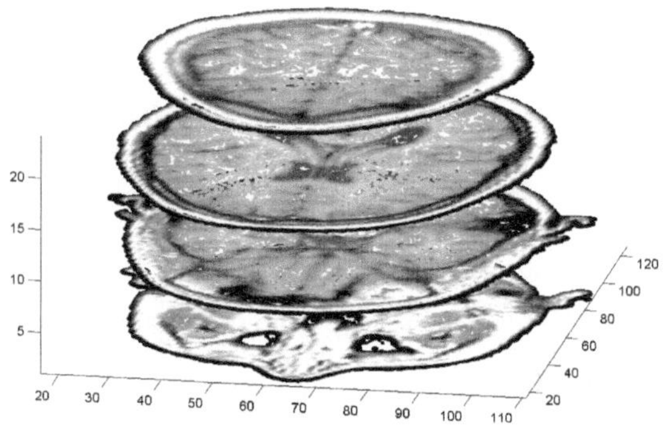

3D View

Fig. 11.1 MRI Images and 3D Visualization

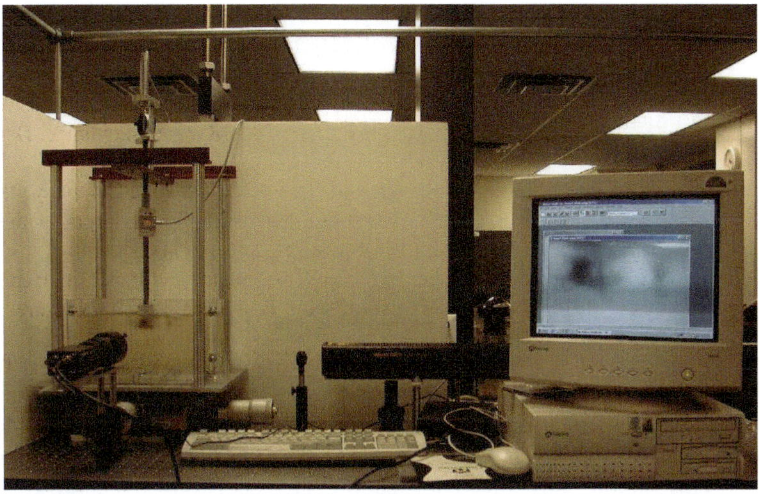

Fig. 11.2 Test Set-up

11.2.3 Digital Image Processing

The relative displacement field was obtained by correlating two consecutive images taken during deformation. The image processing technique used in this study was digital image correlation (DIC), discussed in chapter 8. DIC has been widely used in many engineering fields with different names, such as particle image velocimetry (PIV) in fluid dynamics measurement (Adrian 1991), surface displacement analysis (SDA) in solid mechanics (Chu et al. 1985). An advanced cross-correlation algorithm named adaptive cross-correlation (ACC) from *Dantec Measurement Technology* (2000) was adapted in this research to reduce the prediction errors.

Comparison of conventional DIC and ACC demonstrated that ACC gives a better performance than conventional DIC (see Section 8.7). ACC not only reduces the prediction errors but also relieve the limit of the maximum reliable displacement, which depends on the interrogation window size. ACC uses both variable window sizing and window shifting. ACC permitted measuring deformation with a resolution of 0.1 pixels, which corresponded to a displacement on the order of 0.02 mm based on the CCD camera and the optical settings in the test. This accuracy can be improved by increasing the resolution of the camera or adjusting the optical setting, for example, using a smaller focus area.

The parameters of ACC used in this chapter were the following. The interrogation window sizes were 128×128 pixels, 64×64 pixels, to the final window size of 32×32 pixels. The total window offsets were 6. No window overlapping was used in this chapter.

11.2.4 Test Procedure

A footing on sand overlying soft clay was simulated using silica gel over consolidated silica powder. Silica powder was Flo-Gard SP (FGSP) from *PPG Industry Inc.* with an average aggregate diameter of 25 μm. Silica gel (SG) was angular fine silica gel from *Multisorb Technologies Inc.* with an aggregate diameter of 0.5-1.5 mm. Both materials consisted of small particles on the order of 0.02 μm. More details can be found in Chapter 5 and 6. The bottom layer of FGSP was first consolidated into the mould and then de-aired silica gel was poured onto the silica powder to the designed depth. Detailed descriptions on the sample preparation can be found in section 10.6.2.

After sample preparation, the model was adjusted and placed on the linear stage platform. The loading mechanism was designed to push down the model footing vertically onto the soil. The laser light sheet was used to slice the sample. The sample was sliced at specific cross-sections by controlling the movement of the sample through a linear stage at selected loading stages. The movement was perpendicular to the laser light sheet. Images on different cross-section were captured and imported through the frame grabber to a PC. The slicing procedure started from the surface of the model further into the sample one by one. When further displacement had been applied, slicing started from the previous cross-section location under former deformation stage back to the surface of the model one by one. Slicing moved back and forth until the end of the test. This slicing method reduced the influence of the speckle change with time on the test results.

The load settlement p-z curve of the model footing is shown in Fig. 11.3. Load and settlement were measured using a load cell and an LVDT connected to a load frame system. The deformation steps are labeled on the figure. The load settlement curve represented a punching or a local shear failure, which was reasonable considering the low strength of the "clay" layer relative to the strength of top "sand" layer.

Creep was prominent in the transparent soil model, as evident in Fig. 11.3. For example, after the second deformation steps finished, the load measured at the first cross-section was 6.4 N and reduced with time to 5.4 N. Study of the consolidation properties of transparent amorphous silica powder discussed in section 5.7 showed that the material has two porosities, internal porosity and inter-aggregate porosity. This property is shared with silica gel and could have contributed to the time dependent behavior evident in Fig. 11.3.

The images of the four different cross-sections are shown in Fig. 11.4, where the first slice was taken 1 mm inside the transparent synthetic soil model. The other three planes were 4.18 mm, 7.36 mm, 11.54 mm inside the sample. The distance between each plane was precisely controlled by the linear stage. 3D views of slices under different deformation stages are shown in Fig. 11.5.

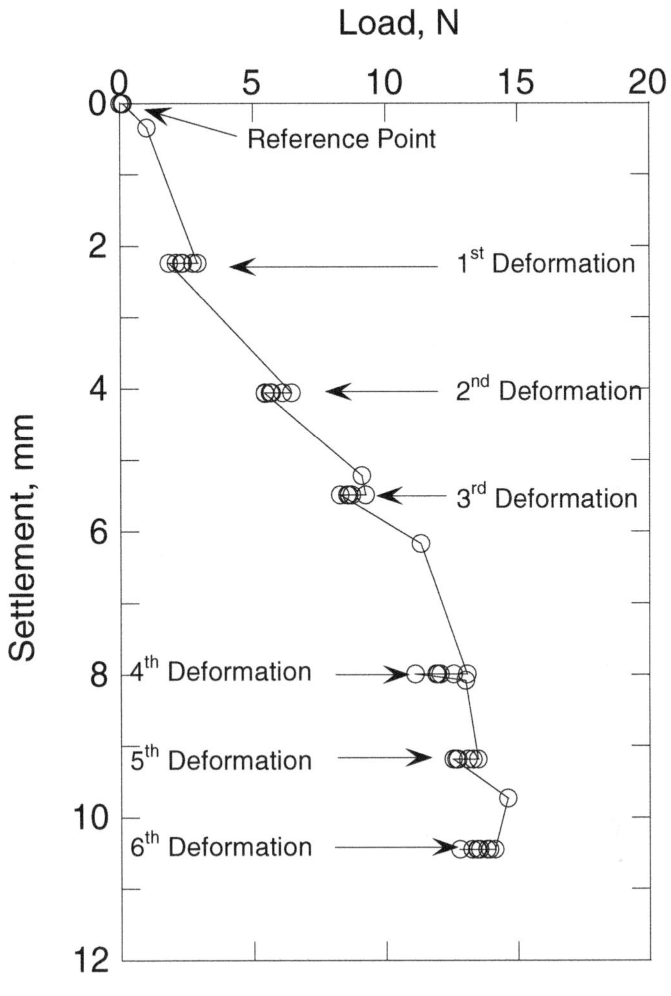

Fig. 11.3 Load-Settlement Curve of Model Footing

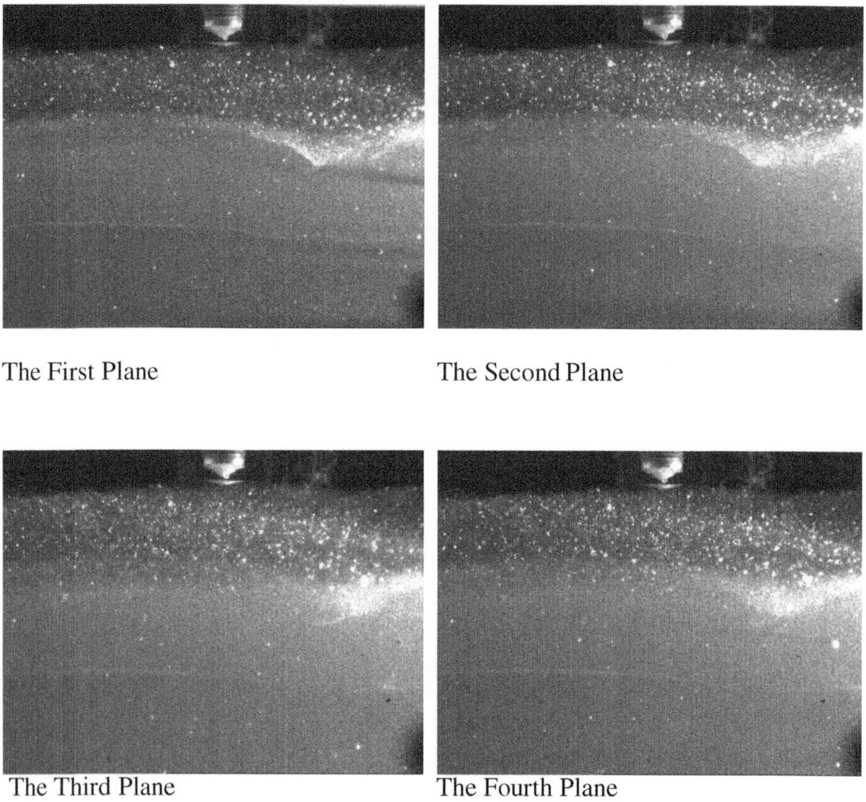

Fig. 11.4 Image of Each Cross Section

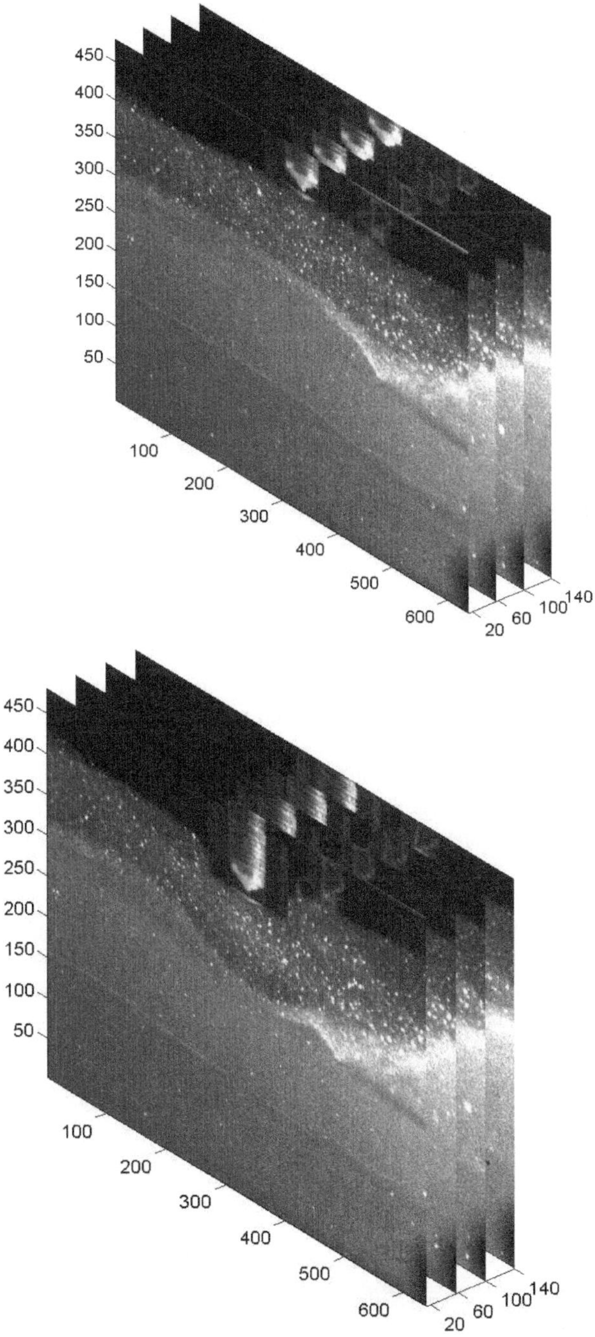

Fig. 11.5 3D View of Image Slices. The First (top), Sixth (bottom) Deformation Stage.

11.3 Displacement Field Analysis

The displacement field was calculated by cross-correlating images from the same cross-section at different deformation stages. Fig. 11.6 and 11.7 show the 2D displacement fields of four different cross-sections. The displacement field was clearer in the first plane than the fourth plane because of the contrast loss due to light diffusion during propagation through transparent soil. The displacement pattern along the footing was consistent.

Streamlines are used mostly in the aerodynamics field. A streamline is a path traced by a particle as it moves with the flow. Velocity is tangent to streamline at every point and mass does not cross streamlines. The concept of streamline is similar to the flow net used in geotechnical engineering. Streamlines of the displacement field were used to visualize the distribution and variation of displacement (Fig. 11.8), which shows the change in the direction of displacement.

The streamlines along the footing are shown in Fig. 11.9, where the sixth deformation stages were used for the illustration.

The magnitude of displacement was calculated as follows:

$$V = \sqrt{u^2 + v^2} \tag{11.1}$$

where u and v were horizontal and vertical displacement components. The image plane was treated as a plane in the x and z axes. The distance between cross-sections was treated as the y axis. The 2D displacement magnitude fields from the four different cross-sections were combined into a three-dimensional displacement magnitude field.

Slicing the 3D displacement magnitude field details the distribution of displacement inside the field as shown in Fig. 11.10 and Fig 11.11, for the first and sixth loading stages respectively, where the unit in y direction was the number of cross-section, the units in the x and z axes were the numbers of the interrogation windows used in image analysis.

The maximum displacement spread out in the sixth loading step more than that in the first loading step. The grayscale shows qualitatively the magnitude. The darker the color, the larger the magnitude of displacement. However, the data can be queried numerically within the Matlab® model.

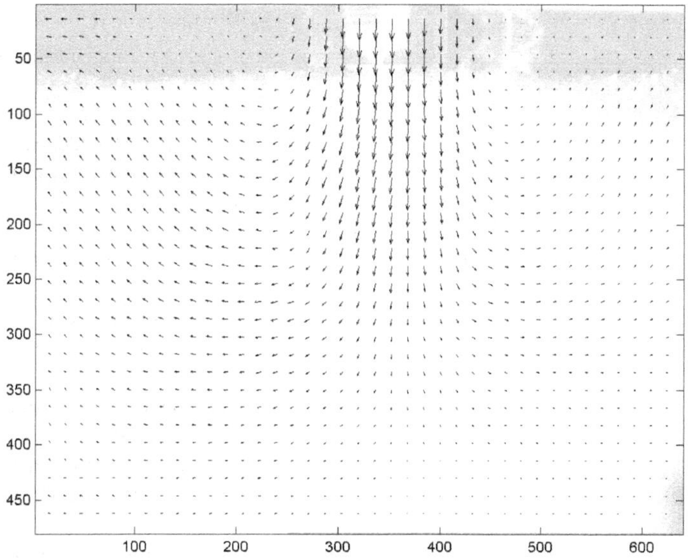

The First Plane

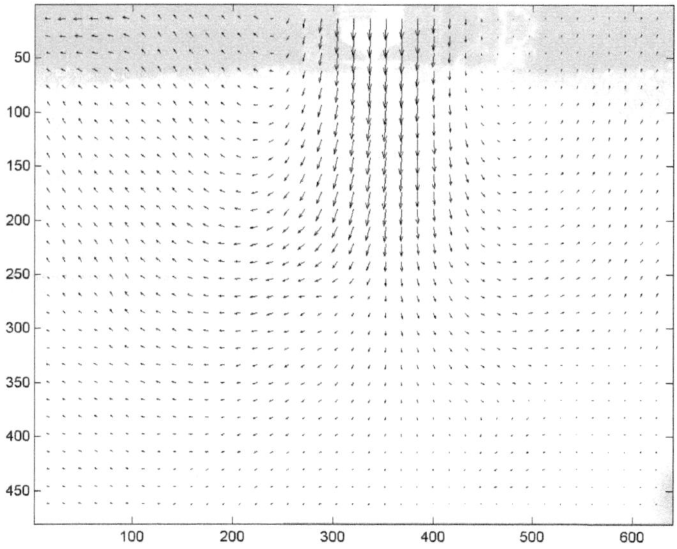

The Second Plane

Fig. 11.6 Displacement Field from The First Deformation Stage (First & Second)

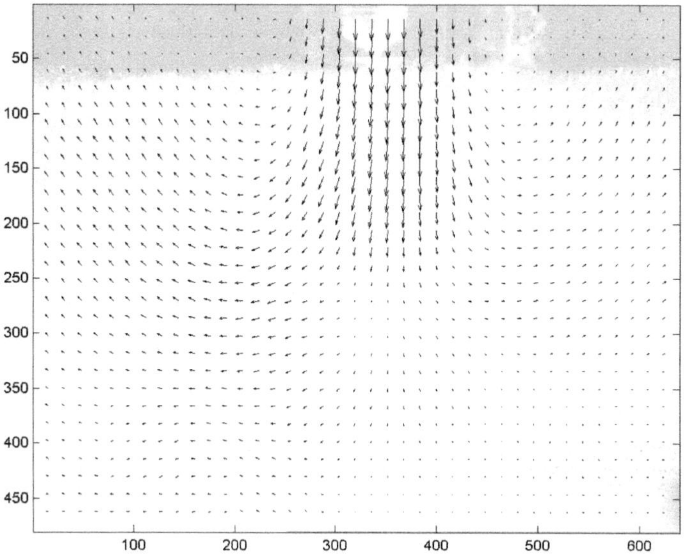

The Third Plane

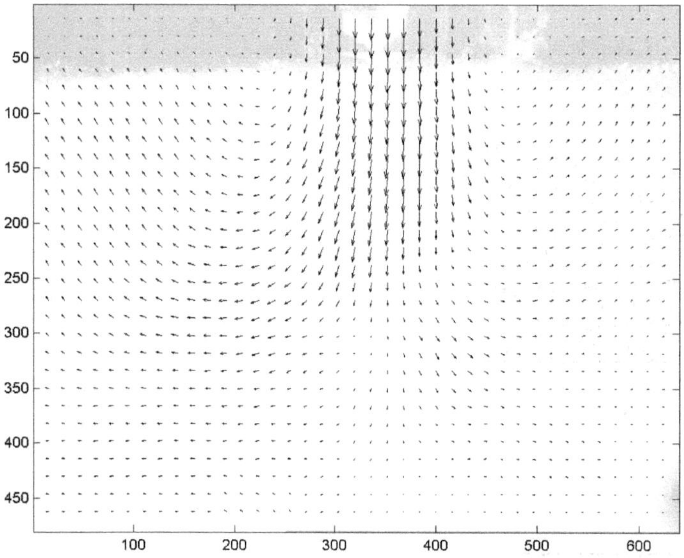

The Fourth Plane

Fig. 11.7 Displacement Field from The First Deformation Stage (Third & Fourth Planes)

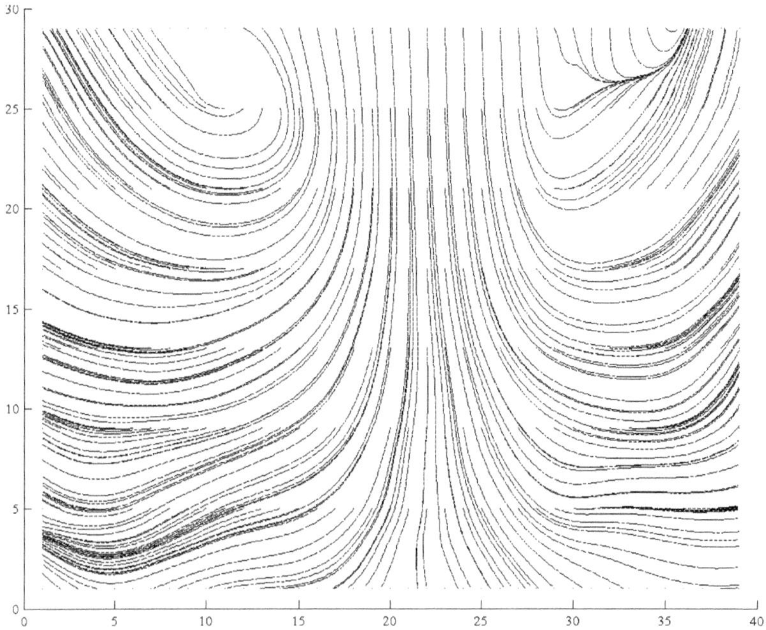

Fig. 11.8 Streamline of Displacement Field from The Sixth Deformation Stage. The second Plane Shown.

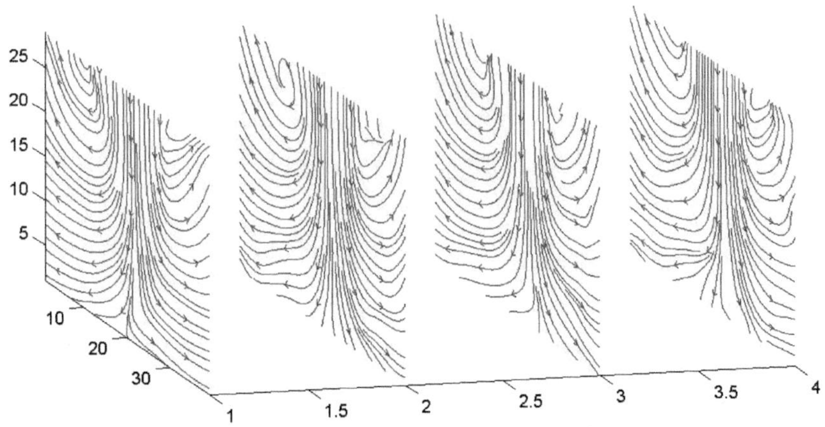

Fig. 11.9 3D View of Stream Line of Displacement Field. (Sixth Loading Stage Shown)

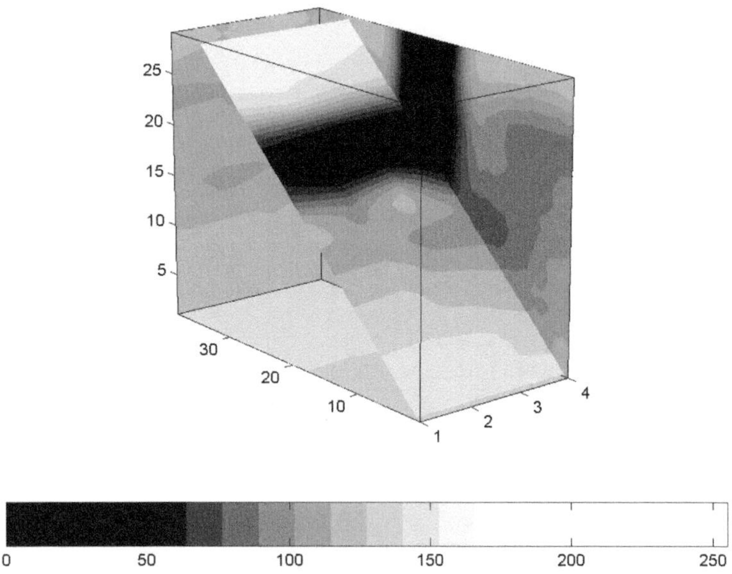

Fig. 11.10 Slice of 3D Displacement Magnitude Field (The First Deformation Stage). Darker color represents larger displacement.

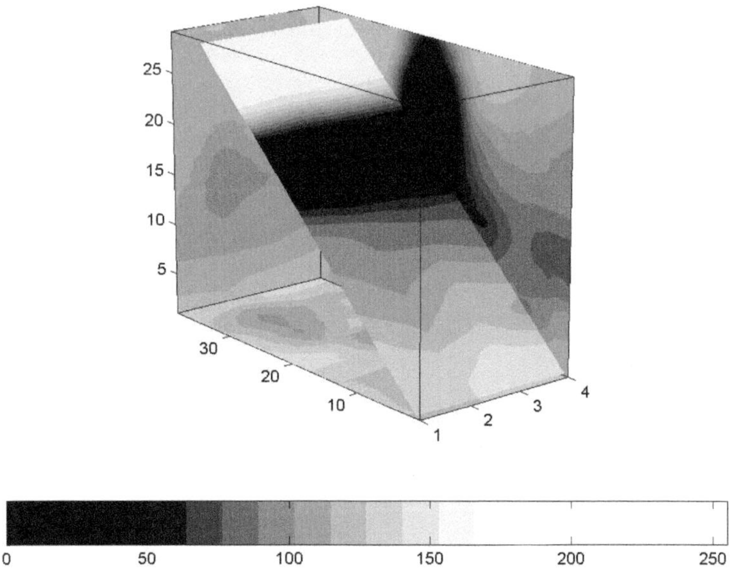

Fig. 11.11 Slice of 3D Displacement Magnitude Field (The Sixth Deformation Stage). Darker color represents larger displacement.

11.4 Strain Field Analysis

Strain fields were obtained by differentiating the predicted displacement field. A similar approach has been used by many researchers, including Roscoe et al (1963) and Yamamoto and Kusuda (2001). Green's strains (Bathe 1995) were used regardless of the magnitude of the displacement, as follows:

$$\varepsilon_x = \frac{\partial u}{\partial x} + \frac{1}{2}\left[\left(\frac{\partial u}{\partial x}\right)^2 + \left(\frac{\partial v}{\partial x}\right)^2\right] \tag{11.2}$$

$$\varepsilon_y = \frac{\partial v}{\partial y} + \frac{1}{2}\left[\left(\frac{\partial u}{\partial y}\right)^2 + \left(\frac{\partial v}{\partial y}\right)^2\right] \tag{11.3}$$

$$\gamma_{xy} = \frac{\partial u}{\partial y} + \frac{\partial v}{\partial x} + \left[\frac{\partial u}{\partial x}\frac{\partial u}{\partial y} + \frac{\partial v}{\partial x}\frac{\partial v}{\partial y}\right] \tag{11.4}$$

The maximum shear strain was calculated using the strains above.

$$\gamma_{\max} = \sqrt{(\varepsilon_x - \varepsilon_y)^2 + \gamma_{xy}{}^2} \tag{11.5}$$

The failure envelope was assumed to coincide with the peak maximum shear strain; the failure development was studied by locating the peak maximum shear strain. The out-of plane displacement component was not obtained by the optical system, because only one camera was used. The out-of plane displacement can be obtained by updating the developed system with two cameras set under a well-defined configuration. The modeled problem in this chapter was a typical plane strain problem. Therefore, it was assumed that the influence on the strain field from out-of plane displacement component was insignificant in this study.

The maximum shear strain from the first and sixth deformation stage is shown in Fig. 11.12 and 11.13, where the peak shear strain regions were shaded. A punching shear failure mode occurred in the top "sand" layer and extended downward to the surface of the lower clay layer. In the bottom "clay" layer, the failure envelope was similar to that under a model footing in a uniform "clay" layer. The failure model was similar to that reported by Meyerhof (1974) for a footing on thin sand overlying clay, which is the problem being simulated.

No continuous failure plane was developed in both the first and the sixth deformation stages (Fig. 11.12 and Fig. 11.13). The peak strain fields were shaded in the figures. The failure envelopes developed on the sixth deformation stages were clearer than the first deformation stage, which was consistent to their loading conditions. The sixth deformation stage exceeded the bearing capacity of model footing.

The distributions of maximum shear strain in different cross sections agreed with each other at the same deformation steps. Small variation existed due to the prediction errors in the displacement field.

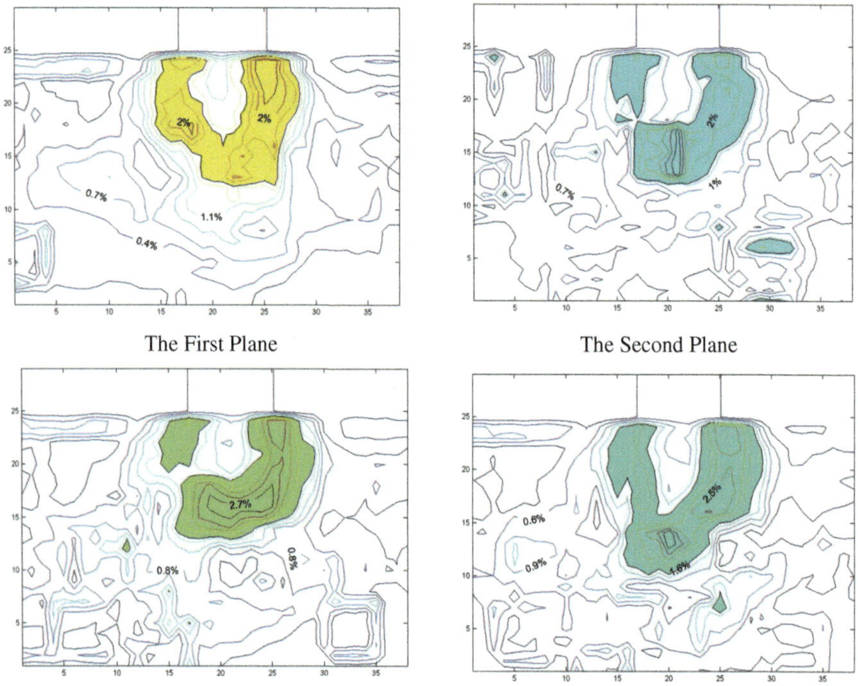

The First Plane The Second Plane

The Third Plane The Fourth Plane

Fig. 11.12 Maximum Shear Strain Field from The First Deformation Stage

A 3D view of maximum shear strain field is shown in Fig. 11.14, where the units in the image plane were the numbers of the interrogation windows used in the image processing, the unit in the direction between images was the number of cross-section. Fig. 11.14 clearly shows that the maximum shear strain is mainly located around the footing in the first deformation stage. At the sixth deformation stage where the loading exceeded the bearing capacity of the model footing, the failure envelope was much clearer than the first deformation stage. The magnitude of shear strain is proportional to the light intensity in the images. The brighter the image, the larger the magnitude of shear strain.

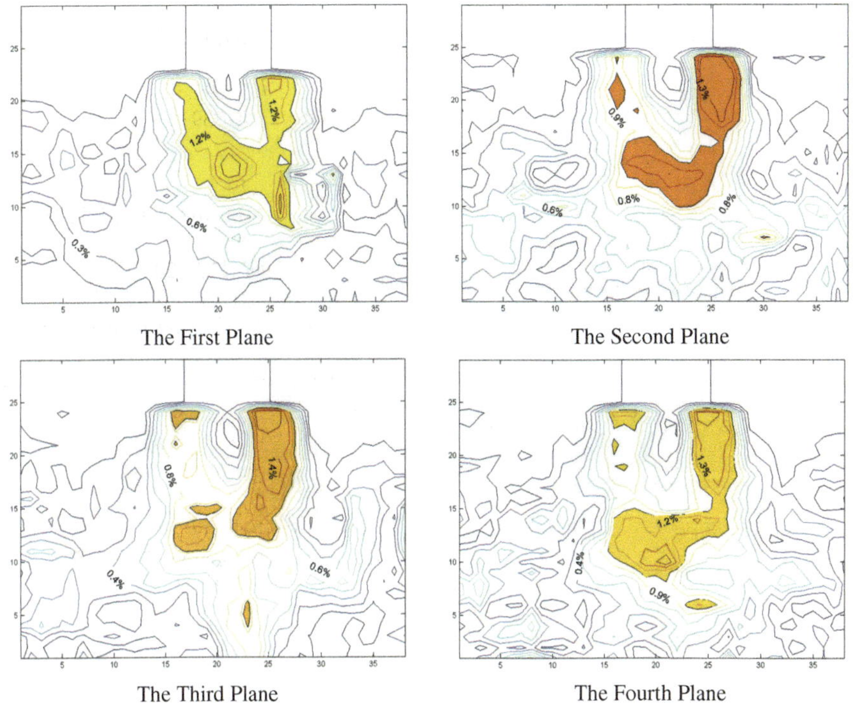

Fig. 11.13 Maximum Shear Strain Field from The Sixth Deformation Stage

A similar approach to the one used in presenting 3D view of displacement magnitude field was used to view the 3D strain field. The 3D views of maximum shear strain field in the first and sixth deformation stages are shown in Fig. 11.15. In order to illustrate the information on each cross-section clearly, the scale along the footing is not in the same scale as the image. The studied case is a typical plane strain problem. Slicing of strain field in both deformation stages showed that the strain field was similar along the footing, which is consistent with the problem being studied.

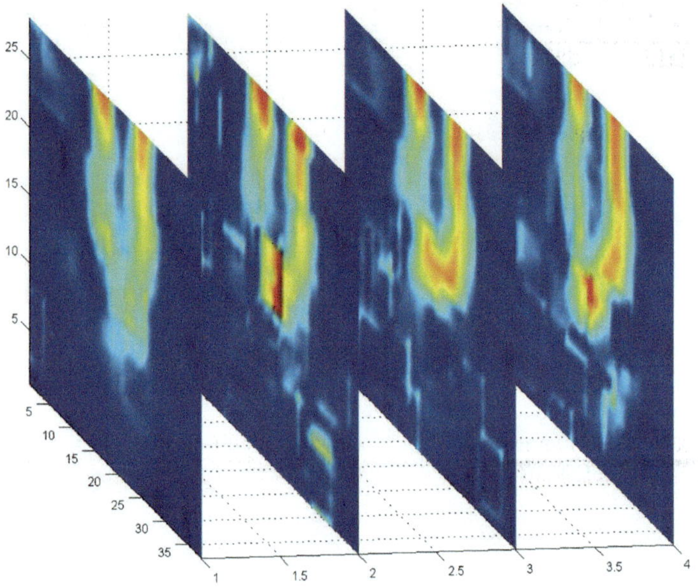

The First Deformation Stage

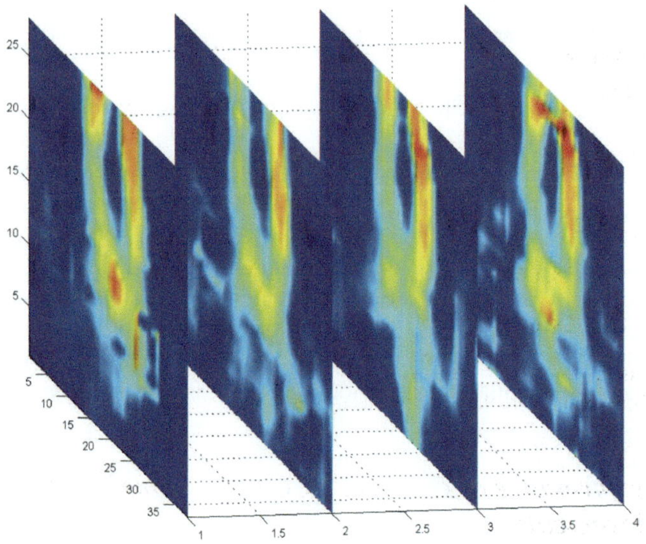

The Sixth Deformation Stage

Fig. 11.14 3D View of Maximum Shear Strain Field

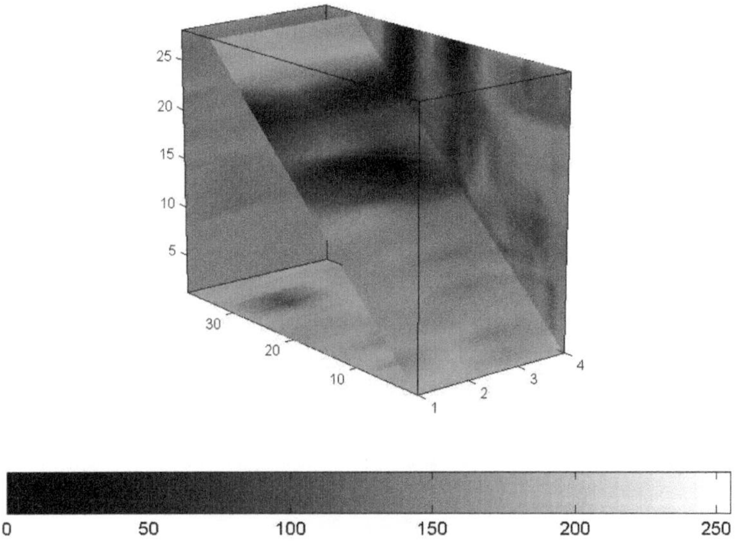

The First Deformation Stage

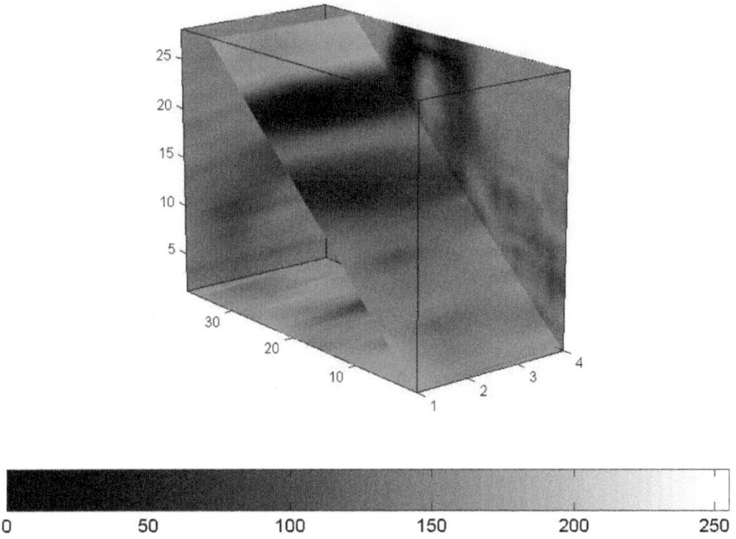

The Sixth Deformation Stage

Fig. 11.15 Slice of 3D Maximum Shear Strain Field. Darker Scale Represent a Higher Magnitude of Strain.

11.5 Displacement and Strain Development

11.5.1 Displacement Field

The proposed mechanism for 3D displacement analysis can also reveal the development of displacement field or strain field with deformation. The 2D displacement magnitude fields of the same plane under different deformation stages are presented in 3D as shown in Fig. 11.16, where the unit along images was the deformation stage (Fig. 11.3); the units in image plane were the numbers of the interrogation windows used in image analysis. It is difficult to give a quantitative analysis on the development (on paper) because of the different magnitudes. However, it qualitatively shows the development of displacement field with deformation. The displacement field can be investigated numerically in Matlab.

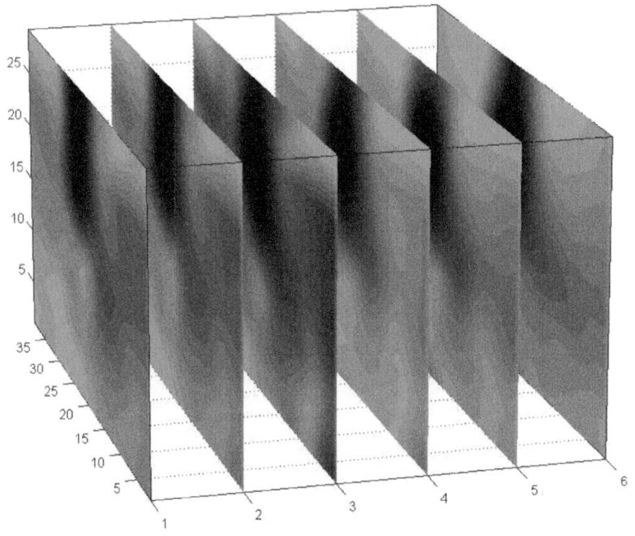

Fig. 11.16 3D View of Development of Displacement Field. (The Third Plane (7.36 mm Inside Sample). Darker Shades Represent Larger Displacements.

Slicing the displacement magnitude field revealed the development better (Fig. 11.17). The unit between image planes was the deformation stage, and the units along the image plane were the number of the interrogation windows. The middle two planes were the displacement magnitude fields at the second and fourth deformation stages. The displacement field extended to a larger area with the deformation. The variation occurred on the third deformation stage was visible, which showed that there were errors on this stage. A smoother transition with deformation is expected by unifying the displacement increment between the stages.

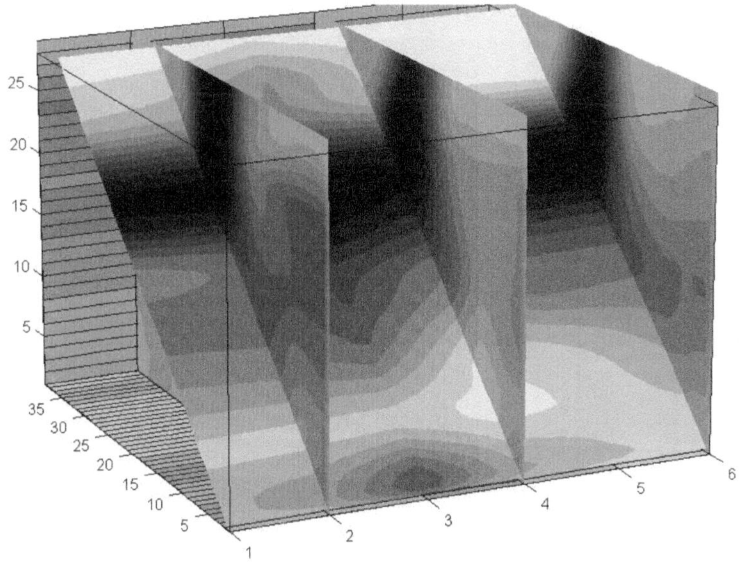

Fig. 11.17 Slicing 3D Displacement Magnitude Development Field. The Third Plane (7.36 mm Inside Sample) is shown.

11.5.2 Strain Field

A 3D view of the shear strain field development was obtained by combing the 2D strain fields at different deformation stages instead of the different cross-sections. Slicing the maximum shear strain fields revealed the development of shear strain with the deformation on the same cross-section, shown in Fig. 11.18. The figure clearly shows that the maximum shear strain at the first deformation stage was not prominent. It became clearer after the third deformation stage. The shear failure became more obvious with deformation, which was indicated by the dark zone.

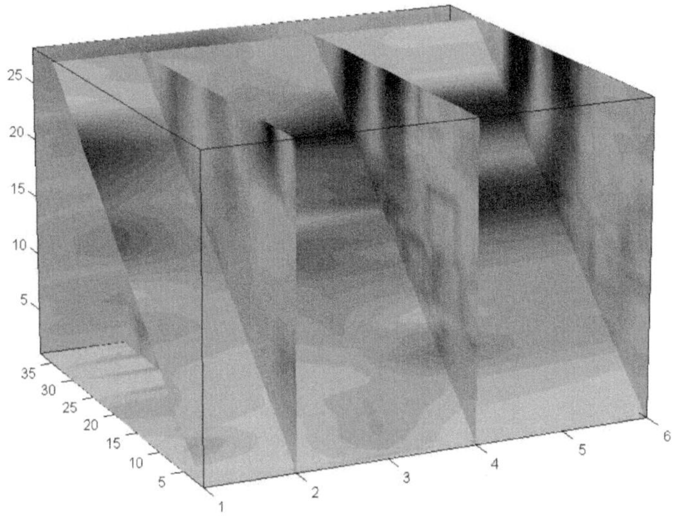

The First Plane (1 mm Inside Sample)

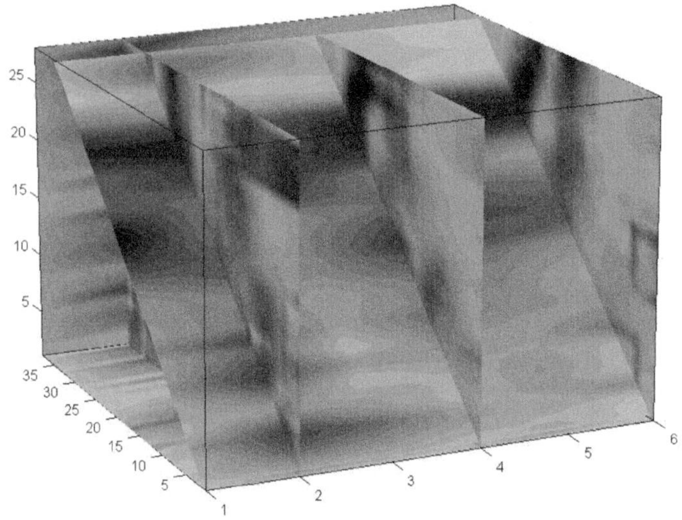

The Third Plane (7.36 mm Inside Sample)

Fig. 11.18 Maximum Shear Strain Development with Loading. Darker Shades Represent Higher Strains.

11.6 Error Analysis

11.6.1 Result Analysis

The studied problem was a typical plane strain problem. Therefore, the displacement field was supposed to be uniform along the footing length. Assuming that the surface friction of the mould was insignificant, the variation in the predicted settlement between cross-sections was used to verify the optical method. Comparisons were conducted on three different horizontal planes, (1) across the footing bottom (in Fig. 11.19), (2) 0.5 B from the footing bottom (in Fig. 11.20), and (3) 1 B from the footing bottom (in Fig. 11.21), where B was the footing width. Six points within 2.5B from the footing center were selected in the study. Comparisons of the settlements between the four different cross-sections are detailed in Table 11.1 to 11.3. The errors in the tables refer to the largest deviation from the average settlement obtained from the predicted values. The maximum deviations obtained from optical measurement were generally smaller than 0.1 mm.

The errors on the points through the footing centerline were specifically tabulated in Table 11.4. The errors on this cross-section were smaller than 10 percent of the corresponding deformation increments. As the points went deeper from the footing bottom, the errors increased because of the increasing influence from non-uniform compaction during sample preparation. For example, in the first deformation stage, the error on the point immediately below the footing bottom was only 3 percent while the error on the point located 1.5 B deep from footing bottom was 9.3 percent. The heterogeneity of the model was believed to contribute the error increments. Considering the magnitude of error on the order of 0.1 mm and all errors are within 10% of the measured value, the proposed optical method is feasible for studying 3D deformation problems.

Table 11.1 Predicted Settlement for the Plane across Footing Bottom

Measuring Points	A	B	C	D	E	F
Distance From Footing Center	2.5 B	2.0 B	1.5 B	1.0 B	0.5 B	0 B
The First Loading						
Average Value, mm	0.31	0.35	0.36	0.19	-0.80	-2.12
Standard Deviation, mm	0.067	0.069	0.079	0.043	0.044	0.067
Error, mm	0.065	0.084	0.101	0.061	0.030	0.072
The Sixth Loading						
Average Value, mm	0.20	0.14	0.05	-0.07	-0.72	-1.23
Standard Deviation, mm	0.028	0.007	0.013	0.006	0.053	0.015
Error, mm	0.026	0.008	0.011	0.008	0.071	0.014

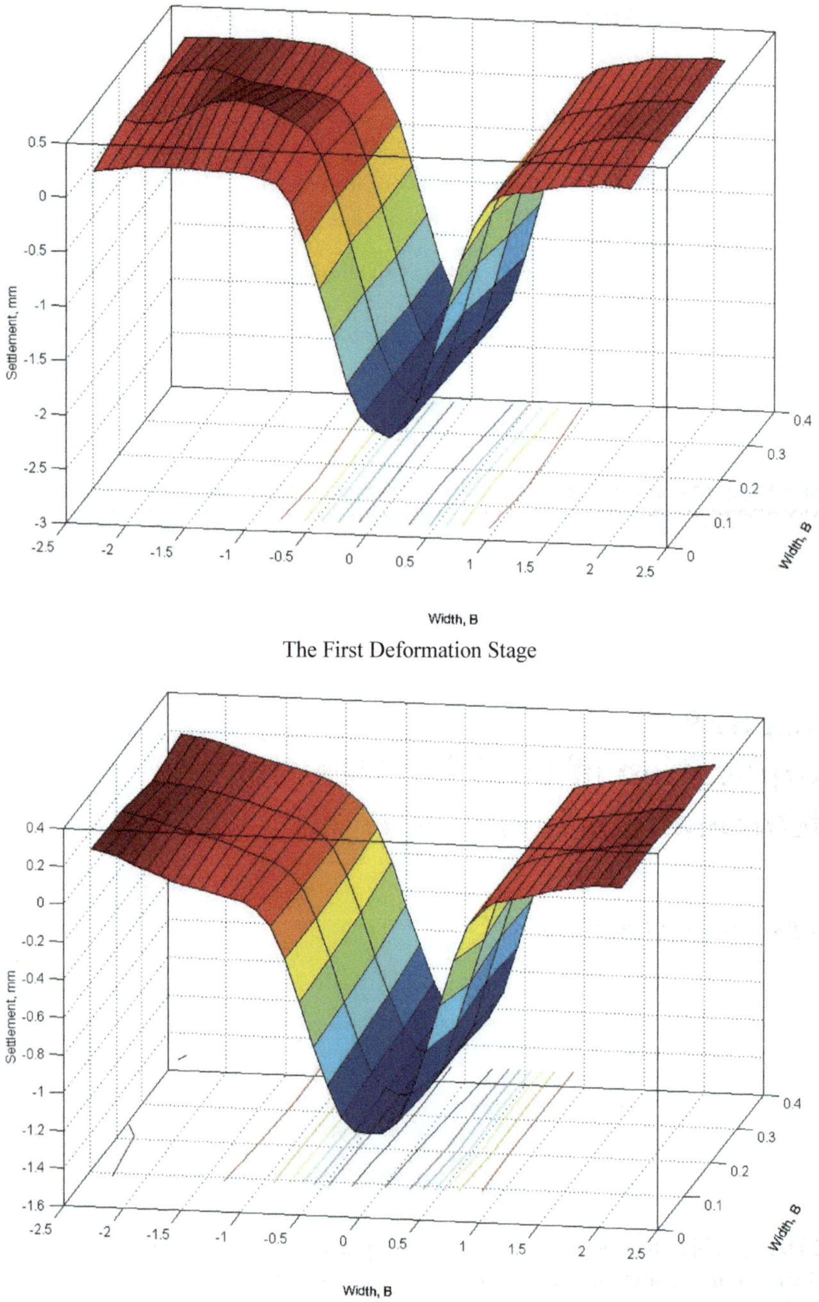

The First Deformation Stage

The Sixth Deformation Stage

Fig. 11.19 Settlement of the Plane across Footing Bottom

Table 11.2 Predicted Settlement for Plane 0.5 B below Footing Bottom

Measuring Points	A	B	C	D	E	F
Distance From Footing Center	2.5 B	2.0 B	1.5 B	1.0 B	0.5 B	0 B
The First Loading						
Average Value, mm	0.52	0.52	0.52	0.16	-0.94	-1.89
Standard Deviation, mm	0.054	0.138	0.044	0.026	0.084	0.127
Error, mm	0.062	0.109	0.042	0.034	0.120	0.176
The Sixth Loading						
Average Value, mm	0.17	0.14	0.04	-0.12	-0.71	-1.19
Standard Deviation, mm	0.040	0.008	0.016	0.021	0.035	0.020
Error, mm	0.051	0.011	0.023	0.027	0.052	0.026

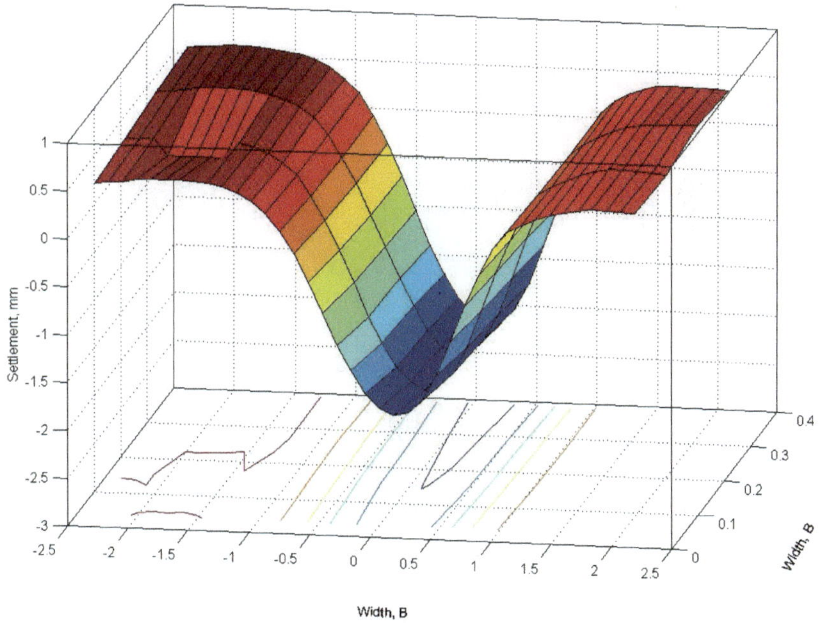

Fig. 11.20 Settlement of Plane 0.5 B below Footing Bottom. The First Deformation Stage Shown.

Table 11.3 Predicted Settlement for Plane 1B below Footing Bottom

Measuring Points	A	B	C	D	E	F
Distance From Footing Center	2.5 B	2.0 B	1.5 B	1.0 B	0.5 B	0 B
The First Loading						
Average Value, mm	0.42	0.51	0.45	0.02	-0.72	-1.50
Standard Deviation, mm	0.062	0.063	0.043	0.090	0.147	0.177
Error, mm	0.058	0.056	0.052	0.110	0.195	0.139
The Sixth Loading						
Average Value, mm	0.15	0.11	-0.02	-0.21	-0.59	-0.98
Standard Deviation, mm	0.049	0.069	0.037	0.064	0.089	0.099
Error, mm	0.037	0.103	0.050	0.072	0.125	0.116

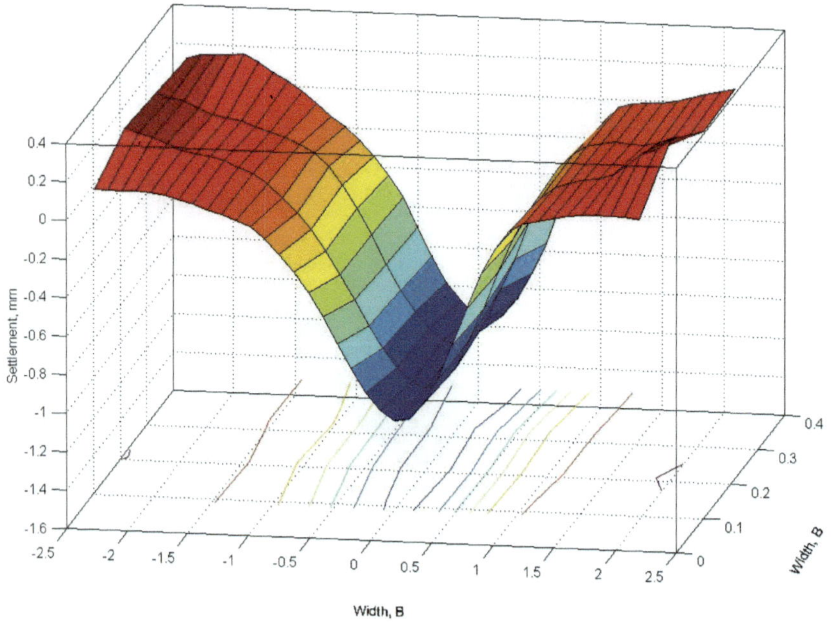

Fig. 11.21 Settlement of Plane 1 B below Footing Bottom. The Sixth Deformation Stage Shown.

Table 11.4 Errors at the Point across Footing Center, F

Depth From Footing	The First Loading			The Sixth Loading		
	0 B	0.5 B	1 B	0 B	0.5 B	1 B
Settlement, mm	2.12	1.89	1.50	1.23	1.19	-0.98
Error, mm	0.072	0.176	0.139	0.014	0.026	0.099
Percentage, %	3.4	9.3	9.3	1.1	2.2	11.6

11.6.2 Image Distortion Analysis

Image distortion was analyzed by using a standard paper ruler taped on the surface of a model box. The image was captured under the same optical setting used in the 3D deformation measurement discussed above. The image was analyzed manually using Microsoft Photo Editor. The locations for measuring distortion were shifted from their original positions, shown in Fig. 11.22. The labeled numbers were the numbers of pixels in the image corresponding to 10 mm in the ruler. No apparent image distortion was found from the measurement for the optical setting employed, where the CCD camera was set 200 mm away from the mold with the maximum zoom condition. The 10 mm distance was measured as 54 pixels everywhere in the image, except one location, where it was measured as 53 pixels. The judgment on the edges and misalignment might contribute to this error.

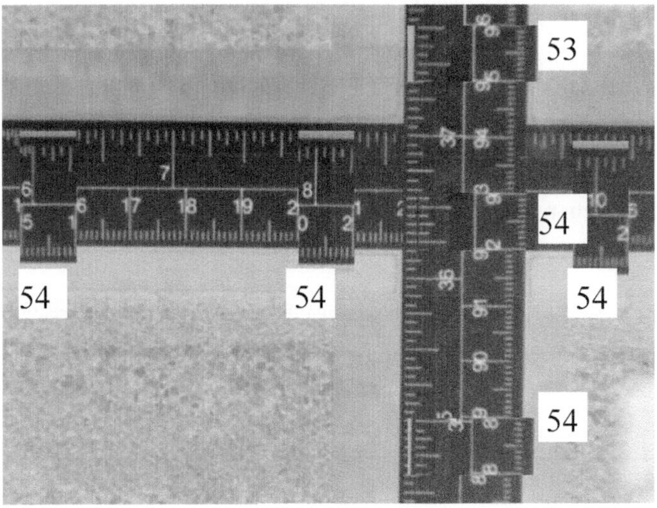

Fig. 11.22 Image for Image Distortion Analysis

11.6.3 Speckle Stability over Time

A fundamental assumption for the proposed optical measurement method was that the statistical and physical distribution of the speckle images is stable over time. The stability of speckle image over time was investigated by correlating two images taken 5 minutes apart. The distribution of the predicted displacement field from the two images is shown in Fig. 11.23. A random distribution of predicted displacement was evident. The average of displacement components, u and v, were 0.003 mm and 0.008 mm. The standard deviations for these two components were 0.056 and 0.067 mm.

The histograms of occurrence plotted against light intensity of two images taken 5 minutes apart are shown in Fig. 11.23. Two curves matched each other over the whole intensity level, which proves that the statistical properties of speckle images were stable over time.

11.6.4 Linear Stage Error

The position of slices was controlled manually by adjusting the linear stage. Mistakes might be made during the test. Analysis of the mechanism for position control was required to verify the proposed method. Besides the number of rolls, the position was also marked on the edge of the mold to minimize the possibility of blunders. The maximum possible error was 0.635 mm (the distance linear stage travels per roll), because of the errors in counting the number of rolls. The deformations from two images 0.635 mm away from each other were compared to the result of two images from the same location (Fig. 11.24), where transparent silica gel was used in this analysis. A fixed horizontal displacement (0.635 mm) was applied. The predicted displacements from two images, 0.635 mm from each other, were comparable to the one from the same location, shown in Table 11.5. The laser beam used in this research was a 35 mW Helium Neon Laser made by Melles Griot. The output beam diameter was 1.24 mm with a beam divergence of 0.68 mrad. The depth of the laser light sheet increases to about 1.5 mm at a penetration depth of 10 cm. The distance between the two images was within the depth of laser sheet, which proves that the proposed mechanism is robust against small alignment errors.

Table 11.5 Comparison on Predicted Displacement Field Due To Position Errors

Distance Between Two Images	Horizontal Displacement		Vertical Displacement	
	$\bar{u}$	σ	$\bar{v}$	σ
The Same Location	0.6364	0.0859	-0.0036	0.0717
0.635 mm From Each Other	0.6416	0.0732	-0.0034	0.0607

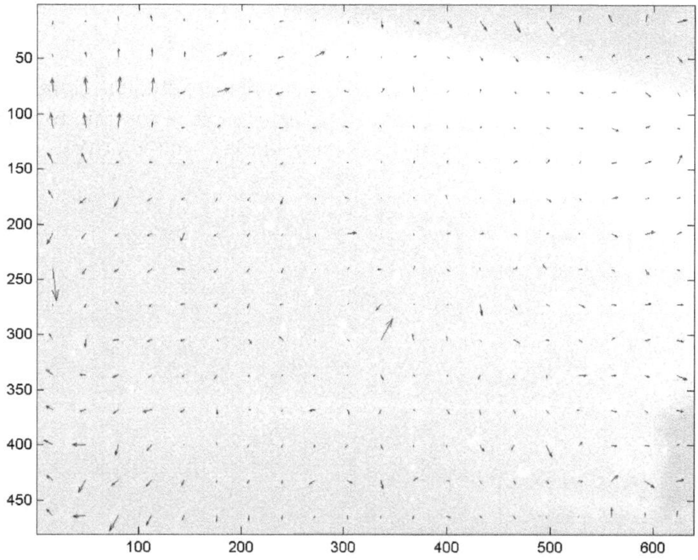

Displacement Field from Two Images

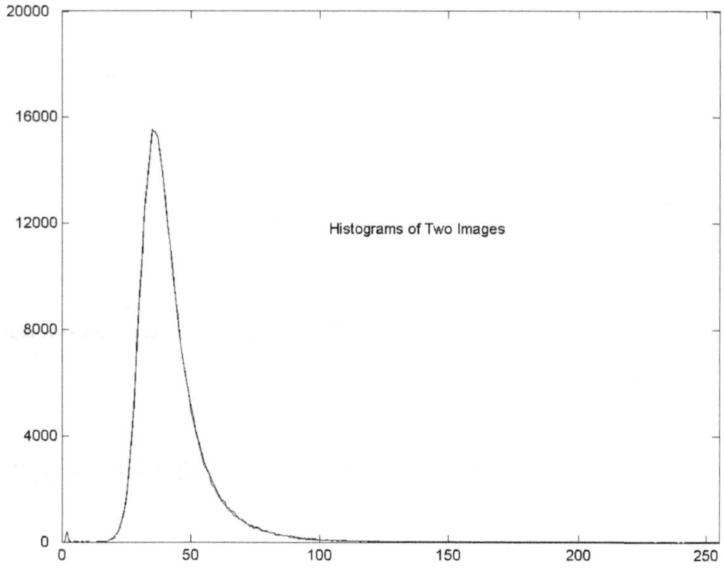

Histograms of Two Images are Identical to Each Other

Fig. 11.23 Stability of Speckle Properties over Time

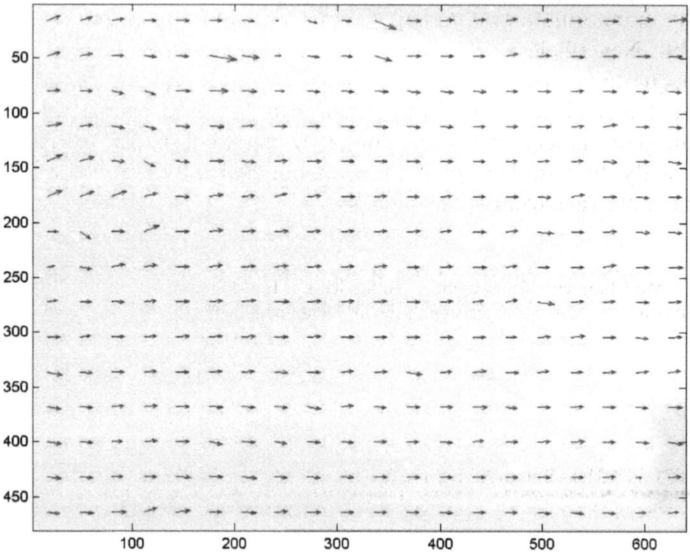

The Same Location

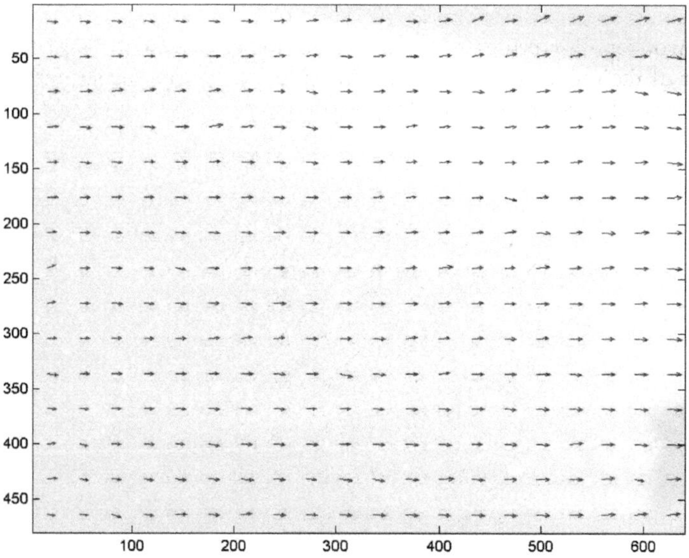

0.635 mm from Each Other

Fig. 11.24 Comparison of Displacement Field with Location Misalignment

11.7 Conclusions and Recommendations

In this chapter, 3D deformations inside a transparent synthetic soil model were measured non-intrusively by cross-correlating the speckle images. The speckles were generated by interaction between laser light and transparent synthetic soils. A soil profile consisting of "sand" overlying soft "clay" was simulated. The deformation and strain field under the model footing were similar to those published. Error analysis of the mechanism and system showed that the proposed method is suitable for measuring spatial deformation non-intrusively.

The out-of plane deformations were not captured in the study. This deficiency, however, can be solved by updating the proposed system with multiple cameras. The load-frame system needs to be connected and synchronized with the optical system to eliminate system errors. A computer-controlled linear stage is needed to minimize the errors caused by manually control of the system. A larger model is recommended for eliminating boundary effects.

References

Adrian, R.: Particle-imaging techniques for experimental fluid mechanics. Annual Review of Fluid Mechanics 23, 261–304 (1991)

Bathe, K.: Finite Element Procedures, p. 1037. Prentice-Hall, Englewood Cliffs (1995)

Bergfelt, A.: Loading tests on clay. Geotechnique 11, 15–31 (1956)

Chu, T., Ranson, W., Sutton, M., Peters, W.: Application of digital image correlation techniques to experimental mechanics. Experimental Mechanics 25(3), 232–244 (1985)

Dantec Measurement Technology A/S, FlowmapÒ Particle Image Velocimetry Instrumentation-Installation & User's Guide, Tonsbakken 16-18, DK-2740 Skovlunde, Denmark (2000)

Dupre, L., Lagarde, A.: Photoelastic analysis of a three-dimensional specimen by optical slicing and digital image processing. Experimental Mechanics 37(4), 393–397 (1997)

Grantham, S., et al.: Study of internal deformation fields in granular materials using 3-D digital speckle X-ray flash photography. In: Proc. of SPIE - The International Society for Optical Engineering, vol. 4101 (B), pp. 321–328 (2000)

Heinz, W.: Discussion of papers in session 1" Stress-Strain Behavior of Soils. In: Parry, R., (ed.) Proc. Roscoe Memorial Symposium, Cambridge University, G.T. Foulis, Henley-on-Thames, England, pp. 120–122 (1971)

James, R., Bransby, P.: Experimental and theoretical investigations of a passive earth pressure problem. Geotechnique 20(1), 17–37 (1970)

Konagai, K., Tamura, C., Rangelow, P., Matsushima, T.: Laser-aided tomography: A tool for visualization of changes in the fabric of granular assemblage. Proc. JSCE, No 455 I-21 in Struct. Eng./Earthquake Eng. 9(3), 193–201 (1992)

Liner, L.: Geophysics and NIH image. Computers & Geosciences 25(4), 403–414 (1999)

McKenna, J., Sherlock, D., Evans, B.: Time-lapse 3-D seismic imaging of shallow subsurface contaminant flow. J. of Contaminant Hydrology 53(1-2), 133–150 (2001)

Meyerhof, G.: Ultimate bearing capacity of footings on sand layer overlying clay. Canadian Geotechnical J. 11(2), 224–229 (1974)

Nemat-Nasser, S., Okada, N.: Radiographic and microscopic observation of shear bands in granular materials. Geotechnique 51(9), 753–765 (2001)

Repin, D., Pavlis, G.: Interactive three-dimensional visualization of seismic array data. Computers & Geosciences 23(10), 1079–1092 (1997)

Roscoe, K., Arthur, J., James, R.: The determination of strains in soils by an X-ray method. Civil Engineering and Public Works Review 58, 873–876, 1009–1012 (1963)

Spiesberger, W., Tasto, M.: Processing of medical image sequences. In: Huang, T.S. (ed.) Image Sequence Analysis, pp. 381–428. Springer, New York (1981)

Udupa, J., Herman, G.: Guest editorial medical image reconstruction, processing, visualization, and analysis: the MIPG perspective. IEEE Transactions on Medical Imaging 21(2), 281–295 (2002)

Weber, L., Schumate, E., Mawer, N.: Experiments on flow at a 90° open-channel junction. J. of Hydraulic Engineering, ASCE 127(5), 340–350 (2001)

Zenina, A., Dupre, J., Lagrade, A.: Separation of isochromatic and isoclinic patterns of a slice optically isolated in a 3-D photoelastic medium. European J. of Mechanics A - Solids 18(4), 633–640 (1999)

Yamamoto, K., Kusuda, K.: Failure mechanism and bearing capacities of reinforced foundations. Geotextiles and Geomembranes 19(3), 127–162 (2001)

Chapter 12
2D Flow in Transparent Synthetic Soils

Magued Iskander, Jinyuan Liu, Hung-Chieh Lo, and Kazunori Tabe

Abstract. In this chapter, 2D multiphase flow and surfactant flushing tests were simulated using a variety of transparent synthetic soils including silica gel, fused silica, and Aquabeads. The developed technology allows for visualizing the contamination concentration profiles and the surfactant's sweeping efficiency. The breakthrough curves of dye through both materials were typical of the bell-shape reported for natural soil. Image analysis shows both the advection and dispersion processes clearly visible in the tests, which demonstrates that transparent synthetic soil can be used to visualize and non-invasively measure pollutant transport in natural soil.

12.1 Introduction

The United States Environmental Protection Agency (EPA) estimates that there are approximately 330,000 hazardous waste sites where ground water may be contaminated (EPA 1993). Estimates by the Nuclear Regulatory Commission (NRC) range from 300,000 to 400,000 sites (NRC 1994). Based on EPA data, estimates are that the United States as a whole will spend $480 billion to 1 trillion over the next 30 years cleaning up these sites (Charbeneau 2000). Many of these pollutants are released as either an aqueous solution or a non-aqueous phase liquid (NAPL).

There are three basic physical mechanisms by which dissolved pollutants are transported in the subsurface environment: advection, diffusion, and mechanical dispersion. The process of pollutant transport with the groundwater bulk fluid movement is advection, which is the most significant mass transport process. Diffusion is the process of mass transport associated with random molecular motions (Bird et al 1960). These molecular motions result in dissipation of pollutants from higher concentration regions to regions with a lower concentration.

Mechanical dispersion is a transport process associated both with bulk fluid movement and with the complexity of pore spaces. When there is a concentration gradient, mechanical dispersion results in the deviation from the bulk water movement. Mechanical dispersion is usually modeled as a diffusion or Fickian

M. Iskander: Modeling with Transparent Soils, SSGG, pp. 259–292.
springerlink.com

type process (Charbeneau 2000). The advection-dispersion is the basic process used in describing pollutant transport in porous media.

In order to clean up a site, one of the most difficult tasks is to identify the location of the contaminants. Currently, core samples (borings) are relied on for both quantifying and locating the contaminants and are often unreliable and costly. The development of better technologies to address this problem requires laboratory testing using physical models.

Transparent materials including glass beads or quartz powder have been used to help in understanding flow processes through the soil (Chen and Wada 1986, Brock and Orr 1991, Matsubayashi et al 1997, and Herrmann et al 2002). However, glass and quartz surrogates are limited by their poor transparency, because air entrapped in the tiny cracks of glass particles reduces the transparency of these materials. These materials are also limited by their inability to represent the geotechnical properties of natural soils. In the last five years, glass bead has been increasingly used in micromodels by many researchers to implement imaging technique in their research (Corapcioglu et al 1997, Corapcioglu and Fedirchuk 1999, Jia et al 1999, and Huang et al 2002). At the present time, large-scale models made of glass bead or quartz powder are limited by their poor transparency.

The transparent synthetic soils described in Chapters 5-7 have been shown to exhibit macro-scale geotechnical properties similar to those of natural soils. These materials are truly transparent — not merely translucent. In this chapter, 2D flow tests are performed to study the potential application of transparent synthetic soils for studying pollutant transport in soils. Visualizing the advection-dispersion process of pollutant transport is demonstrated by using image processing techniques to give a better understanding of the process.

12.2 Flow Tests Using Silica Gel

12.2.1 Materials

Amorphous silica gel (silica gel) was obtained from *Multisorb Technologies, Inc.* with an average aggregate size of 1 mm. Two pore fluids were used in the flow tests. The first was mineral oil blend used in the previous chapters, which is a 1:1 blend of *Drakeol* 35 mineral oil and *Norpar®* 12 paraffinic solvent, by weight. The refractive index, viscosity, and density of the oil blend at room temperature (24°C) were 1.447, 5.0 cP, and 800 kg/m^3, respectively. The second pore fluid was calcium bromide brine, which is a mixture of calcium bromide and water to have a refractive index of 1.448 at 25°C to match the refractive index of mineral oil blend. The viscosity and density of this mixture at room temperature were 3.6 cP and 1572 kg/m^3, respectively.

12.2.2 Sample Preparation

Sample preparation proceeded as follows. First, amorphous silica gel was immersed in the oil blend. Second, vacuum was applied to de-air the mix until the mixture became transparent. The time required for vacuum depends on the depth of silica gel and the power of the vacuum pump. Approximately 24 hours was required to fully saturate the silica gel. Third, silica gel was packed into a Plexiglas model tank to form a transparent amorphous silica gel sample. Before packing, two filters consisting of four layers of steel screens with the opening size ranging from 2 mm to 0.5 mm were placed at the both ends of the mold in order to distribute the flow vertically along the mold depth.

The model tank was partially filled with oil blend before packing. Silica gel was slowly poured into the tank using a spoon. At the same time, the silica gel sample was stirred to release air bubbles entrapped during pouring. The tank was vibrated while being packed and silica gel was placed in 30 mm lifts. A piezoelectric vibrator was used to compact silica gel during placement in the model tank. The magnitude of the vibration was 0.1 mm and duration for vibration was 2 minutes for each lift. Once silica gel filled the model tank, a cover plate was screwed tightly onto the tank with a gasket between the plate and the tank.

12.2.3 Flow Equipment Setup

The schematic layout of the experimental setup is shown in Fig. 12.1. A Plexiglas model tank with an internal dimensions of 25×152×406 mm was used to perform the flow test. The model tank was glued to prevent leakage everywhere except the top. Besides the precisely cut gasket used between the cover plate and box, additional gasket pads with holes were used between the screw cap and the cover plate. This technique was successful in preventing leakage. Pore fluid containing an oil-soluble dye was injected through one end of Plexiglas mold. Two reservoirs containing clear and dyed oil blend were connected to the pressure panel board to keep the pressure head constant. A manual switching valve was used to control the tracer injection. Flow was collected at the other end using a graduated cylinder.

12.2.4 Optical Measurements

The optical system shown in Fig. 12.2 was used to measure the dye movement through transparent synthetic soil model. The camera was set 1.20 m from the model with its optical axis perpendicular to the Plexiglas mold. Images were captured and imported to the PC through a frame grabber. A fluorescent lamp was used to illuminate the experiment. The lamp was kept in the same position throughout the test so as not to change the lighting conditions. A black curtain was hung to prevent light scattering. Images were captured during the test and simultaneously imported to the computer.

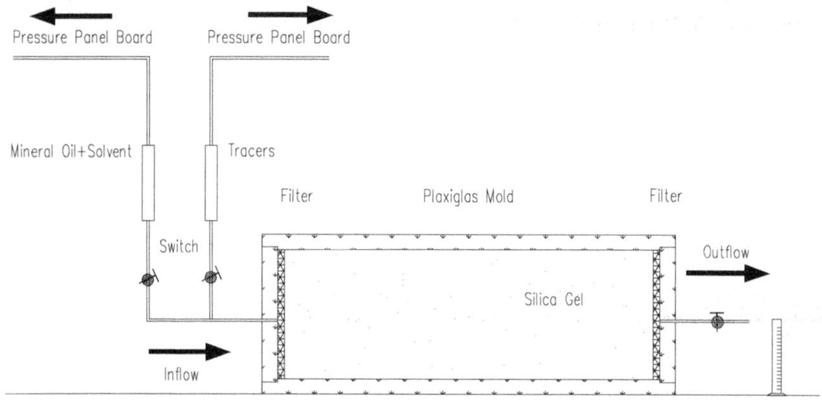

Fig. 12.1 Schematic of Experimental Setup

12.2.5 Calibration of Concentration

Oil Red O from *Fisher Scientific Company*, Atlanta, GA. was used to dye the pore fluid in these tests. The concentration of the oil-soluble dye was measured using a fluorometer. A fluorometer measures the luminescence of a fluorescent substance when it is subjected to a light source of a given wavelength. The higher the concentration of the fluorescent substance, the more emitted light, which the fluorometer will detect. The fluorometer used in this research was Spectronic® 20 Genesys™ (SG) from *Spectronic Instruments* (Fig. 12.3). A calibration curve was made using ASTM Standard D5613. The luminescence of a series of precisely diluted Oil Red O solutions with known concentrations was measured. A 4.5mL four-sided clear methacrylate cuvette from *Fisher Scientific Company* was used to perform the luminescence measurements in the fluorometer.

The frequency of light was selected based on the dye used and kept the same throughout the test. A frequency of 581 nm was used. Next, the reading of the empty cuvette was measured as zero. Finally, a filled cuvette was recorded as the reading for the sample. The calibration curve for Oil Red O is shown in Fig. 12.4. Fluorometer readings were linear up to a concentration of 0.15g/l.

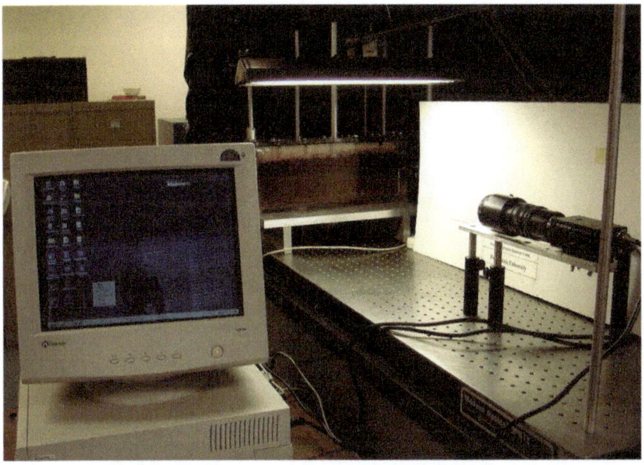

Fig. 12.2 Optical Setting for Flow Measurement

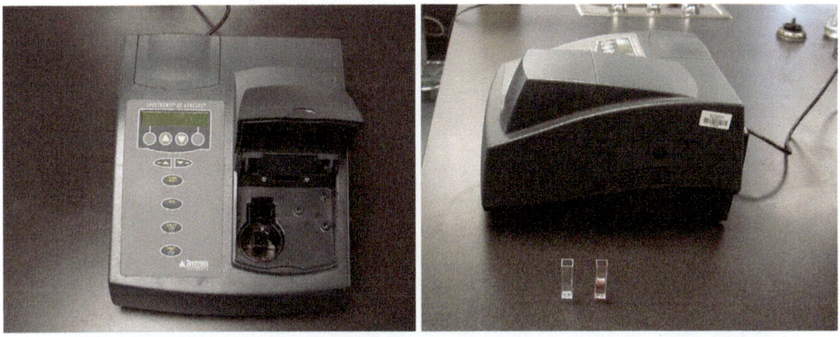

Fig. 12.3 Fluorometer and Cuvettes

12.2.6 Chromatographic Separation

There is a chromatographic separation in silica gel when using an oil-soluble dye for flow test, shown in Fig. 12.5, where the dye was adsorbed by silica gel and became separated from the moving oil blend.

Flooding transparent synthetic soil specimen with calcium bromide brine having a matching refractive index prior to flow tests with dyed mineral oil was recommended by Mannheimer and Oswald (1993), and later used by Welker (1999) to prevent chromatographic separation in transparent silica powder. The effectiveness of this method for use in silica gels is demonstrated by Fig. 12.6. Two silica gel samples obtained from the same saturated silica gel sample were immersed in dyed mineral oil blend first. Then, the calcium bromide brine was

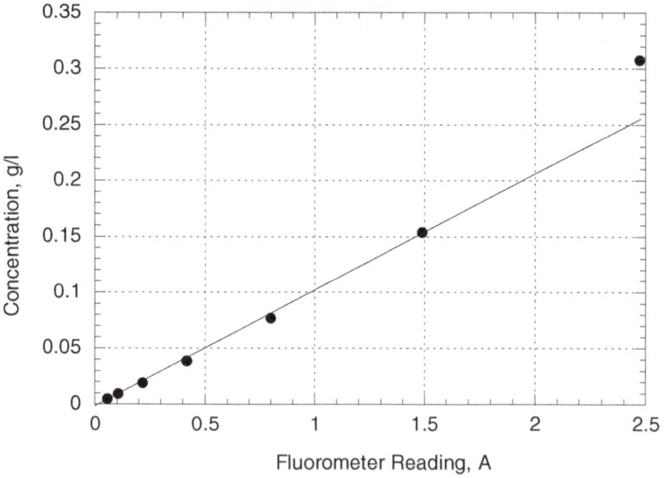

Fig. 12.4 Calibration Curve for Oil Red O at 581 nm

Fig. 12.5 Chromatographic Separation in Silica Gel

added to the sample on the left hand side (LHS), because the density of brine is heavier than mineral oil blend and silica gel is water wet, mineral oil blend in silica gel was replaced by calcium bromide brine. Another sample was left without any further processing (RHS). After both samples were left for 8 hours, the difference in the color of silica gel became very prominent. Silica gel sample without processing (RHS) adsorbed the dye from the dyed mineral oil blend and became darker. The color of dyed mineral oil became lighter due to the loss of dye. The sample flooded with calcium bromide brine remained the same as the beginning.

Fig. 12.6 Calcium Bromide Flooding to Prevent Chromatographic Separation in Silica Gel. Specimen on LHS was flooded with Calcium bromide. Specimen on RHS was not.

After the sample preparation described in Section 12.2.2, two additional steps were followed.

- The sample was flooded with calcium bromide brine until the oil blend in the inter-aggregate pore volume was displaced by calcium bromide brine
- The sample was flooded with the oil blend again until the brine in the inter-aggregate pore volume was replaced by mineral oil blend.

12.2.7 Image Analysis

When using a linear detector, the intensity of the image is proportional to the concentration of tracer. A similar method has been used by many researchers, including Aeby (1998), Corapcioglu and Fedirchuk (1999), and Huang et al (2002). In a solution containing a dye with a concentration, *C*, the fluorescence intensity *F* is given by Slavik (1994) as,

$$F = \phi_f I_0 (1 - e^{-2.303\varepsilon Cl}) \tag{12.1}$$

where I_0 is the intensity of the incident light, l is the path length of the light, ε is the molar extinction coefficient of the dye, and ϕ_f, named the fluorescence efficiency, is the fraction of the absorbed light which is re-emitted as fluorescence.

At low concentrations, where $2.303\varepsilon Cl < 0.05$, Eq. 12.1 can be approximated as follows,

$$F = 2.303\phi_f I_0 \varepsilon Cl \tag{12.2}$$

In the tests, a linear detector (a CCD camera) was used. Therefore, the intensity of fluorescence in the image, *I*, is proportional to the fluorescence intensity *F*.

$$I \propto F \tag{12.3}$$

The concentration of the fluorescent dye is directly proportional to the intensity of the fluorescence. Background correction is required to reduce non-uniform lighting or camera response in different areas of the image. The correction was obtained by using an image processing toolbox of Matlab®. The background correction was performed as follows.

- Image captured before the test was stored as the background image. In order to eliminate image noise, several images can be averaged as the background image.
- The background image and images during the test were imported into Matlab®.
- The correction was performed by subtracting the background image from the images captured during the test and imported as background corrected images.

Movement of the tracer through silica gel is shown in Fig. 12.7, where the intensity shown in the images are background corrected images. In order to have a

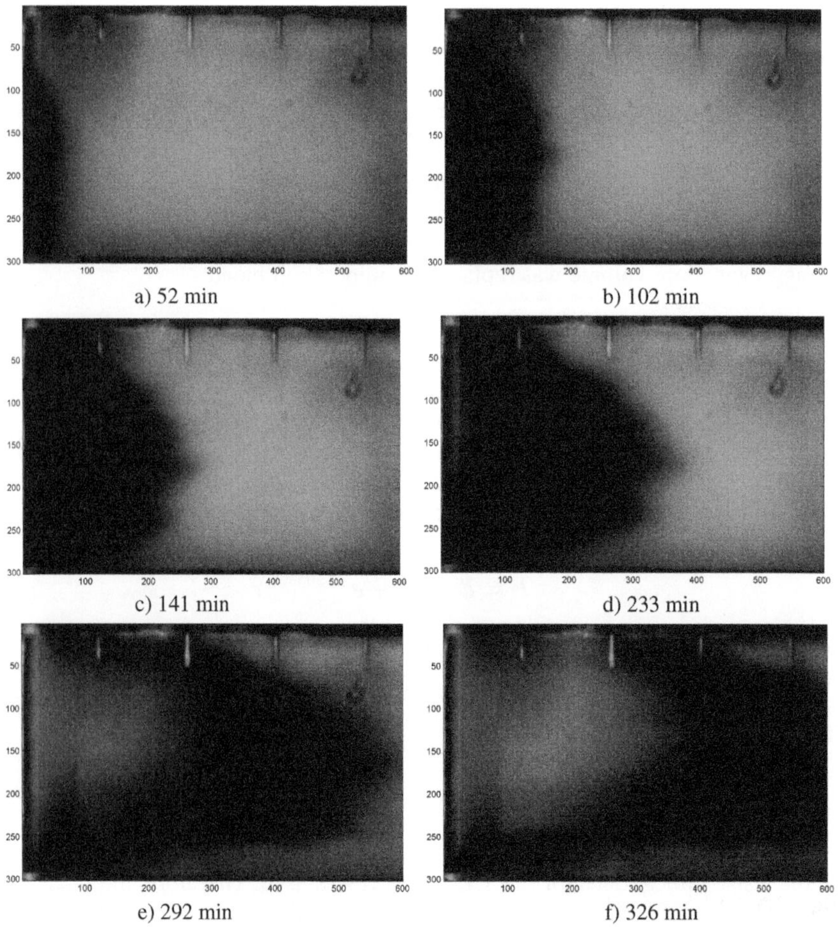

Fig. 12.7 Corrected Images Showing Flow in Transparent Silica Gel

clearer view of flow movement, a three-dimensional view is shown in Fig. 12.8, where the intensity value at each point was projected along the z-axis. The intensity of peak dye concentration was reduced gradually with time as expected and the region of same dye concentration expanded with time. Both figures show that both advection and dispersion processes occurred in the flow transport through transparent silica gel.

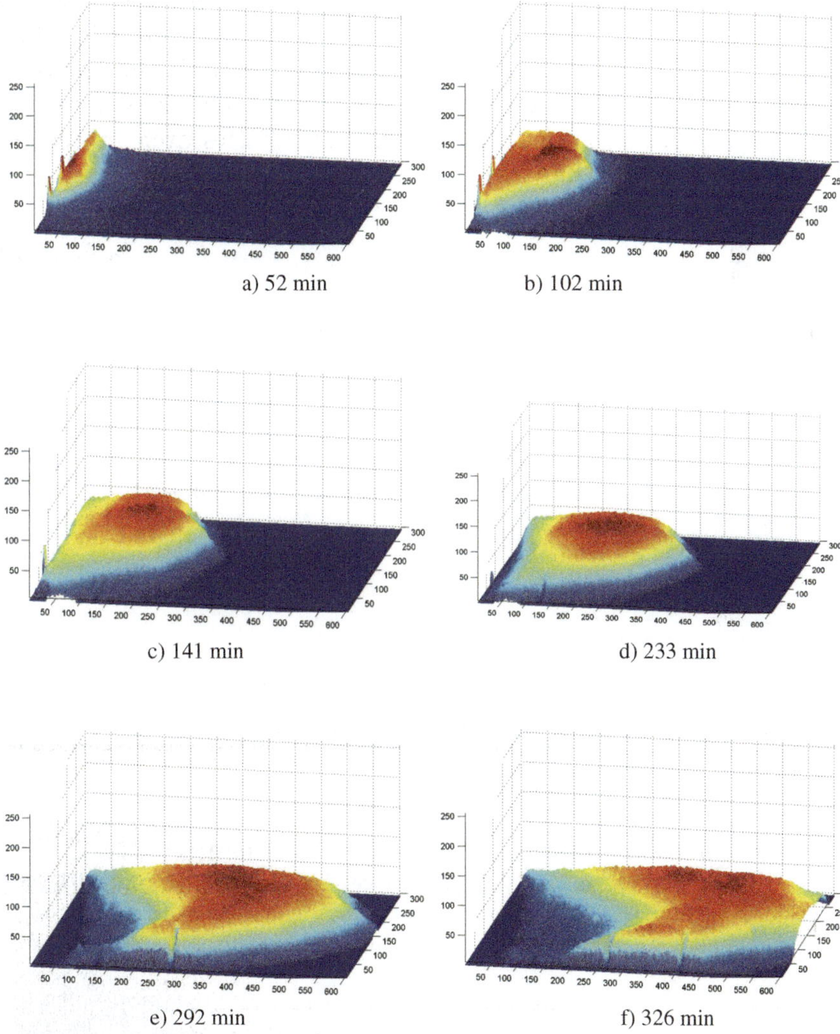

Fig. 12.8 3D View of Solute Transport during Test (The image intensity at each point is projected along the vertical (z)-axis. Note Image intensity is linearly proportional to concentration).

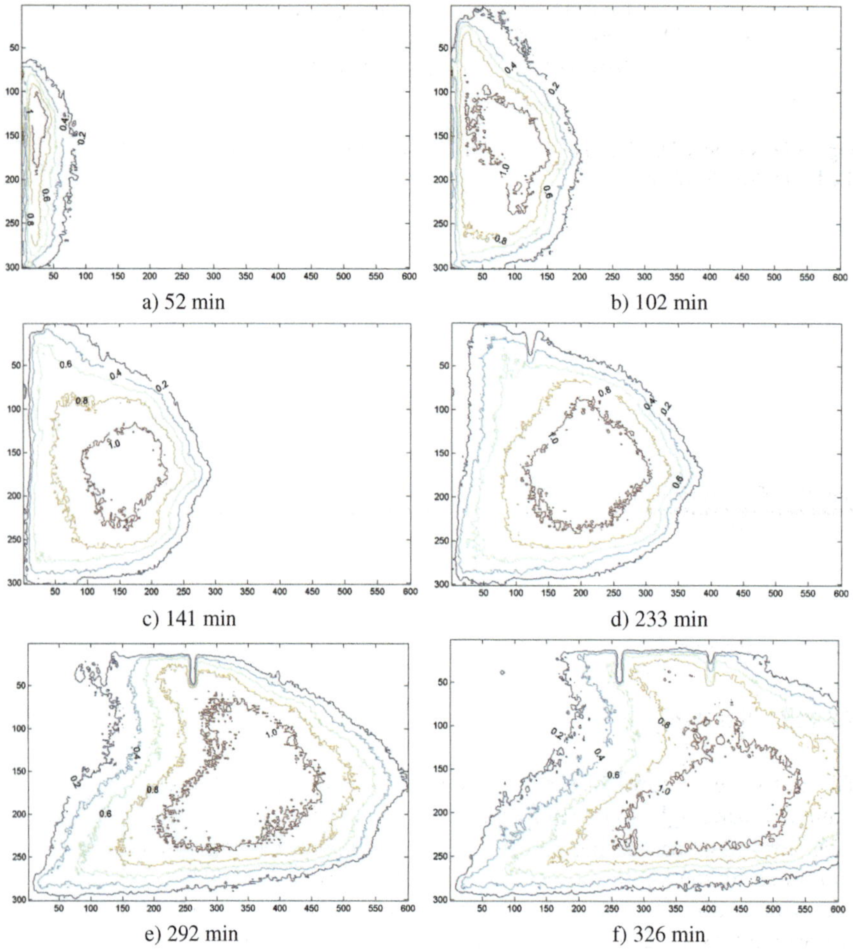

Fig. 12.9 Contour of Dye Concentration (contours represent 5 image intensities scaled linearly, 20%, 40%, 60%, 80%, and 100%)

In order to illustrate dispersion with time, the contours of relative intensity are shown in Fig. 12.9. The intensity of image was scaled to 5 levels based on the grayscale intensity between the darkest points to brightest point of the image itself. The regions were labeled to illustrate qualitatively the dispersion process with time.

Because of the linear relation between dye concentrations and image intensity, the volumes contained by the distribution curves should be proportional to the mass of dye injected. The volume should theoretically be same over time based on the assumption of mass balance. The volumes obtained directly from Fig. 12.8 are given in Table 12.1. The areas under the distribution curves were increasing with

Table 12.1 Volume under the Intensity Surface

Time, min	Volume	Change	Time, min	Volume	Change
163	3644647	---	233	4376765	---
180	3648481	1%	261	4457327	1.9%
216	4086105	12%	280	4678450	6.7%

time. The volumes were only comparable within a certain period of time. The volume of dyed mineral oil blend injected was 51 ml. The concentration used in this experiment, 400 mg/l, was expected to exceed the linear range of the relation between dye concentration and image intensity. A lighter concentration is recommended for future research. A detailed calibration between the image intensity and dye concentration is also required to eliminate this error.

12.2.8 Breakthrough Curve

The concentration of dye measured in the effluent is plotted against the produced mineral oil blend volume or the time from injecting. This curve of the concentration against time or fluid volume is called the breakthrough curve. The breakthrough curve of the dye through silica gel is shown in Fig. 12.10, where the flow volume was measured in the effluent using a graduated cylinder. Times for the measurements were also recorded at the same time. The injected fluid volume was recorded from the beginning of tracer injection.

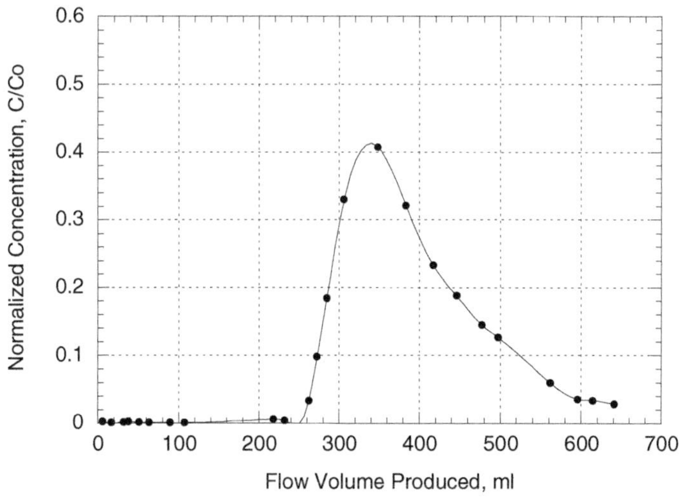

Fig. 12.10 Breakthrough Curve of Tracer through Transparent Silica Gel

The breakthrough curve resembled the typical bell shape and was slightly steeper on the rising limb than on the falling limb. The tail was much longer and flatter than the leading stage. Dye was not detected in the first 250 ml. The concentration increased sharply after this until the peak of concentration was measured at the flow volume of approximate 340 ml. Concentration of dye reduced gradually after the peak concentration was reached and still could be detected after the flow volume of 650 ml, where the test was stopped.

12.2.9 Characterizing the Properties of Silica Gel

Porosity and Pore Volume: Silica gel consists of ultra fine particles with individual diameters on the order of 0.02 μm – same as silica powder (Fig. 5.2). These particles combine to form larger porous aggregates. So silica gel has a two-pore structure, internal porosity and inter-aggregate porosity. Inter-aggregate porosity is of more interest for the geotechnical and geoenvironmental engineering applications. It is very difficult to precisely determine the inter-aggregate porosity or the internal porosity.

The total porosities under loosest and densest test conditions were estimated by measuring the volume of dry silica gel. Based on an assumption of specific gravity of silica gel, the volume was indirectly estimated by measuring the unit weight of sample. The loosest test was conducted by pouring silica gel through a funnel from a height of 6 cm. The densest sample was obtained by vibrating the mold on a vibrator a layer of 30 mm at a time. The largest density determined under dry condition is believed to be higher than that for saturated condition because of the energy loss due to damping in the saturated condition. The total volume of the test mold was 1495 mm^3. The measured densities of dry silica gel ranged from 745 to 850 kg/m^3. Assuming the specific gravity of silica gel is 2.1, the pore volumes of the loosest and densest sample were estimated as 960 and 888 mm^3. The total void ratios of the loosest and densest conditions were 1.79 and 1.46 corresponding to total porosities of 0.64 and 0.59.

Two methods have been used to estimate the inter-aggregate porosity of the silica gel. First, natural sand with a similar grain size distribution (Fig. 10.2) was used to estimate the inter-aggregate porosity of silica gel. The pore volume measured by natural sand represents the inter-aggregate pore volume of silica gel under the assumption that the influence from the particle eccentricity on the pore volume is ignorable. The sand sample was compacted using vibrator under the same condition as that for densest silica gel model. The pore volume was estimated as 558 mm^3 based on the specific gravity of 2.65. The inter-aggregate void ratio was 0.596 corresponding to a porosity of 0.37.

Pore volume can also be estimated by performing a tracer test and studying the resulting concentration history measured at the out flow. For a single-phase, non-reactive flow in a packed bed, the pore volume corresponds to the dimensionless mean residence time (Himmelblau and Bischoff 1968, Jin 1995), $\overline{t_D}$, which is calculated from the tracer breakthrough curves resulting from an idealized instantaneous tracer pulse.

$$\overline{t_D} = \frac{\int_0^\infty t_D C_D dt_D}{\int_0^\infty C_D dt_D} \tag{12.4}$$

where $C_D = \dfrac{C - C_{initial}}{C_{injected} - C_{initial}}$ is a dimensionless tracer concentration, which is a function of t_D, and $t_D = \dfrac{\int_0^t q dt}{V_p}$ is dimensionless time. V_p is the pore volume of vessel and q is the volumetric flow rate. The dimensionless pore volume is the difference between the mean residence time of output and input tracer curves for a finite-volume. Corrected pore volume is given by

$$\overline{t_D} = \frac{\int_0^\infty t_D C_D dt_D}{\int_0^\infty C_D dt_D} - \frac{t_s}{2} \tag{12.5}$$

where t_S is duration of slug injected.

The pore volume estimated from breakthrough curve was 353 ml, which was 40 % smaller than that estimated by using dry sand. The mass of slug predicted from breakthrough curve was 63.7 ml with an initial concentration, C_0, of 0.39 g/l, which was 25 % larger than the injected slug, 51 ml at the same concentration. The difference was mainly caused by the errors in the estimation of the inter-aggregate pore volume. The pore fluid volume for manual sampling was not accounted in the pore volume, which introduced an error in pore volume of around 40 ml just before the peak. Manual sampling and a low sampling frequency were believed to introduce some errors in the test. A better result is expected by using an auto-sampler, sampling at a higher sampling frequency.

Hydraulic Conductivity: Flexible wall permeability tests were performed according to ASTM test method D5084. The samples were permeated with the oil blend during the triaxial tests. Hydraulic conductivity around 1.5×10^{-4} cm/sec was measured for the silica gel, which corresponds to intrinsic permeability of 1 Darcy.

Dispersivity: The longitudinal dispersivity, D_l , is a characteristic of soil used to describe the dispersion in the direction of flow. The longitudinal dispersivity was calculated using the method mentioned by Domenico and Schwartz (1990) based on the measured breakthrough curve. The same method was also tried by Corapcioglu and Fedirchuk (1999).

$$D_l = \frac{v^2 \sigma_t^2}{2t} \tag{12.6}$$

where σ_t is the standard deviation derived graphically from breakthrough curve form, which can be estimated as the following.

$$\sigma_t = \frac{t_{84} - t_{16}}{2} \tag{12.7}$$

where t_{84} and t_{16} are the breakthrough times corresponding to relative concentration of 0.84 and 0.16, respectively, and t is the time corresponding to relative concentration of 0.5. The breakthrough curve is shown in Fig. 12.11. An average flow rate of 0.008 mm/sec was used. Based on the method mentioned above, a longitudinal dispersivity around 0.60 mm was calculated, which is within the measured laboratory range of 0.1 to 1mm for natural soil (Charbeneau 2000).

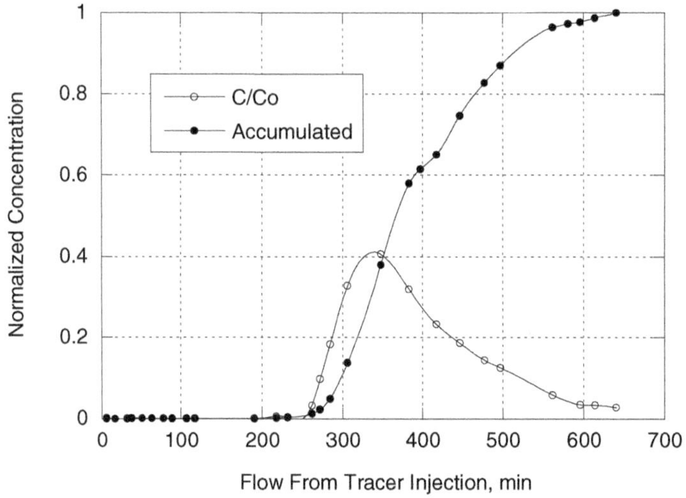

Fig. 12.11 Relative and Accumulated Concentration Curves

12.3 Flow Test Using Fused Silica

12.3.1 Transparency Degradation in Silica Gel

In the experiments using silica gel in Section 12.2, samples were flooded by calcium bromide brine with a matching refractive index to prevent chromatographic separation. The results show that this method is efficient (Fig. 12.6). However, a second problem arises with flooding, which is demonstrated in Fig. 12.12. The transparent silica gel model on the left hand side was the sample before flooding with calcium bromide, where high quality of transparency is demonstrated by the advertisement clearly visible through the 25 mm sample. The sample on the right hand side was after flooding by calcium bromide brine, where transparency degradation is evident, even though the same refractive index was used for both liquids. It is believed that a new chemical material is generated by the chemical reactions between calcium bromide and silica gel, which influences the sample transparency. Similar transparency degradation in silica powder was reported by Welker et al (1999) for amorphous silica powder.

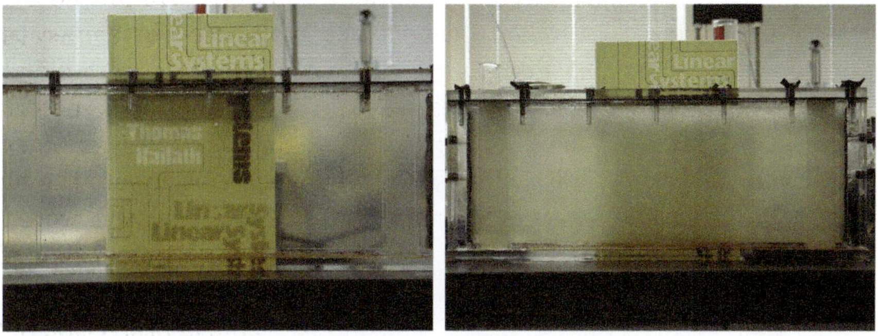

Fig. 12.12 Transparency Degradation due to Flooding with Calcium Bromide. Before Flooding (LHS), and After Flooding with Calcium Bromide (RHS).

12.3.2 Fused Silica

Fused silica (Fig 12.13) is another transparent material having the same refractive index of silica gel. Fused silica was obtained from *Accumet Materials Co., Ossining, NY*. The grain distribution of the fused silica used is shown in Fig. 12.14, along with that of amorphous silica gel.

Fig. 12.13 Fused Silica

The surface charge of silica gel may contribute to the chromatographic separation. The surface force was measured to increase sharply as the distance between the particles reduces (Santamarina and Fam 1995). Fused silica is solid and exhibits porosity similar to natural soil. The distance between particles is larger than that in silica gel. Fused silica will have surface charge smaller than silica gel. Because it lacks an internal porosity chromatographic separation is less in this material (Fig. 12.15). The test procedure was similar to that used in Fig. 12.6. The picture was taken after the samples were left for 8 hours. There is no difference between the sample flooded with calcium bromide brine (LHS) and the sample without any processing (RHS).

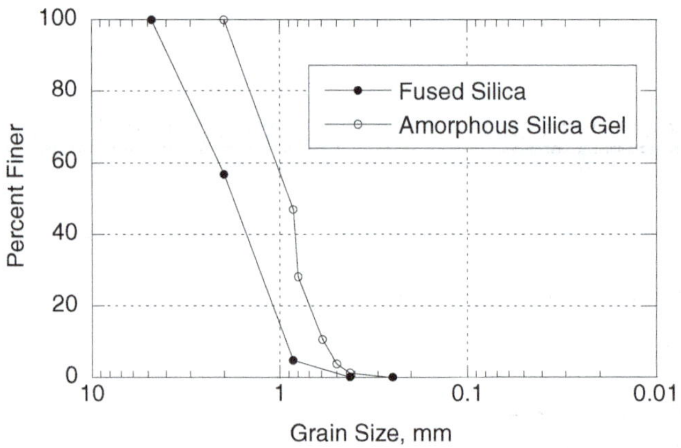

Fig. 12.14 Fused Silica and Its Grain Size Distribution

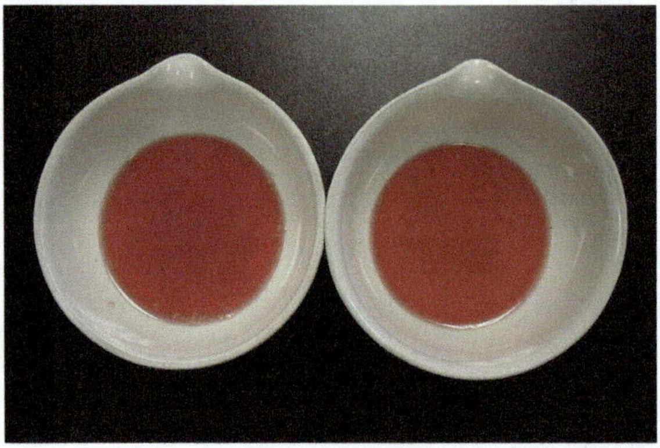

Fig. 12.15 No Chromatographic Separation in Fused Silica. Flooded by Calcium Bromide (LHS) and Without Any Processing (RHS).

Transparency of fused silica is presented in Fig. 12.16(a), where the left hand side sample was filled with dry fused silica and on the right hand side was the saturated sample with the oil blend. The transparency is lower than that of silica gel because of the impurities in the sample. Samples with fewer impurities were difficult to obtain, but they are required for better transparent of the model.

A typical image of flow through transparent fused silica is shown in Fig. 12.16(b), where the advection and dispersion with time are visible through the variation of intensity of the dye cloud.

Fig. 12.16 (a) Fused Silica Gel for Flow Test. LHS is Dry specimen. RHS is saturated with a matched Refractive Index fluid. Note transparency in RHS is lower than in Silica Gel.

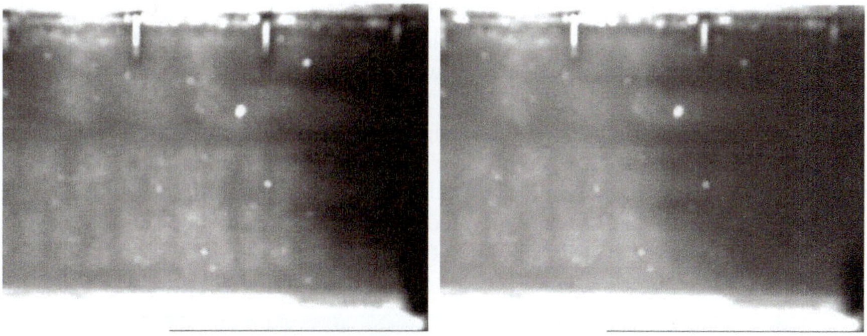

Fig. 12.16 (b) Flow in Fused Silica

The breakthrough curve through transparent fused silica with pore fluid is shown in Fig. 12.17. The flow volume vs. concentration curve was a typical bell shaped similar to that of amorphous silica.

The pore volume was determined by measuring the weight of dry fused silica to fill the box. The estimated pore volume was 615 ml based on an assumed specific gravity of 2.1. The measured pore volume from breakthrough curve was 603 ml, which was only 2 % smaller than the calculated one. The mass of the slug estimated from the breakthrough curve was 120 ml with a concentration, C_0, of

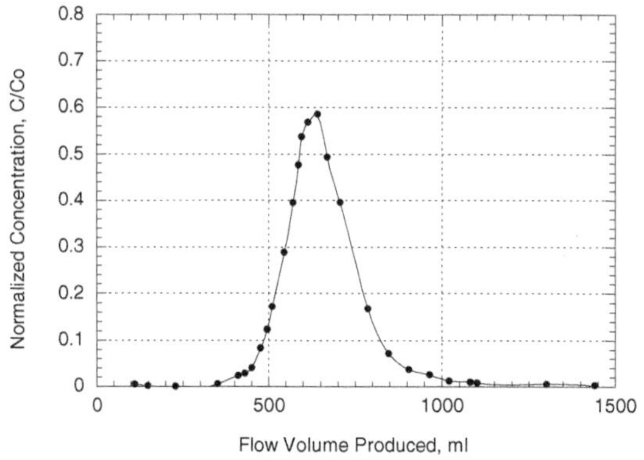

Fig. 12.17 Breakthrough Curve of Tracer through Transparent Fused Silica Gel

0.40 g/l, which was 7% larger than the actual slug volume, 112 ml. These errors are much smaller than that used in amorphous silica gel under the same manual sampling conditions.

12.3.3 Immiscible Flow Test

The potential of studying immiscible flow through transparent synthetic soils was demonstrated using fused silica, where calcium bromide brine was flowed through a sample saturated with dyed mineral oil blend. In order to give a better visual effect, the image was reversed (negative image) where the grayscale changed from 0 is black and 255 is white to 0 is white and 255 is black. The negative image of typical interface in immiscible flow test is shown in Fig. 12.18, where the small darker part on the bottom of the mold was calcium bromide blend and the rest in the sample was dyed oil blend. The difficulty arises in immiscible flow tests because of the differences in both the viscosity and density of two liquids. A clear interface slope demonstrates the potential application of transparent synthetic soil in immiscible flow study.

Fig. 12.18 Immiscible Flow through Fused Silica Gel

12.4 Flow Tests Using Aquabeads

12.4.1 Materials

Three types of Aquabeads were used for modeling soils with different relative hydraulic conductivities: (1) NC100 ($K = 2.8\sim7.3\times10^{-2}$ cm/sec), (2) #200–0.5% ($K = 4.2 \sim 5.6\times10^{-4}$ cm/sec) were used to model permeable porous media; and (3) C50 ($K = 1\sim2.4\times10^{-5}$ cm/sec) was used to model the less permeable media.

A dyed fluid made of de-aired water and red food coloring, was used to conduct miscible flow tests. The red food coloring (Amazon.com Standard Identification Number (ASIN): B00085FA84) is a fine red powder which easily dissolved in water. This dye was selected since its dye properties were established by Kostarelos (1998) and used with success as a tracer, despite having a slightly higher density than water. In addition, two kinds of oil, white mineral oil (CAS# 8042-47-5), and motor oil (Castrol 10w40), were dyed using Oil Red O (CAS#1320-06-5) and utilized as contaminants in this study. Because oil does not dissolve in water, multi-phase flow transport process could be observed.

12.4.2 Concentration Calibration

The purpose of concentration calibration of dyed fluid is to show that there is a linear relationship between concentration and fluorescence, which in turn is proportional to image intensity. A *Spectronic 20 Genesys* fluorometer (Fig 12.3) and many 4.5ml four-sided clear methacrylate cuvettes were used for calibration according to ASTM D5613. The concentration of red food coloring and Oil Red O is linear up to 0.5 g/L and 0.0625 g/L, respectively (Fig. 12.19). The calibration of dyed motor oil was difficult due to its dark color; therefore a calibration was developed using a surrogate made of mineral oil and 0.25 g/l Red O dye, which was approximately as dark as the motor oil mixture.

12.4.3 Packing of Aquabeads for 2D Flow Test

A Plexiglas model tank with internal dimension 40.0×2.8×14.7 cm was used to perform flow tests. Although a 2.8 cm wide model is used in this study, larger models up to 8.0 cm wide that can be used for 3D modeling were also transparent. In order to generate a uniform horizontal flow through the model tank, two filters consisting of different sizes of steel screens were placed at each end of the tank. First, Aquabeads were immersed in de-aired water until the desired water absorption was achieved. Next, Aquabeads were poured into the model tank. During the pouring process, it was necessary to stir the Aquabeads to release air bubbles entrapped in the pore space. Finally, a cover plate with a gasket was screwed tightly

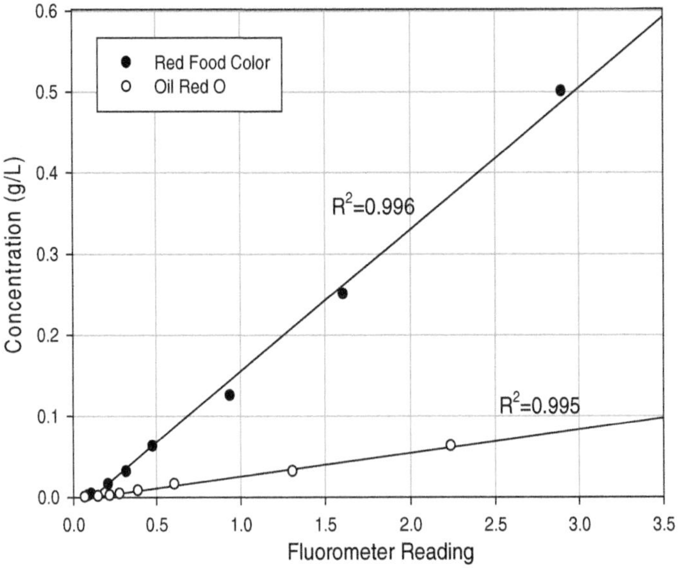

Fig. 12.19 Linear Range of Calibration Curve of red food coloring and oil red O

onto the tank to prevent leakage. The effective porosity of various tank models was obtained using miscible flow tests. The effective porosity was 345.3 ml for NC100 and 569.07 ml for #200-0.5% models. For a layered model made of 75% NC100 over 25% C50 (by height) the effective porosity were 258.71 ml and 450.71 ml for 75% of #200-0.5% over 25% C50.

12.4.4 Flow System Setup and Effluent Collection

All transparent soil models require saturation of the system since presence of air with a different refractive index is a major source of opacity. After packing, deaired water, dyed fluid, and surfactant solutions, placed in three reservoirs were introduced to the model tank through a manual switching valve similar to the one shown in Fig 12.1. A stable pressure panel board was connected to the reservoirs in order to provide a constant head (and flow rate). Initially, water was injected into the model tank to ensure good transparency and that constant flow rate was achieved. Next, dyed fluid was injected, and the effluent was collected using cuvettes at the other side of the tank. The concentration of each effluent sample was measured using the fluorometer. The measured concentrations were used to obtain the volume of NAPL collected. The testing criteria of the flow tests are summarized in Table 12.2. It is assumed that changes in the size and shape of Aquabeads during flow are negligible because (1) the model tank has a constant volume, and (2) the effective stress is maintained constant.

Table 12.2 Criteria of Miscible Flow Tests

Testing Criteria	Single Layer		Double Layer	
	100% NC-100	100% #200-0.5%	75% NC-100 over 25% C-50	75% #200-0.5% over 25% C-50
Dye concentration (g/L)	0.5	0.5	0.5	
Dye volume (ml)	29.43	30.73	19.64	21.75
Hydraulic gradient	0.038	0.075	0.038	0.038
Flow rate (ml/min)	0.92	0.8	0.98	0.87

12.4.5 Optical System and Image Analysis

The optical system described in Section 12.2.4 was used (Fig 12.2). The movement of dyed fluid in the middle of the tank was captured and imported to a computer during the test. A background correction is required in order to reduce the image noise caused by non-uniform lighting and camera response. In this study, the background correction was applied using Adobe Photoshop CS2 filter. The corrected images were imported and analyzed by the image processing toolbox of Matlab®.

The image analysis approach described in Section 12.2.7 was used again, where a linear relation is assumed between concentration and image intensity.

12.4.6 Miscible Flow Tests

Movements of the red dye through a single layer of NC100 is shown in Fig. 12.20, for two time intervals after a 0.1 PV slug was injected using the switching valve and entered the tank from the left hand side traveling in a predominantly horizontal direction. The original (side view) images are shown on the top. Three dimensional views of the plume shape are shown in the middle where the intensity value of each point of the plume is projected along the z-axis which clearly indicates the concentration profile. The 2D intensity contours obtained using Matlab are also shown in the bottom. The contours were scaled to five levels, based on the grayscale intensity from the darkest point to brightest point of the corrected captured image. Both images show the advection-dispersion process as the dyed fluid migrated thorough the model.

For a single layer of #200-0.5% the images indicate that flow lacks uniformity (Fig. 12.21). Because of its relatively higher hydraulic conductivity and transmissivity, NC100 possess a stronger ability to transmit water compared with #200–0.5%, resulting in a wider plume when similar amount of dye migrate through the model. The flow through #200-0.5% was also slower than NC100.

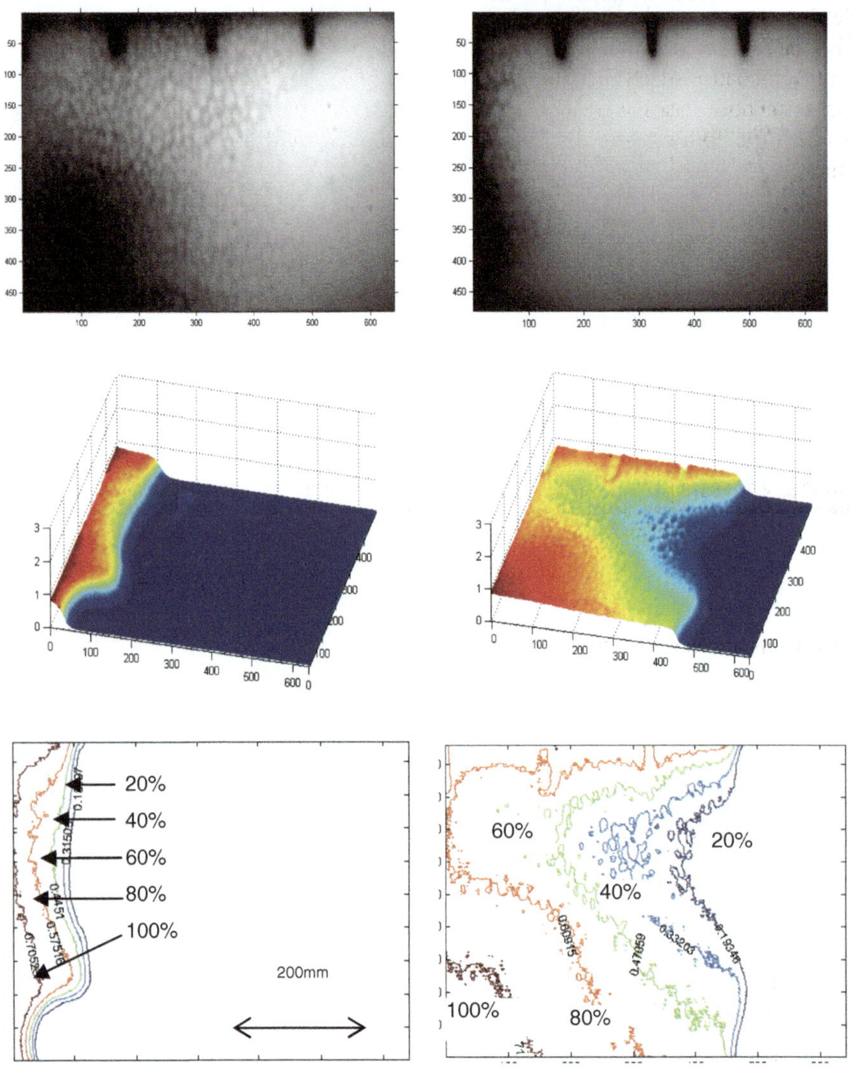

Fig. 12.20 Modeling of 2-D Miscible Flow in a Single Layer of Aquabeads NC100

For layered systems of NC100 and #200-0.5 over C50 (double layer) shown in Fig. 12.22 and 12.23, a 'tail' is evident between the two layers due to the difference in hydraulic conductivity. Fingering in the #200 model (Fig. 12.23) is visible, perhaps due to inadequate sample packing.

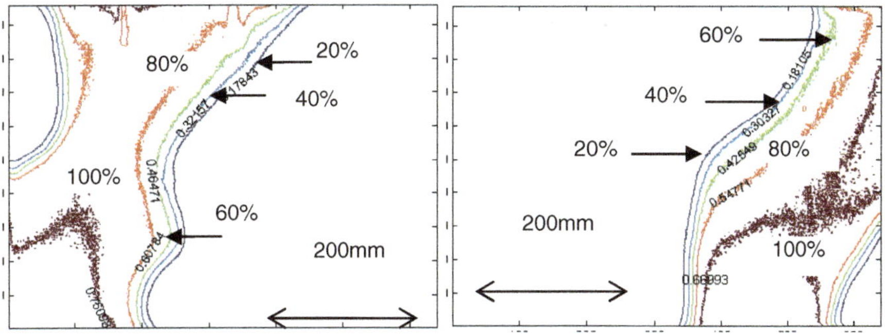

Fig. 12.21 Modeling of 2-D Miscible Flow in a Single Layer of Aquabeads #200-0.5%

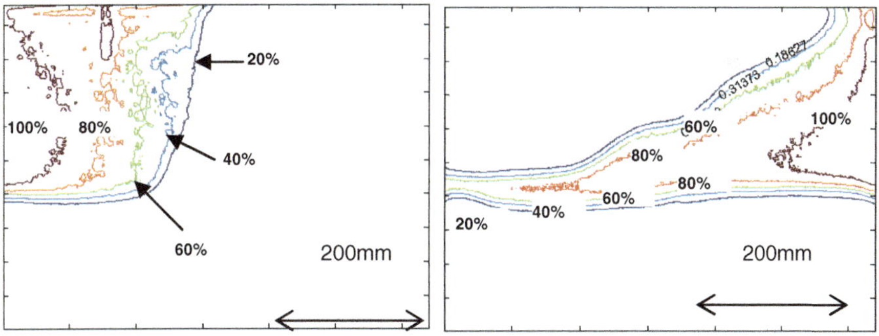

Fig. 12.22 Modeling of 2D Miscible Flow in a Double Layer of NC100 Over C50

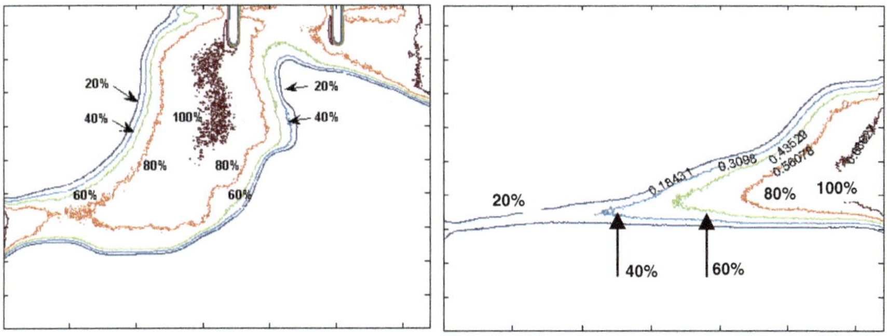

Fig. 12.23 Modeling of 2D Miscible Flow of a Double Layer of #200–0.5% Over C50

12.4.7 Multi-phase Flow

Immiscible flow images (of the middle of the tank) after the injection of 0.1 PV of motor oil are shown in Fig. 12.24. An upward slower migration is observed due to oil's light density and high viscosity can be observed from the image analysis. Flushing 20 PV of water, did not remove the oil, and merely spreads it even when a higher flow rate of water on the order of 2.2-2.5 ml/min was flushed. A mass balance was used to calculate the amount of oil recovered. The recovery after flushing 20 PV was less than 5% of total oil volume injected in the tank.

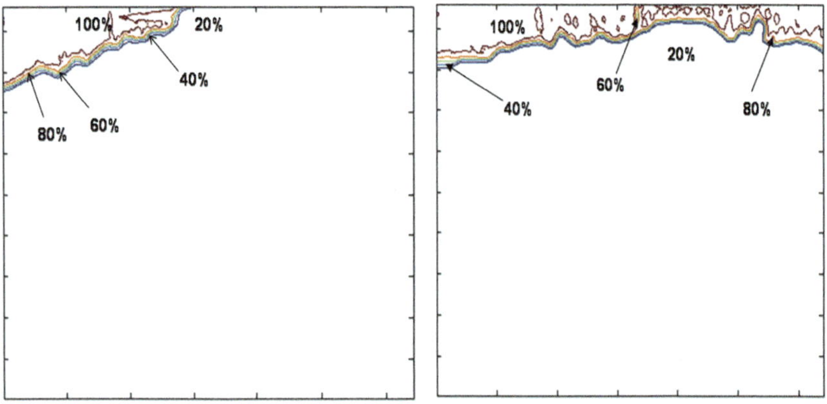

Fig. 12.24 Modeling of 2D Multiphase Flow – Motor Oil (Left: after 0.4 PV water flushed, Right: after 20.3 PV water flushed) Note that flushing with 20 PV of water merely spreads the NAPL.

12.4.8 Hydraulic Characteristics of Aquabeads

Hydraulic Conductivity measurements were performed during single layer flow tests, and were found to be 1.12×10^{-2} and 4.56×10^{-3}cm/sec for NC100 and #200-0.5%, respectively. These values are consistent with the conductivities obtained from fixed wall permeability tests. The conductivity of #200-0.5% was one order of magnitude higher than the results of fixed wall hydraulic conductivity test, perhaps due to inadequate packing.

The Intrinsic Permeability of NC100 and #200–0.5% was found to be 1.15×10^{-7}cm^2 (11.65 Darcy) and 4.67×10^{-8}cm^2 (4.67 Darcy), respectively, which is consistent with that of sand.

Transmissivity of NC100 and #200–0.5% are 0.164 cm^2/sec and 0.067 cm^2/sec, respectively in the conducted flow tests.

Effective Porosity is the pore space of the porous media which allow the water to pass through. It was estimated by performing a tracer test, where the concentration of the effluent is measured to plot a breakthrough curve. From the breakthrough curve, the swept pore volume can be computed using the method of first temporal moment, which corresponds to an instantaneous tracer pulse (Himmelblau and Bischoff 1968). When a conservative non-reactive tracer is used, the equations for calculating swept pore volume, *Pv*, and effective porosity, n_e , of 1D flow are:

$$Pv = \frac{\sum_{i=1}^{n} C_i V_i \Delta V}{\sum_{i=1}^{n} C_i \Delta V} - \frac{1}{2} V_s \tag{12.8}$$

$$n_e = \frac{P_v}{V} \tag{12.9}$$

where: C_i: concentration of sample, V_i: cumulative pore volume, ΔV: volume of each effluent specimen, V_s: volume of the injected tracer slug, and *V*: bulk volume. It is believed that Eq. 12.8 and 12.9 provide a reasonable approximation of test conditions, and can be used. For single layer flow tests, the center of mass of the tracer breakthrough curve (Fig. 12.25) corresponds to the pore volume of 345.3 cm^3 and 569.07 cm^3 for NC100 and #200–0.5%, respectively. For layered (double) system tests, the pore volume was estimated to be 258.71cm^3 for NC100, and 450.71cm^3 for #200–0.5%. The effective porosity of NC100 determined from the double layer modeling was 75 % of the single layer model, because the double layer model consisted of 75% NC100 (top) and 25% C50 (bottom). In addition, the pore volume estimated from the double layer modeling of 200–0.5% was 79 % of the single layer model, a 4% overestimation. In the double layer tests on #200-0.5%, a small breakthrough curve appeared before the main breakthrough curve due to insufficient sample collection and non uniform model packing. Better result can be expected with better packing and by using an auto-collector with a higher sampling frequency. Fingering observed in the captured images can also be seen from the breakthrough curves shown in Fig. 12.25(b) where falling limbs are much longer and flatter than the rising limbs.

The Effective Porosity of NC100 and #200–0.5% were found to be 0.22 and 0.37, respectively. The porosity seems rather low for the NC100 compared to typical natural media of this size, probably reflecting the effect of packing flexible particles.

The Hydrodynamic Dispersion Coefficient is a mass transport parameter that combines the mechanical dispersion with molecular diffusion. In this research, the

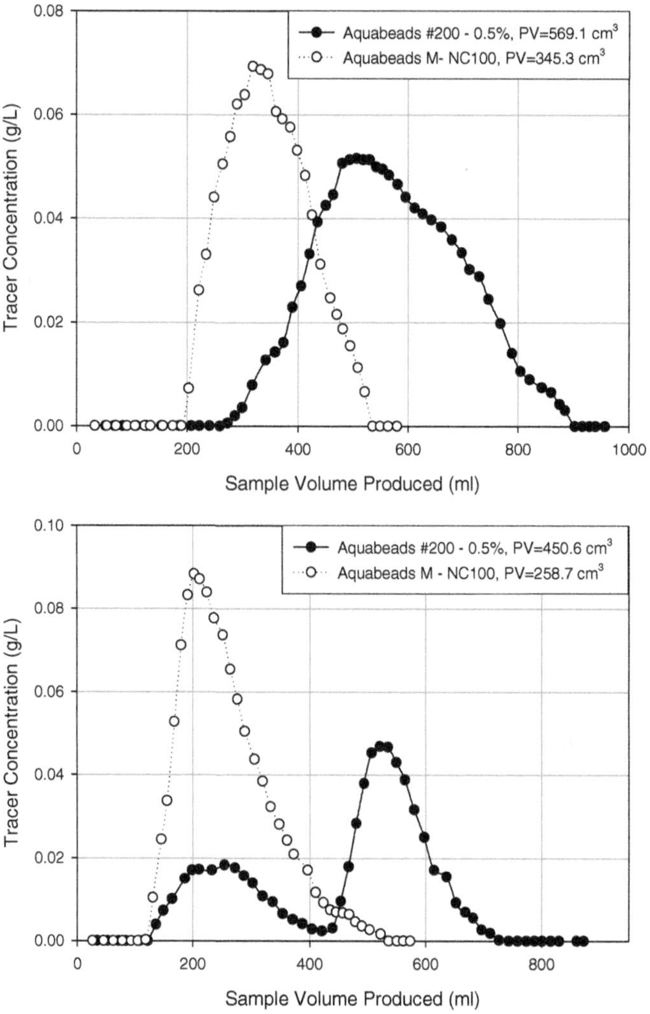

Fig. 12.25 Breakthrough Curve Tracer Test – (a) Single layer Modeling –Top and (b) Double Layer Modeling – Bottom

accumulated breakthrough curve made from the tracer test is shown in Fig. 12.26, which resulted in hydrodynamic dispersion coefficient of 3.29×10^{-3} cm^2/sec and 1.25×10^{-3} cm^2/sec for NC100 and #200–0.5%, respectively.

The Longitudinal Dispersion can be estimated from two points on the breakthrough curve (Fetter 1999). The results show that the longitudinal dispersivity of NC100 is 17.3 mm, and that of #200–0.5% is 13.4 mm. This is slightly higher than the expected laboratory range for natural soils, which typically range from 0.1 to 10 mm (Freeze and Cherry 1979). The hydraulic characteristics of Aquabeads NC100 and #200–0.5% are summarized in Table 12.3.

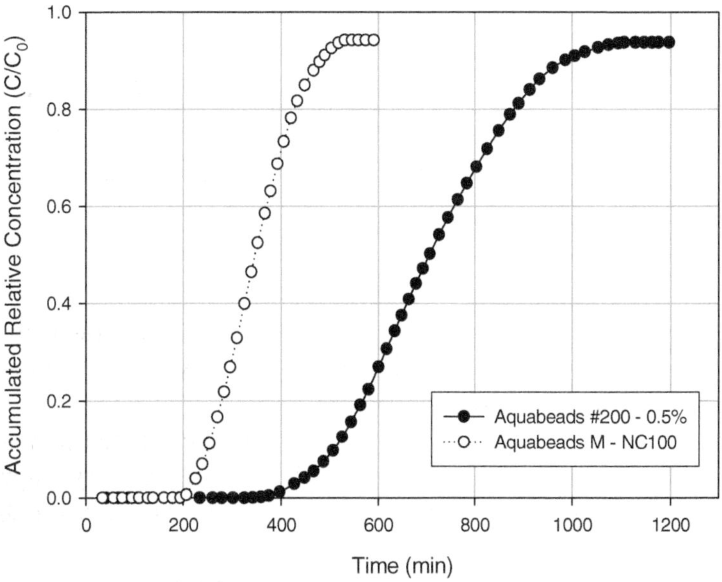

Fig. 12.26 Accumulated Breakthrough Curve of Mono Layer Flow Test]

Table 12.3 Hydraulic Characteristics of Selected Aquabeads

Hydraulic Characteristic	NC100	#200 – 0.5%
Hydraulic conductivity (cm/sec)	1.12×10^{-2}	4.56×10^{-3}
Intrinsic permeability (cm^2)	1.15×10^{-7}	4.67×10^{-8}
Transmissivity (cm^2/sec)	0.164	0.067
*Porosity	0.22	0.37
*Dispersion coefficient (cm^2/sec)	3.29×10^{-3}	1.25×10^{-3}
*Longitudinal dispersivities (mm)	17.3	13.4

* determined by tracer tests.

12.5 Modeling of 2D Surfactant Flushing Using Aquabeads

A limited surfactant flushing test is presented to (1) illustrate that Aquabeads can be used to model geoenvironmental contamination problems, (2) the compatibility of Aquabeads and some commonly used surfactants, and (3) the usefulness of visualizing remediation processes.

12.5.1 Phase Behavior Tests

Conducting phase behavior test is essential in order to understand the reactions between contaminants and a given surfactant solution. Phase behavior provides a basis for surfactant selection, and assures the effectiveness of additives (electrolytes, co-solvents and polymer) and their best surfactant formation. The expected results, such as high contaminant solubilization, fast equilibrium, and minimal liquid crystallization and gel formation, can be observed through these tests (Dwarakanath et al. 1999). Phase behavior tests were performed as follows. First, the surfactant solution and oil contaminants are mixed in a test tube with a 1:1 ratio, by volume. Second, the tubes are sealed. Next, the tubes are tumbled 10 times to obtain a good mixing of the liquids. Finally, the separation of the liquid phases inside of the tube is observed until equilibrium is reached (Fig. 12.27), and the time to reach equilibrium is recorded.

Several hundred phase behavior tests were performed to select appropriate surfactant solutions which are compatible with Aquabeads. All phase tests were performed at 25°C to simulate the temperature during the modeling conditions. Several criteria were used for finding the optimal surfactant solution including: (1) fast equilibrium time, (2) flat and mobile oil/aqueous interface, (3) no liquid crystallization or gels forming, and (4) chemical compatibility with Aquabeads. The commercial surfactant Triton X100 composed of octylphenol ethylene oxide

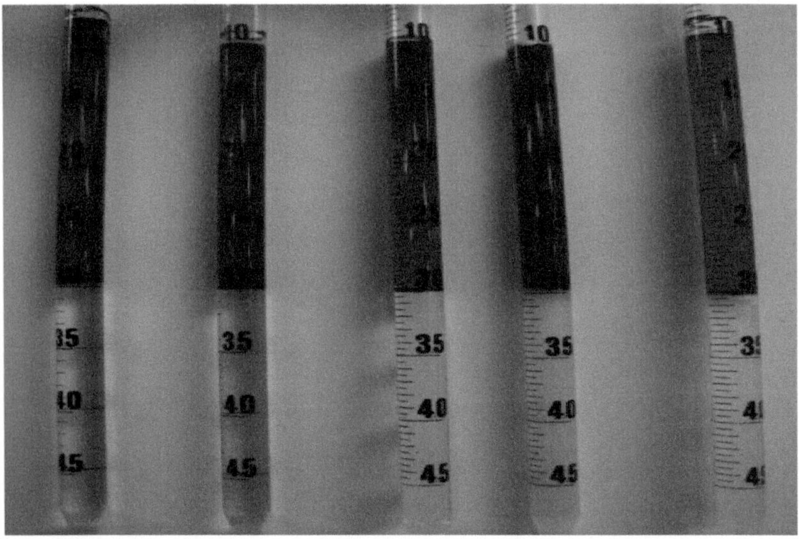

Fig. 12.27 Phase Behavior Test after Equilibrium – Red portion of the tube: Dyed Mineral Oil; Clear portion of the tube: surfactant solution with different concentration (from left to right: Triton X-100 with 0.1, 0.05, 0.025, 0.0125 and 0.00625%, by weight).

consendate (CAS# 9002-93-1) diluted with de-ionized water was selected. In addition, in order to eliminate gel/liquid crystals, Sec-Butanol (CAS# 78-92-2) was used as the co-solvent to mix with X100. Finally, the polymer Xanthan Gum was used as an additive to increase the viscosity of the surfactant solution, which improved the performance of the surfactant in layered systems. No liquid crystallization or gels was observed during testing with the selected surfactant solutions shown in Table 12.4.

Table 12.4 Properties of Selected Surfactant Solutions (% by Weight)

	NAPL	Triton X-100	Sec-Butanol	Xanthan Gum
1	Mineral oil	0.025	0.25	0
2	Mineral oil	0.025	0.25	0.03
3	Motor oil	0.025	2.5	0
4	Motor oil	0.025	2.5	0.03

12.5.2 Test Setup

A layered model, consisting of 25% C50 on the bottom and 75% NC100 on the top, was used. Two kinds of oil, mineral oil and motor oil were utilized as NAPLS in this study. After the system was set up and a constant flow condition was established, dyed oil was injected into the tank, and its residual saturation status was obtained after flushing with a known amount of water. Next, the selected surfactant solution was subsequently injected. The effluent was collected using a glass separatory funnel. Therefore, the amount of water/surfactant solution that had been flushed, and the volume of oil that had been recovered were measured. The testing criteria are summarized in Table 12.5.

Table 12.5 Criteria of Multi-phase Flow Tests

Testing Criteria	Multi-phase Flow Modeling	
Oil	Mineral Oil	Motor Oil
Dyed concentration (g/L)	0.0625	0.25
Volume of dyed fluid (ml)	24.69	25.06
Hydraulic gradient	0.167	0.167
Flow rate (ml/min)	2.21	2.47

12.5.3 Recovery of Mineral Oil

The surfactant solution mobilized the mineral oil trapped in the porous space due to the reduction of interfacial tension. It broke the oil blobs into globules and brought those globules out of the original trapping space. A downward migration of the oil globules was seen during the flushing. It caused the globules to relocate and finally they were trapped in another porous space. 47.6% of mineral oil was removed by the surfactant flushing. In contrast, once Xanthan Gum was added to the surfactant, the oil's downward migration was completely prevented. Xanthan Gum significantly increased viscosity, which resulted in a good oil recovery (Fig. 12.28). A 95.8% oil recovery was achieved after 20 pore volumes of surfactant solution were flushed.

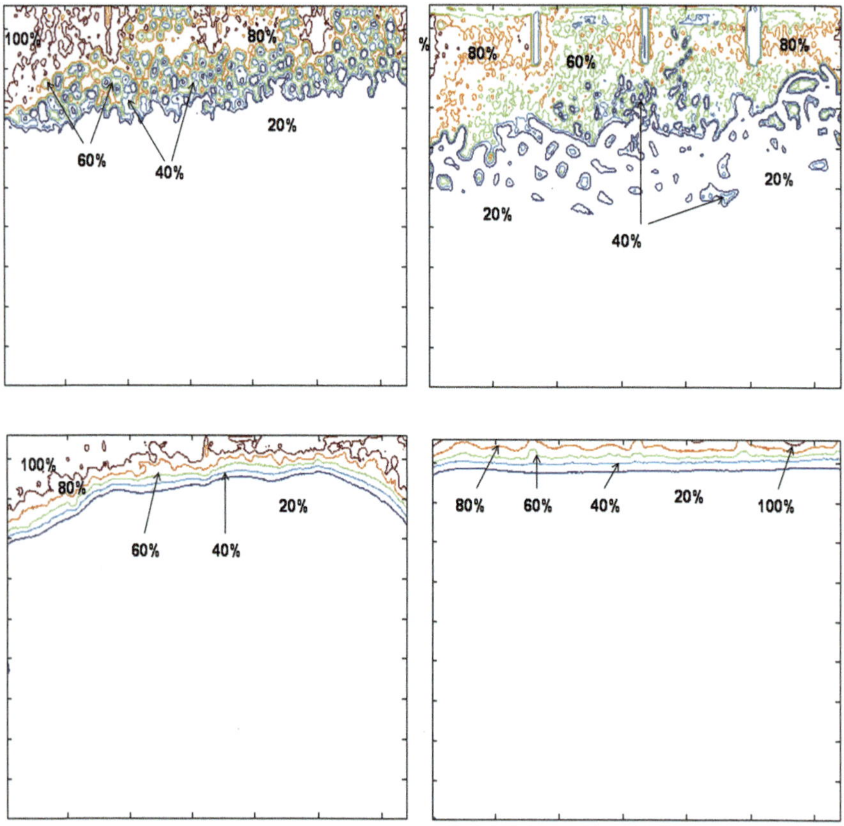

Fig. 12.28 2D Surfactant Flushing – TOP: Mineral Oil without Xanthan Gum (Left: before flushing, Right: after 20.9 PV of surfactant flushed); BOTTOM: Mineral Oil with Xanthan Gum (Left: before flushing, Right: after 20 PV surfactant flushed)

12.5.4 Recovery of Motor Oil

Although no downward migration of oil blobs was observed when flushing the surfactant without adding polymer, only 4.4% of the oil was recovered. Since the motor oil entirely occupied the upper porous space, the surfactant solution traveled

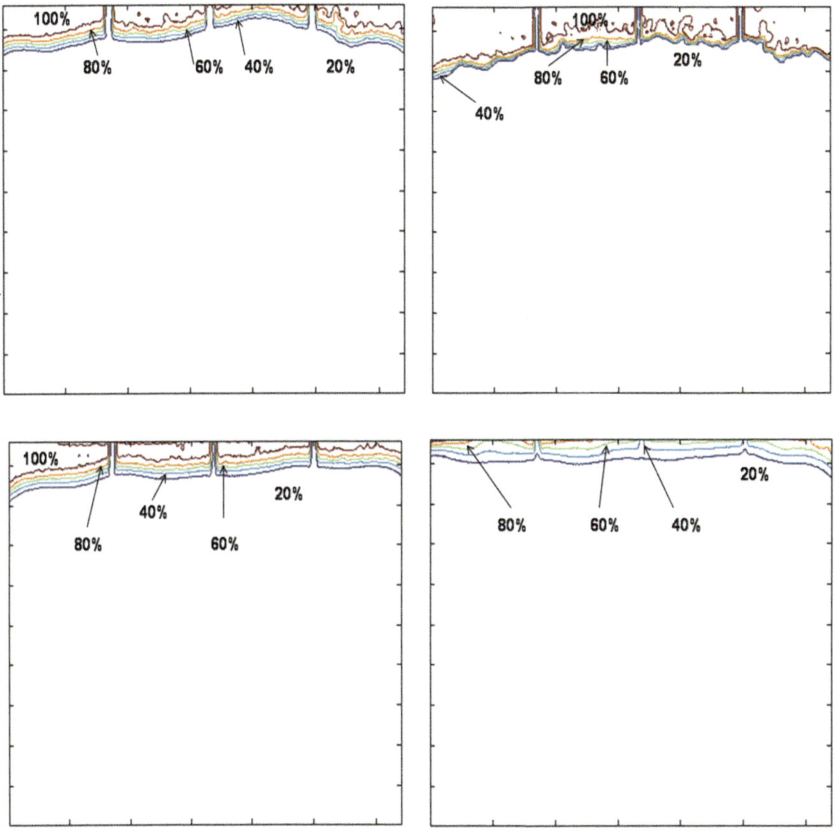

Fig. 12.29 2D Surfactant Flushing – TOP: Motor Oil without Xanthan Gum (Left: before flushing, Right: after 11.9 PV surfactant flushed); BOTTOM: Motor Oil with Xanthan Gum (Left: before flushing, Right: after 14.1 PV surfactant flushed)

primarily through the uncontaminated porous space, thus bypassing the contamination. Therefore, mobilization was reduced due to the limited contact of the surfactant and oil. Inversely, the oil recovery was substantially enhanced when Xanthan Gum was added. 88.5% of the oil was recovered after flushing 14.1 pore volumes of the surfactant solution, which is approximately 20 times the recovery rate achieved without adding polymer (Fig. 12.29). The recovery rate of both mineral oil and motor oil is shown in Fig. 12.30.

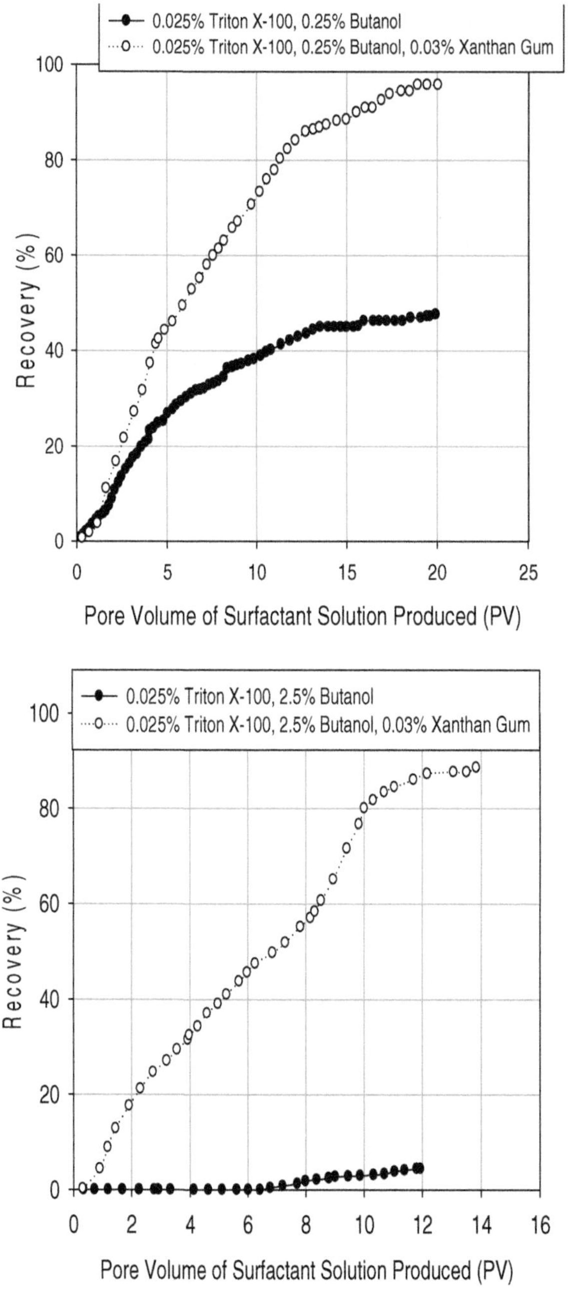

Fig. 12.30 Percent Recovery – (a) Mineral Oil – Top and (b) Motor Oil - Bottom

12.6 Conclusions

In this chapter, multiphase 2D flow of an oil-soluble dye through three different transparent soils was performed. The breakthrough curves of dye through both materials were typical of the bell-shape reported for natural soil. Image analysis shows both the advection and dispersion processes clearly visible in the test. It demonstrated that transparent synthetic soils can be used to visualize and non-invasively measure pollutant transport in natural soil. This study also demonstrates the utility of Aquabeads in evaluating flushing technologies in bench scale model tests prior to conducting in-situ surfactant flushing.

In the short term, Aquabeads are ideal for simulating 2D NAPL transport characteristics, particularly in educational settings including undergraduate labs and high schools. In the long term, Aquabeads may permit 3D modeling of multiphase flow using laser excited fluorescent dyes, thus permitting the study of complex 3D geoenvironmental contamination problems.

More research and improvement in the test set-up and analysis are needed as follows. First, a more accurate relationship between tracer concentration and light intensity is required. Second, a more accurate method is required in estimating the pore volumes in silica gel. Third, transparency of synthetic soil can be improved by using alternative soil-like materials. Finally, comparisons with numerical solutions for well-defined test conditions are needed for the application of transparent soil in pollutant control.

References

Aeby, P.: Quantitative Fluorescence Imaging of Tracer Distributions in Soil Profiles, Ph.D. Dissertation, Swiss Federal Institute Of Technology, Zurich (1998)

ASTM Standard D5084. Standard test method for measurement of hydraulic conductivity of saturated porous materials using a flexible wall permeameter, 1993 Annual Book of ASTM Standards, vol. 04.08, pp. 1199–1206 (1993)

ASTM Standard D5613. Standard test method for open-channel measurement of time of travel using dye tracers, 1994 Annual Book of ASTM Standards, vol. 04.11, pp. 725–742 (1994)

Bird, R., Stewart, W., Lightfoot, E.: Transport Phenomena. Wiley (1960)

Brock, D., Orr, F.: Flow visualization of viscous fingering in heterogeneous porous media. In: SPE Annual Technical Conf. & Exhibition, pp. 211–222 (1991)

Chen, J., Wada, N.: Visualization of immiscible displacement in a three-dimensional transparent porous medium. Experiments in Fluids 4(6), 336–338 (1986)

Charbeneau, R.: Groundwater Hydraulics and Pollutant Transport. Prentice-Hall, NJ (2000)

Corapcioglu, M., Chowdhury, S., Roosevelt, S.: Micromodel visualization and quantification of solute transport in porous media. Water Resource Research 33(11), 2547–2558 (1997)

Corapcioglu, M., Fedirchuk, P.: Glass bead micromodel study of solute transport. J. of Contaminant Hydrology 36(2), 209–230 (1999)

Dwarakanath, V., Kostarelos, K., Pope, G., Shotts, D., Wade, W.: Anionic surfactant remediation of soil columns contaminated by non-aqueous phase liquids. J. of Contaminant Hydrology 38(4), 465–488 (1999)

Domenico, P., Schwartz, F.: Physical and Chemical Hydrogeology. Wiley, NY (1990)

Fetter, C.: Contaminant Hydrogeology. Prentice-Hall, NJ (1999)

Freeze, R., Cherry, J.: Groundwater. Prentice-Hall, Englewood Cliffs, NJ (1979)

Herrmann, K., Pohlmeier, A., Gembris, D., Vereecken, H.: Three-dimensional imaging of pore water diffusion and motion in porous media by nuclear magnetic imaging. J. of Hydrology 267, 244–257 (2002)

Himmelblau, D., Bischoff, K.: Process Analysis and Simulation. Deterministic Systems. John Wiley & Sons, New York (1968)

Huang, W., Smith, C., Lerner, D., Thornton, S., Oram, A.: Physical modeling of solute transport in porous media: evaluation of an imaging techniques using UV excited fluorescent dye. Water Research 36(7), 1843–1853 (2002)

Iskander, M., Lai, J., Oswald, C., Mannheimer, R.: Development of a transparent material to model the geotechnical properties of soils. ASTM Geotechnical Testing J., GTJODJ 17(4), 425–433 (1994)

Iskander, M., Liu, J., Sadek, S.: Transparent amorphous silica to model clay. ASCE J. of Geot. & Geoenv. Eng. 128(3), 262–273 (2002)

Jia, C., Shing, K., Yortsos, Y.: Visualization and simulation of non-aqueous phase liquids solubilization in pore networks. J. of Contaminant Hydrology 35, 363–387 (1999)

Jin, M.: A Study of Nonaqueous Phase Liquid Characterization and Surfactant Remediation, Ph.D. Dissertation, University of Texas at Austin, Austin, USA (1995)

Kostarelos, K.: A new concept: natural buoyancy for surfactant flooding, Ph.D. Dissertation, University of Texas at Austin, Austin, USA (1998)

Mannheimer, R., Oswald, C.: Development of transparent porous media with permeabilities and porosities comparable to soils, aquifers, and petroleum reservoirs. Ground Water 31(5), 781–788 (1993)

Matsubayashi, U., Devkota, L., Takagi, F.: Characteristics of the dispersion coefficient in miscible displacement through a glass beads medium. J. of Hydrology 192, 51–64 (1997)

Multisorb Technologies Inc., 325 Harlem Road Buffalo, NY 14224-1893 USA (2001), `http://www.multisorb.com`

Sadek, S., Iskander, M., Liu, J.: Geotechnical properties of transparent material to model soils. Canadian Geotechnical J. 39(1), 111–124 (2002)

Santamarina, J., Fam, M.: Changes in dielectric permittivity and shear wave velocity during concentration diffusion. Canadian Geotechnical J. 32(4), 647–659 (1995)

Slavik, J.: Fluorescent Probes in Cellular and Molecular Biology. CRC Press, Boca Raton (1994)

Welker, A., Bowders, J., Gilbert, R.: Applied research using transparent material with hydraulic properties similar to soil. Geotechnical Testing J., GTJODJ 22(3), 266–270 (1999)

Epilogue

Magued Iskander

Abstract. This epilogue presents the transparent soil story, along with a summary of the state of the art and current limitations of transparent soil modeling. Areas of potential improvement and suggestions for applications and future research are also presented.

The Transparent Soil Story

In fall 1991, I was a graduate student at The University of Texas at Austin. One of my professors, Ken Stokoe, introduced me to Dr. Richard Mannheimer a chemist at South West Research Institute (SWRI) in San Antonio, Texas. Dr. Mannheimer had recently invented a transparent slurry and published an article titled *Slurries you can see through [1]*. The slurry was made of precipitated amorphous silica and a matched refractive index fluid. The slurry was invented as a tool for research in solid transport using non-Newtonian liquids, such as Ketchup. Dr. Mannheimer also consolidated a specimen of the slurry that appeared to represent a soil like substance. At the time, the consolidated specimens was referred to as cores because they were manufactured using a piston consolidometer and the resulting samples resembled a rock core, in shape.

The project manager at SWRI was Dr. Chuck Oswald, a civil engineer and an expert in blast resistant structural design. I met with Chuck and Richard in San Antonio to address the question of to what extent do the cores represent natural soils? That meeting in fall 1991 represented the birth of transparent soil modeling.

Chuck hired me as a consultant to SWRI. I partnered with my, then, office mate Jiunnren Lai to conduct an investigation of the geotechnical properties of the cores. We used equipment owned by our doctoral adviser, Professor Roy E. Olson, to determine the strength and consolidation properties of cores made of precipitated silica powders and matched refractive index fluids. Simultaneously Chuck and Richard were looking into the flow properties of the same cores. Our preliminary studies indicated that the material was suitable for representing the macroscopic geotechnical properties of a generic low plasticity clay. Two journal articles were published [2, 3].

Transparent soils made of precipitated silica were later categorized as the first family of transparent soils. Two pioneering studies were conducted using this family. The first was Andrea Welker's doctoral dissertation [4] where she studied the

M. Iskander: Modeling with Transparent Soils, SSGG, pp. 293–298.
springerlink.com

flow properties around wick drains under the supervision of Professors Robert Gilbert and John Bowders at The University of Texas at Austin. The second was David Gills' doctoral dissertation [5] where he studied deformations due to pile penetrations under the supervision of Professor Barry Lehane at Trinity College.

The next chapter in the story is in hindsight fortuitous. American geotechnical reviewers are, for lack of a better term, *conservative. Very Conservative!* As a result, not too many new ideas survive peer review. This explains why so many geotechnical innovations originate in Europe. An unusually enlightened group of National Science Foundation peer reviewers and program director (Dr. Priscilla Nelson) funded my proposal titled *CAREER: Modeling 3D Flow and Soil Structure Interaction Using Optical Tomography*. This funding allowed me to support my first two doctoral students, Samer Sadek and Jinyuan Liu.

I worked with Samer and Liu to develop the technology (equipment and methodology) to conduct transparent soil modeling. Samer's main contributions were (a) identifying digital image correlation as the appropriate method for measuring deformations in transparent soils, and (b) discovering the second family of transparent materials suitable for representing sand. The second family which is made of silica gel has the same refractive index like precipitated silica so the two can be combined in the same model. This discovery can be credited to a discussion that Samer and I had with Dr. Bruce Garetz of Polytechnic's chemistry department, when he casually mentioned that most silicas have the same refractive index. That day Samer was wearing a new leather jacket, and one of the pockets contained a packet with silica gel as a desiccant, as is common in many leather goods. So he immersed the desiccant in the same pore fluid we have been using for amorphous silica and the desiccant was transparent. Later we tested many other silica gels to obtain a better refractive index match.

Liu's main contributions were to (1) greatly improve the analysis algorithm, (2) increasing the size of the database of transparent materials suitable for representing natural soils by determining the geotechnical properties of many varieties of silica powders and gels, (3) better documenting the similarity in properties between natural and transparent soils, and (4) adapting transparent soils for simulating flow.

Coincidence plays an important role in research. One of my doctoral students, Kazunori Tabe, is an active member in Japanese American United Church in New York. There, he met Mr. Hiroaki Yoshino, President of Kuraray America, a multinational conglomerate that produces a variety of polymers. One of Kuraray's products is a water based transparent polymeric gel called KI-GEL 201K-F2. Tabe studied its geotechnical properties and showed that it can be used to model flow in many soils and the strength of very weak clays. We named that material Aquabeads and classified it as the third family of transparent soils. Aquabeads are particularly suited for use in educational settings because it is water based.

The developed technology generated considerable interest. The author has been contacted by researchers from approximately 20 universities and is aware of at least 7 published doctoral dissertations and two master theses at 6 universities that are based on the work presented in this monograph.

No one knows what the future holds. Nevertheless, the author is confident that the developed materials provide the necessary foundation for utilizing optical techniques to study spatial deformation patterns and flow characteristics of soil models, non-intrusively. In the short term, the technology can be readily adjusted and applied in geotechnical engineering research. In the long term, transparent synthetic soils are expected to become a powerful tool in solving many geotechnical and geoenvironmental engineering problems.

Capabilities of Available Transparent Soils

A database of transparent materials, which can be customized to meet model test requirements in terms of strength, deformation, and permeability has been developed. Transparency is achieved by matching the refractive index of synthetic particles and pore fluids. Three families of transparent synthetic soils have been presented in this book. The first two families have the same refractive index and can be combined in the same model. The first family is made of amorphous silica powder to model the macroscopic geotechnical properties of natural clays (Chapter 5). The second family which models sand was made of silica gel (Chapter 6). These two families constitute the only transparent system capable in representing a wide range of natural soils, which permits the simulation of different geological profiles using transparent soils. A third family is made of a water based polymer can be used for modeling flow and strength of very weak soils (Chapter 7).

An optical set-up and image processing technique were developed and investigated to measure two-dimensional deformation inside transparent synthetic soil models, non-intrusively. The deformation was obtained by correlating two speckle images resulting from the interaction between laser light and transparent synthetic soils (Chapter 8 and 10). Comparisons of deformation under a model footing in sand and silica gel proved that transparent synthetic soil can be used to model natural soils (Chapter 9). The 3D deformation field under a model footing was obtained by combining 2D deformation fields of several slices in Matlab® (Chapter 11). 2D flow tests were performed in transparent soils and results demonstrate that they can be used to study advection-dispersion process of pollutant transport in natural soils (Chapter 12).

The developed models can potentially be used to study a wide array of geotechnical problems using optical techniques. For example, the developed models can be used to study interaction between soils and foundations, while modeling site-specific geologic features. Several transparent immiscible pore fluids have also been identified, thereby permitting the study of multi-phase flow problems such as decontamination of aquifers.

Limitations of the Technology

- **Material Limitation**: Transparent synthetic soils are able to model natural soils in terms of deformation and strength. However, the lower modulus and high secondary consolidation cause a difficulty in comparing model results with those from natural soils.

- **Model Limitation:** Transparency is reduced as the dimensions of the model increase. The main causes of transparency degradation are (1) entrapped air, (2) changes in the refractive index of the pore fluid with time, and (3) presence of impurities. These difficulties can be overcome using (a) a powerful vacuum pump to de-air the soil, (b) a high- power laser to penetrate the sample deeper without losing contrast, and (c) using purer forms of silica.
- **Algorithm Limitation:** Digital image cross-correlation compares two images to obtain the spatial deformation between them. The out-of plane deformations were not obtained in the study. However, these deformation components can be obtained by using an optical system consisting of two perpendicular lasers and cameras.

Recommendations for Future Research

The ideas presented in this book represent the initial step towards using transparent materials in modeling the behavior of natural soils. The following represents areas of potential future research to advance the modeling technology:

- **Transparent Materials:** Other sources of transparent soils should be investigated. These can include (1) amorphous silica including those used for chemical chromatography analysis, (2) other types of silica such as fused silica (3) specialized glass and matched refractive index fluids and (3) fluoride salts, such as Cryolite. Any of these materials may prove to be useful surrogates for specific applications.
- **Model:** A bigger model size is desirable to minimize scale and boundary effects. Also a container with a surcharge system is required to control the test stresses similar to a calibration chamber and to consolidate clay layers at different consolidation pressures to simulate more complicated soil condition with the desired strength parameters.
- **System Integration:** The present system requires significant manual interaction in order to capture images and analyze results. A more integrated system is needed for better efficiency. The load-frame system should be synchronized with the optical system to eliminate corresponding errors. A computer-controlled linear stage is also necessary to minimize errors caused by manual control of the system.
- **Image Processing Technique:** A multi-camera vision system is needed to capture the out-of plane deformation. Filter lenses should be used to eliminate the influence of ambient light. A higher resolution camera can be implemented to capture larger area without loss of resolutions. A more detailed calibration between light intensity and dye concentration is also required for studying pollutant transport in porous media.

- **Speckle Investigation:** A fundamental understanding of light propagation through transparent media and a comprehensive investigation of the statistical behavior of speckle images are required to fully explore modeling with transparent soils.

- **Study of Similitude:** A system of non-dimensional parameters can be developed to reduce the influence of material properties on model results. Also constitutive relationships between the strength and strain of transparent soils are needed.

Potential Applications

This research addresses an important and critical need in geotechnical engineering. Previously, there were no established methods to measure three-dimensional deformation patterns and flow characteristics within a soil mass. As a result, it was difficult to analyze complex soil-structure interaction and ground flow problems. High speed computing has made it possible to simulate large, three-dimensional problems numerically; however, the validity of many codes can only be verified at the boundaries. Measurement of three-dimensional deformation patterns and flow characteristics within a soil continuum are usually limited by the fact that sensors provide data at discrete locations and therefore do not provide a continuous image of the measurement domain. Additionally, soil sensors exhibit static and dynamic characteristics that are different from those of the surrounding soils, and therefore can change the response of the measured continuum.

Transparent soils present an elegant solution to many of the preceding difficulties. It is also better than other non-intrusive measurement methods, like radiography, computerized axial tomography and nuclear magnetic resonance imaging because (1) their low cost permits their routine application, and (2) the developed method can be easily adjusted to meet modeling requirements.

The presented transparent soil surrogates can be used to solve many geotechnical and geo-environmental problems. A few potential applications are listed as follows:

- **Soil-structure interaction problems:** including three-dimensional soil deformation during tunneling, excavation, and pile penetrations.

- **Pollutant transport simulation and site remediation techniques:** A laser excited fluorescence tracer can be used to study 3D flow problem using transparent synthetic soils. Osmotic phenomena can be investigated using the transparency of this material. Comparisons of well-known numerical solutions with results from transparent soil model tests are also needed.

- **Advanced modeling:** Investigation of complex processes such as liquefaction using transparent synthetic soils is possible. Wave propagation through transparent synthetic soil at the particle level can be studied.

- **Coupled measurements:** Measurement of strain and flow of coupled processes can be investigated using transparent synthetic soils.
- **Use in a centrifuge:** Many geotechnical processes are best studied in a centrifuge in order to reproduce the influence of stress in small model tests. The installation of the modeling system in a centrifuge presents a number of difficulties in terms of camera isolation from vibrations. Nevertheless, it is the next frontier!

The author sincerely hopes that readers are convinced of the merits of transparent soil modeling, and attempt to use it in their research. He would be glad to advise a reader regarding the specific details of their experiment setup.

References

1. Technology Today, p. 2 (1990)
2. Mannheimer, R., Oswald, C.: Development of transparent porous media with permeabilities and porosities comparable to soils, aquifers, and petroleum reservoirs. Ground Water 31(5), 781–788 (1993)
3. Iskander, M., Lai, J., Oswald, C., Mannheimer, R.: Development of a transparent material to model the geotechnical properties of soils. ASTM Geotechnical Testing Journal, GTJODJ 17(4), 425–433 (1994)
4. Welker, A.: In situ remediation of contaminated soils using prefabricated vertical drains. Dissertation, The University of Texas at Austin, USA (1998)
5. Gill, D.: Experimental and theoretical investigations of pile and penetrometer installation in clay. Ph.D. Thesis, Trinity College, Dublin, UK (1999)

Appendix
Camera Calibration Using Neural Networks

Honghua Zhao and Louis Ge

Abstract. A neural network camera calibration algorithm has been adapted for image-based soil deformation measurement systems. This calibration algorithm provides a highly accurate prediction of object data points from their corresponding image points. The experimental setup for this camera calibration algorithm is rather easy, and can be integrated into digital image correlation (DIC) to obtain the full-field deformation of a soil model. The performance of this image-based measurement system was illustrated with a small-scale rectangular footing model. This fast and accurate calibration method will greatly facilitate the application of an image-based measurement system into geotechnical experiments.

A.1 Background

With the rapid development in computer vision and information technology, image-based displacement/strain measurement in geotechnical testing has become feasible and popular. It provides full-field deformation data during a test instead of measuring displacements at a given location through a conventional instrumentation. Donohoe et al (1991) developed a system to measure soil displacements by calculating each particle's centroid during the test. The key component of their study was the edge detector technique which estimates the area and orientation for each soil particle. Hryciw and Irsyam (1993) investigated the load transfer mechanisms along the interface between sand and a ribbed plate. A direct shear box was constructed with Plexiglas walls so that a video camera could be used to observe and measure sand particle displacements. Guler et al (1999) adapted two techniques (MATCH and BMAD) to determine soil particle displacement vectors using the images from direct shear tests. Paikowsky and Xi (2000) used a high-resolution digital CCD camera to track an assembly of 2-dimensional disks in an interfacial friction shear box. Image-based deformation measurement systems have also been applied to monitor and analyze shear band formation. Alshibli and Sture (2000) conducted a series of plane strain experiments on sands. A grid pattern was imprinted on the latex membrane for each specimen and images were taken to analyze a shear band formation analyses. Without using markers on the latex membrane, Rechenmacher and Fino (2003) also performed plane strain experiments on dense sands and used Digital Image Correlation to quantify shear band inclination and thickness. Aydilek et al (2004) implemented a block-based matching image algorithm to obtain the strain distribution in geosynthetics during tensile testing. In geotechnical centrifuge testing, Allersma (1991) demonstrated the use of an image processing technique to monitor the progressive failure of a clay slope model during flight. Taylor et al (1998) developed a system for measuring planar and 3-dimensional surface movement in centrifuge tunnel heading experiments to investigate stability and collapse mechanisms. White et al (2003, 2005) adapted particle image velocimetry (PIV) and close-range photogrammetry to create a soil deformation measurement system which was developed for centrifuge testing. The illustrated examples include the installation of a displacement pile, the progressive failure of a model clay embankment, and the deformation patterns beneath an eccentrically-loaded plane strain strip foundation on clay.

In the application of image-based deformation measurement to geotechnical experiments, the performance of the system depends on the image analysis method that is used to determine the correlation between the image and object coordinates or that tracks soil particles during a test (White et al 2003). Camera calibration is an important component of an image-based measurement system, particularly when the camera is not perpendicular to the measurement field, as is the case in centrifuge experiments. Its goal is to link a point $P(X, Y)$ in the object coordinate system to the corresponding point $p(u, v)$ in the image plane coordinate system, as depicted in Fig A.1.

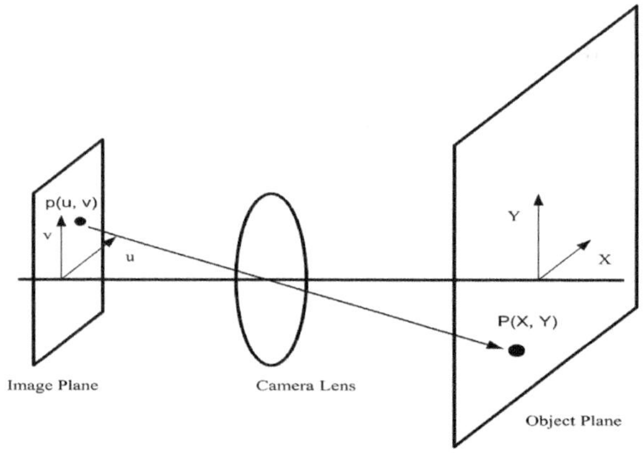

Fig. A.1 Concept of Camera Calibration

The simplest way to do camera calibration is through the pinhole camera method, which is based on the principle of collinearity (Heikkila and Silven, 1997). A linear relation between an object point and its image point is assumed. Due to the fact that camera lens distort images and also because the object is commonly behind a refractive window, a simple relation between the object points and the image points does not normally exist. Taylor et al (1998) first applied close range photogrammetry in centrifuge modeling. In their camera calibration model, nine parameters are needed, including 2 orthogonal offsets between the center of the CCD array and the intersection of the optical axis of the camera with the array, the focal length of the camera, 3 polynomial coefficients representing radial geometric lens distortions, 2 polynomial coefficients corresponding to tangential geometric lens distortions, and a differential scale parameter in the image to account for the non-square pixels. This nine-parameter model can correct image distortions such as fisheye and barreling. Paikowsky and Xi (2000) used a steel plate with 121 evenly distributed 4.76-mm diameter points, 50 mm apart on as the calibration object points. A polynomial function was used to fit points in the object plane to the image plane. The absolute error could reach 0.4 mm. Recently, White et al (2003) have implemented the 14-parameter model of Heikkila and Silven (1997) to perform camera calibration for their image-based deformation measurement system in centrifuge experiments.

The Camera Calibration Toolbox for Matlab developed by Bouguet (2006) at Caltech was used as a framework for the current research (Fig. A.2). The features of the toolbox include extracting image corners, running the calibration engine, displaying the results, controlling accuracies, etc. A neural network-based calibration method was proposed and implemented into the Toolbox. The method was then examined by the calibration data for measuring the planar deformation of soils. An example of this calibration algorithm integrated with PIV code MatPIV (Sveen, 2004) was illustrated with a rectangular footing model. The image acquiring system used in this research is the Light Wise LW-1.3-G-1394 digital camera from Image Solution Group, which has a resolution of 1280 × 1024 pixels. It is noted, however, that this method does not give images containing both the soil under observation and also the calibration points. If the camera or test model container moves between calibration and the experiment, the calibration is invalid. This can be improved by using a method with concurrent soil and control point data (Taylor et al 1998, and White et al 2003).

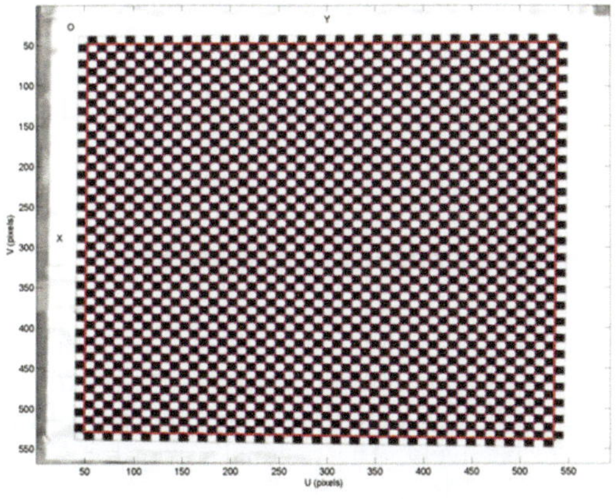

Fig. A.2 Coordinate System with 2.5 mm Spacing Grids for Corner Points

A.2 Neural Network

The neural network trained in this study is a three-layer, feed-forward neural network (2–15–2 NN). The input layer consists of n1 = 2 input nodes, and the output layer consists of n2 = 2 output nodes. The hidden layer consists of 15 neurons. Input data are the image points $p(u, v)$ obtained from the extraction of the corner points. The target data are the corner points $P(X,Y)$ from the calibration plane, and the training data are from half of the calibration points. The other half of the calibration data is used as testing data to test the neural network (Zhao 2007). A t otal of 2500 data points were used in this study with 2.5 mm separation distance between two neighbor points. This gives a field of view (FOV) of 145 mm × 155 mm.

The neural network training algorithm used was a back-propagation function, which updates the weight and bias values according to Levenberg-Marquardt optimization method. In the training process, the input layer takes an input vector of size n1 = 2 and passes it to the hidden layer. Next, the hidden layer maps this input vector to a vector of size n3 = 2 by a tangent sigmoid transfer function tansig (ζ):

$$a = \frac{2}{1+e^{-2\zeta}} - 1 \qquad \text{(A.1)}$$

where ζ is an activation function, the sum of the weighted inputs and biases. This vector from the hidden layer is then mapped by a pure linear transfer function to an output layer. The performance of the neural network model is defined by the normalized square error MSE, expressed as

$$MSE = \frac{1}{2QN} \sum_{q=1}^{Q} \sum_{h=1}^{N} (d_{qh} - x_{out,qh})^2 \qquad \text{(A.2)}$$

where Q is the number of outputs, N is the total number of data sets, d_{qh} is the set of target data; $x_{out,qh}$ is the set of network output data. To determine the appropriate number of neurons in the hidden layer, different numbers of neurons (5, 10, 15, and 20) were tested. It was found that 15 neurons in the hidden layer gave the best performance.

A.3 Neural Network Calibration Model

The proposed neural network calibration model was verified through the testing data, which were not used in the training process. The maximum absolute residual errors for X and Y coordinates were 0.011 mm (1/13182 of FOV) and 0.014 mm (1/11071 of FOV), respectively. The residual errors for X and Y are shown in Fig A.3 and Fig A.4, respectively. It is clear from Fig A.3 and Fig A.4 that the residual error is small and distributed uniformly throughout the entire calibration area.

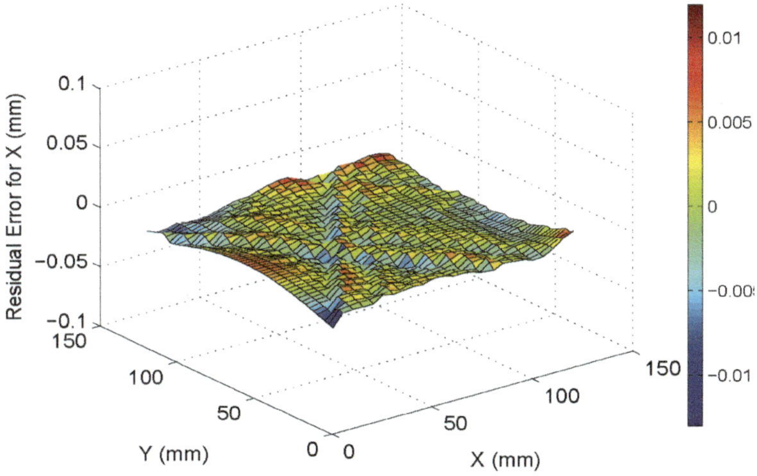

Fig. A.3 Residual Error for *X* from the Neural Network Model

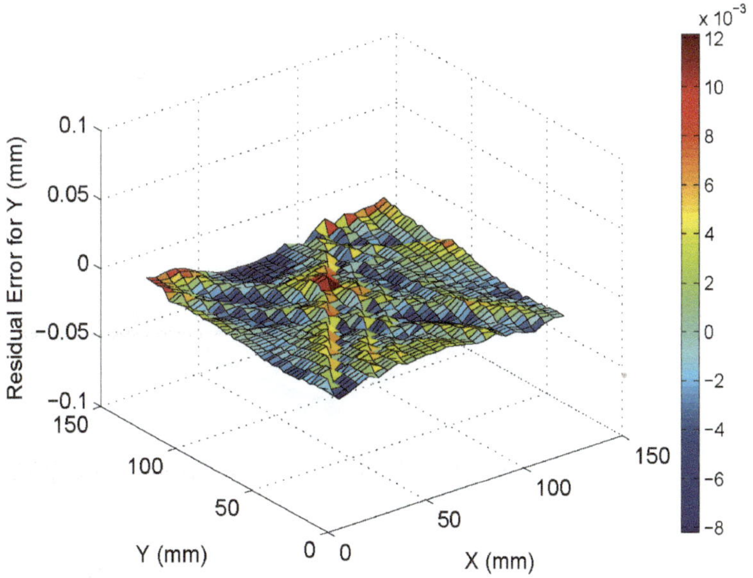

Fig. A.4 Residual Error for *Y* from the Neural Network Model

A.3.1 Calibration Algorithm Comparison

To examine the effectiveness of this neural network model, the linear and second-order polynomial calibration algorithms were also studied in this research, which are expressed as in Equation (A.3) and Equation (A.4):

$$\begin{cases} X = a_1 + a_2 u + a_3 v \\ Y = b_1 + b_2 u + b_3 v \end{cases} \tag{A.3}$$

$$\begin{cases} X = a_1 + a_2 u + a_3 v + a_4 u^2 + a_5 v^2 + a_6 uv \\ Y = b_1 + b_2 u + b_3 v + b_4 u^2 + b_5 v^2 + b_6 uv \end{cases} \tag{A.4}$$

where *X*, *Y* are the coordinates of point *P* in the object plane; *u*, *v* are the coordinates of corresponding image point *p* in the image plane; and a_i, b_i are the coefficient for the model. The coefficients a_i, b_i were determined from the normalized least square method using the same training data as used for the neural network model. These models were also tested using the same testing data to examine their prediction accuracy.

Fig A.5 shows the residual error for X from the linear calibration model. The residual error is small in the center of the calibration plane and becomes large towards the edges of the calibration plane. Fig A.6 shows the residual error for Y from the linear calibration model. The residual error for Y has a convex distribution.

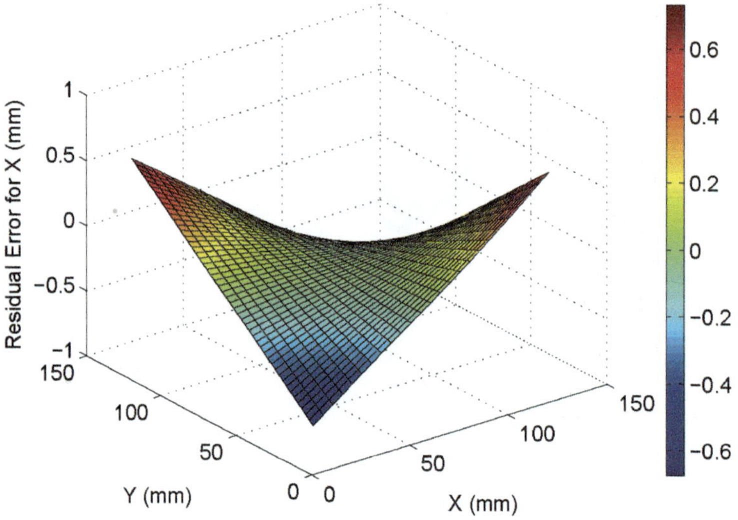

Fig. A.5 Residual Error for *X* from the Linear Calibration Model

Along X = 50 mm and 100 mm, the residual error is small. The residual error increases as it moves away from these two lines with the maximum residual error appearing at the edge and center area parallel to the X axis. Fig A.7 and Fig A.8 present the residual error distribution for X and Y over the calibration plane from

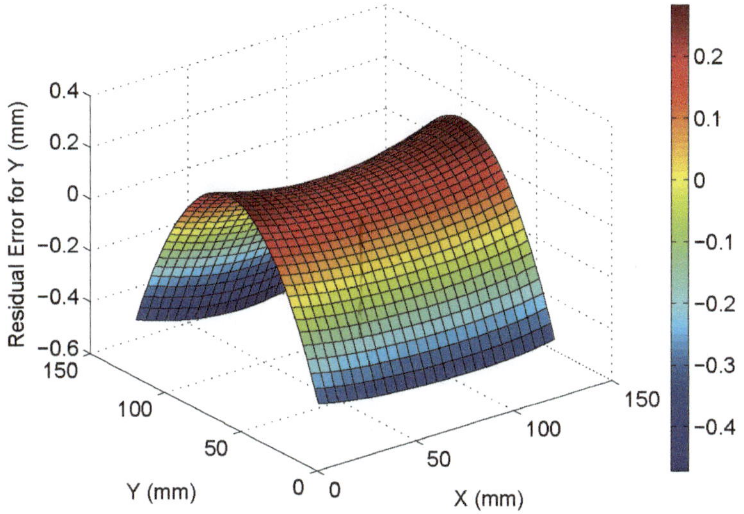

Fig. A.6 Residual Error for *Y* from the Linear Calibration Model

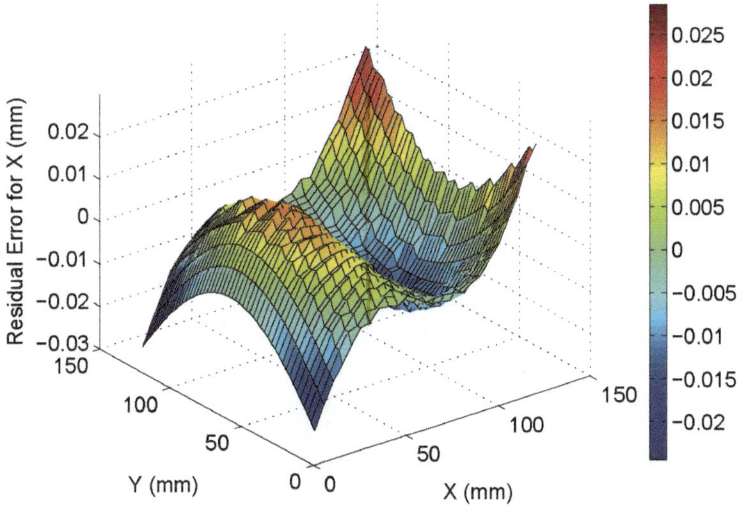

Fig. A.7 Residual Error for *X* from the Second-Order Polynomial Model

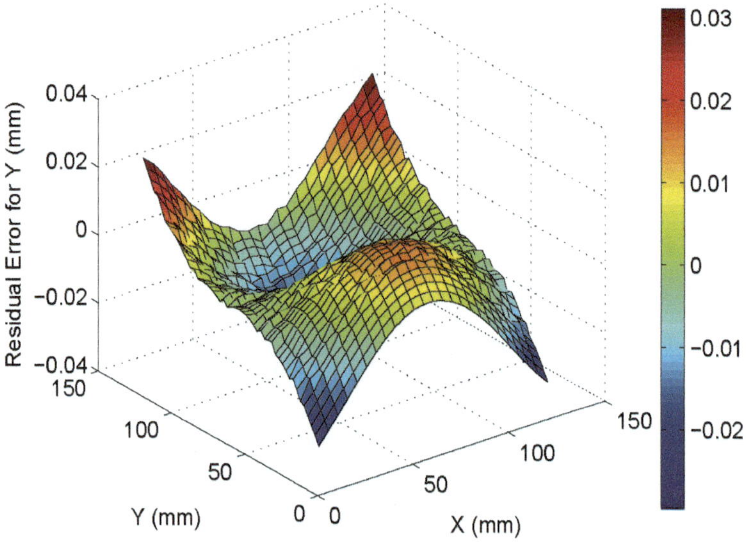

Fig. A.8 Residual Error for *Y* from the Second-Order Polynomial Model

the second-order polynomial model. The residual error for X is the same as that for Y when rotating by 90°. Both have a larger error at the four corners of the plane and a smaller error at the center area of the calibration plane. The maximum absolute residual error is 0.672 mm (1/215 of FOV) for X and 0.685 mm (1/266 of FOV) for Y from the linear calibration algorithm. The maximum absolute residual error is 0.030 mm (1/4833 of FOV) for X and 0.032 mm (1/4844 of FOV) for Y from the second-order polynomial algorithm. These values are much higher than those errors from the neural network calibration algorithm.

A.4 Angle Error Analysis

The effects of the angle (θ) between the calibration plane and the image sensor plane on the residual error were also investigated. In the setup for this experiment, first the calibration plane was closely aligned to the image sensor plane; then, the calibration plane was rotated to different angles (θ), as shown in Fig A.9. Images of the calibration plane were taken by the camera while keeping the position of the camera fixed during this process. Table A.1 shows the residual errors obtained from the neural network model, the linear calibration model, and the second-order polynomial model for each calibration plane. From Table A.1, it is evident that the linear algorithm gives the largest absolute residual error among the three different calibration methods.

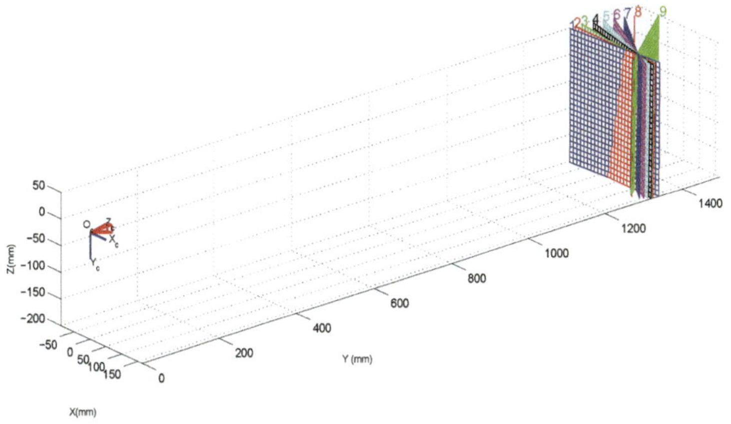

Fig. A.9 Different Calibration Planes for Angle Error Analysis

Table A.1 Absolute Error for Nine Planes from the Three Calibration Algorithms

Plane No.	Angle (°)	Maximum Absolute Residual Error					
		Linear Model		Second-Order Polynomial Model		Neural Network Model	
		mm	Fraction Of FOV	mm	Fraction of FOV	mm	Fraction of FOV
1	7.62	0.672	1/215	0.030	1/4833	0.014	1/11071
2	13.61	1.499	1/97	0.044	1/3295	0.025	1/5800
3	17.97	1.932	1/75	0.057	1/2544	0.016	1/9063
4	25.60	1.932	1/75	0.095	1/1526	0.039	1/3718
5	30.87	3.348	1/43	0.130	1/1115	0.038	1/3816
6	36.49	3.944	1/37	0.199	1/729	0.014	1/10357
7	39.43	4.427	1/33	0.202	1/718	0.016	1/9063
8	50.66	4.482	1/32	0.274	1/529	0.056	1/2589
9	53.93	4.722	1/31	0.303	1/479	0.062	1/2339

When the angle, θ, is small, the second-order polynomial model can predict the data well. However, the error increases as the angle becomes larger. The neural network calibration algorithm was found to perform the best, because it gives the minimum absolute residual error, and also the error does not vary much as the angle increases.

Fig. A.10 Experiment Setup for the Rectangular Footing Model

A.5 Application in Digital Image Correlation (DIC) and Particle Image Velocimetry (PIV)

A small-scale wood block (6.35 cm × 7.62 cm × 15.88 cm) was placed on top of the sand in a model container (35.56 cm × 6.60 cm × 35.56 cm), as shown in Fig A.10. The height of the sand was 18.92 cm (7.45 in.), and a loading frame was used to apply the vertical load to the foundation. The camera was set up to take images of the soil below the footing during the loading. The loading rate was set to 2.54 mm/min (0.1 in/min). The test setup is illustrated in Fig A.11. A long steel plate was put on the loading platen of the frame, the sand box rested on the steel plate, and the digital camera was set about 2 m from the sand box at the other end of the steel plate. This setup ensured that the digital camera was moving together with the loading platen during the loading process. Images were taken every 30 seconds. The proposed neural network camera calibration algorithm was implemented into MatPIV, a computer code initially developed by Sveen (2004) for PIV, to process the images. The principles behind DIC (PIV) image pattern matching techniques are presented in Chapter 8.

Assume two images (I1 and I2) are obtained, and these two images are divided into smaller regions (sub-windows, interrogation-windows, or interrogation-regions). Then, each sub-window in the image, I1, is compared with the corresponding sub-window in the image, I2. The cross-correlation between the two sub-windows is defined as

(a)

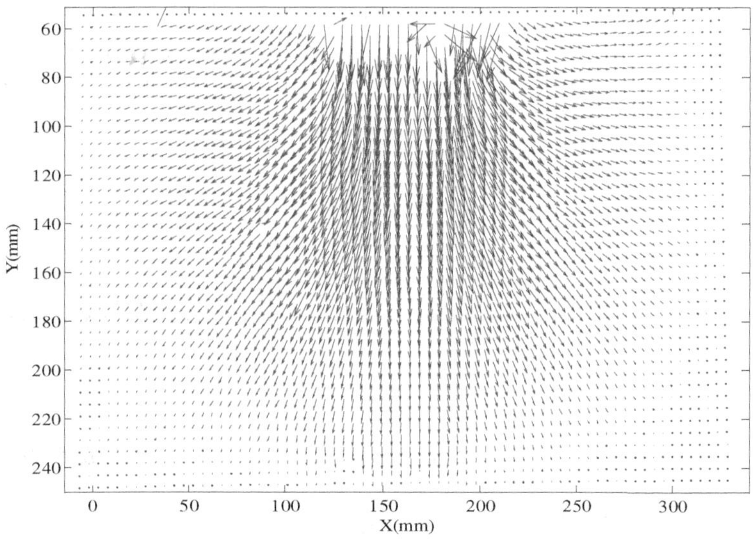

(b)

Fig. A.11 (a). Deformation Vector Field of Sand under the Rectangular Footing in the Image Space (t = 6 min, P = 0.92 kN). **(b).** Deformation Vector Field of Sand under the Rectangular Footing in the Object Space (t = 6 min, P = 0.92 kN).

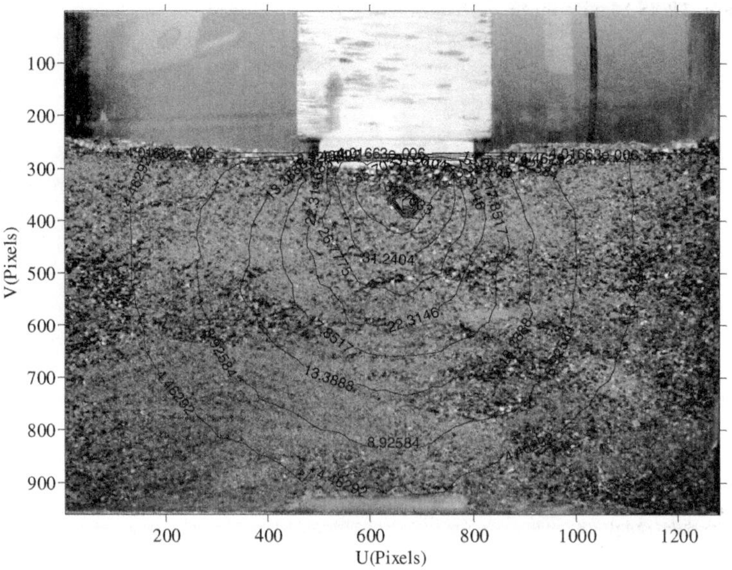

(a)

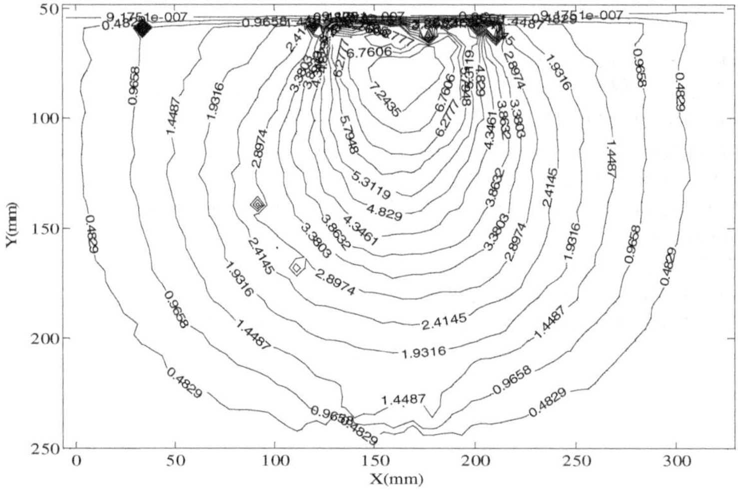

(b)

Fig. A.12 (a). Magnitude Contour of Displacement Vector for Sand under the Rectangular Footing in the Image Space (contour label in unit of pixels, t = 6 min, P = 0.92 kN). **(b).** Magnitude Contour of Displacement Vector for Sand under the Rectangular Footing in the Object Space (contour label in unit of pixels, t = 6 min, P = 0.92 kN)

$$R(s,t) = \sum_{m=0}^{M-1}\sum_{n=0}^{N-1} I_1^{\,i,j}(m,n) \times I_2^{\,i,j}(m-s,n-t) \tag{A.5}$$

where is $I_1^{i,j}$ the sub-window number (i, j) in the first image, I_1; and $I_2^{i,j}$ is the sub-window number (i, j) in the second image, I_2. When this expression reaches its peak value, the two sub-windows will most likely match each other. The sub-window has moved s and t units in i and j directions, respectively. By a centroid fit, a Gaussian fit, or a parabolic fit to the region close to the peak, the system can achieve sub-pixel resolution. The smallest window size used in this analysis was 36 pixels × 36 pixels in order to cover the large displacement of sand particles.

The digital image correlation method was applied to analyze the two images to obtain the full-field deformation of the soil surrounding the footing. Fig A.11 (a) and (b) show the displacement vector field of the soil under the footing at t = 6 min, and P = 0.92 kN (20.6 lb), which are similar to cylindrical cavity expansion. Therefore, Fig A.12 (a) and (b) show circular contours of displacement. From the experimental results, no Terzaghi's failure mechanism was observed, i.e. the vertical movement extends to too great a depth to match Terzaghi's mechanism, and there is no heave beside the footing as predicted by Terzaghi.

A.6 Summary

A three-layer back-propagation neural network calibration algorithm was developed for camera calibration in an image-based measurement system. This algorithm was compared with the linear calibration and second-order polynomial calibration algorithms. When the angle between the image plane and object plane became large, the linear calibration algorithm and second-order polynomial calibration algorithm gave a considerable error. It was found that the neural network calibration model will give a very accurate result independent of the angle between the image plane and the object plane. This algorithm was integrated into MatPIV software. This fast and accurate calibration method should facilitate image correction in situations when the camera is not aligned with the model.

References

Allersma, H.: Using image processing in centrifuge research. In: Proc. Centrifuge 1991, Colorado, Balkema, pp. 551–559 (1991)

Alshibli, K., Sture, S.: Shear band formation in plane strain experiments of sand. ASCE J. of Geotechnical and Geoenvironmental Engineering 126(6), 495–503 (2000)

Aydilek, A., Guler, M., Edil, T.: Measurement of particle movement in granular soils using image analysis. ASCE J. of Computing in Civil Engineering 18(1), 65–74 (2004)

Bouguet, J.: Camera Calibration Toolbox for Matlab, Computer Vision Research Group, Department of Electrical Engineering, California Institute of Technology (2006), `http://www.vision.caltech.edu/bouguetj/calib_doc/index.html`

Donohoe, G., Bacobella, M., Gill, J.: Segmentation and object tracking for the microstructure analysis of soil. IEEE 1-2(2), 1215–1219 (1991)

Guler, M., Edil, T., Bosscher, P.: Measurement of particle movement in granular soils using image analysis. ASCE J. of Computing in Civil Engineering 13(2), 116–122 (1999)

Heikkila, J., Silven, O.: A four-step camera calibration procedure with implicit image correction. In: IEEE Computer Society Conference on Computer Vision and Pattern Recognition (CVPR 1997), San Juan, Puerto Rico, pp. 1106–1112 (1997)

Hryciw, R., Irsyam, M.: Behavior of sand particles around rigid ribbed inclusions during shear. Soils and Foundations 33(3), 1–13 (1993)

Paikowsky, S., Xi, F.: Particle motions tracking utilizing a high-resolution digital CCD camera. ASTM Geotechnical Testing J. 23(1), 123–134 (2000)

Rechenmacher, A., Finno, R.: digital image corelation to evaluate shear banding in dilative sands. ASTM Geotechnical Testing J. 27(1), 1–10 (2004)

Sveen J.: An Introduction to MatPIV, v.1.6.1. Eprint No.2, Dept. of Math. University of Oslo (2004) ISSN 0809-4403

Taylor, R., Grant, R., Robson, S., Kuwano, J.: An image analysis system for determining plane and 3-D displacements in soil models. In: Proc. Centrifuge 1998, Tokyo, pp. 73–78 (1998)

White, D., Take, W., Bolton, M.: Soil deformation measurement using particle image velocimetry (PIV) and photogrammetry. Geotechnique 53(7), 619–631 (2003)

White, D., Randolph, M., Thompson, B.: An image-based deformation measurement system for the geotechnical centrifuge. Int. J. of Physical Modeling in Geotechnics 5(3), 1–12 (2005)

References

1. Adams, J.: The Engr. behavior of a Canadian Muskeg. In: Proc. 6th Int. Conf. Soil Mech. & Found. Engr., vol. 1, pp. 3–7 (1965)
2. Adeli, H., Hung, S.: A fuzzy neural network learning model for image recognition. Integrated Computer-Aided Engr. 1(1), 43–55 (1993)
3. Adrian, R.: Particle-imaging techniques for experimental fluid mechanics. Annual Review of Fluid Mechanics 23, 261–304 (1991)
4. Aeby, P.: Quantitative fluorescence imaging of tracer distributions in soil profiles, Ph.D. Dissertation, Swiss Federal Institute of Technology, Zurich (1998)
5. Ajlouni, M.: Geotechnical properties of peat and related Engr. problems, Ph.D. Dissertation, University of Illinois at Urbana-Champaign (2000)
6. Allersma, H.: Photo-elastic stress analysis and strains in simple shear. In: Vermeer, P.A., Luger, H.J. (eds.) Proc. IUTAM Symposium on Deformation and Failure of Granular Materials, Delft, pp. 345–353 (1982)
7. Allersma, H.: Using imaging technologies in experimental geotechnics. In: Proc. 2nd Int. Conf. on Imaging Technologies, Davos, Switzerland, pp. 1–9. ASCE (1998)
8. Allersma, H.: Using image processing in centrifuge research. In: Proc. Centrifuge 1991, Colorado, Balkema, pp. 551–559 (1991)
9. Aloufi, M., Santamarina, C.: Low and high strain behavior of grain masses - the effect of particle eccentricity. Transactions of ASAE 38, 877–887 (1995)
10. Alshibi, K., Sture, S.: Sand shear band thickness measurement by digital imaging techniques. J. of Computing in Civil Engr. 13(2), 103–109 (1999)
11. Alshibli, K., Sture, S.: Shear band formation in plane strain experiments of sand. ASCE J. of Geotechnical and Geoenvironmental Engr. 126(6), 495–503 (2000)
12. Ammouche, A., Breysse, D., Hornain, H., Didry, O., Marchand, J.: A new image analysis technique for the quantitative assessment of microcracks in cement-based materials. Cement and Concrete Research 30(1), 25–35 (2000)
13. Anandan, P., Bergen, J., Hanna, K., Hingorani, R.: Motion analysis and image sequence processing: Hierarchical model-based motion estimation. In: Sezan, M., Lagendijk, R. (eds.) Motion Analysis and Image Sequence Processing. Kluwer (1993)
14. Anon: Video technology in traffic Engr. and transportation planning. J. of Transportation Engr. 125(3), 169–175 (1999)
15. ASTM D698, Test method for laboratory compaction characteristics of soil using standard effort (12,400 ft-lbf/ft3(600 kN-m/m3)), Annual Book of ASTM Standards, vol. 04.08, doi: 10.1520/D0698-07E01
16. ASTM D2434, Standard test method For permeability of granular soils (constant head). Annual Book of ASTM Standards, vol. 4.08, doi: 10.1520/D2434-68R06
17. ASTM D2435, Standard test method for one-dimensional consolidation properties of soils. Annual Book of ASTM Standards, vol. 4.08, doi: 10.1520/D2435-04
18. ASTM D3080, Standard Test Method for Direct Shear Test of Soils under Consolidated Drained Conditions. Annual Book of ASTM Standards, vol. 4.08, doi: 10.1520/D3080-04
19. ASTM D4015 Standard test methods for modulus and damping of soils by resonant-column method. Annual Book of ASTM Standards, doi: 10.1520/D4015-07

20. ASTM D4254, Standard test method for minimum index density and unit weight of soils and calculation of relative density. Annual Book of ASTM Standards, vol. 04.08, doi: 10.1520/D4254-00R06E01
21. ASTM D4767, Standard test method for consolidated-undrained triaxial compression test on cohesive soils. Annual Book of ASTM Standards, vol. 04.08, doi: 10.1520/D4767-04
22. ASTM D5084, Standard test method for measurement of hydraulic conductivity of saturated porous materials using a flexible wall permeameter. Annual Book of ASTM Standards, doi: 10.1520/D5084-03
23. ASTM D5613, Standard test method for open-channel measurement of time of travel using dye tracers. Annual Book of ASTM Standards, vol. 04.11, doi: 10.1520/D5613-94R08
24. Aydilek, A., Guler, M., Edil, T.: Measurement of particle movement in granular soils using image analysis. J. of Computing in Civil Engr. 18(1), 65–74 (2004)
25. Bai, X., Smart, P.: Change in microstructure of kaolin in consolidation and undrained shear. Geotechnique 47(5), 1009–1017 (1997)
26. Balasubramaniam, A., Chaudhry, A.: Deformation and strength characteristics of soft Bangkok clay. J. Geotech. Eng. Div. 104(GT9), 1153–1167 (1978)
27. Barron, J., Fleet, D., Beauchemin, S., Burkitt, T.: Performance of optical flow techniques. In: Proc. IEEE Computer Vision and Pattern Recognition 1992, pp. 236–242 (1992)
28. Barron, J., Fleet, D., Beauchemin, S.: System and experiment: performance of optical flow techniques. Int. J. of Computer Vision 12(1), 43–77 (1994)
29. Bathe, K.: Finite Element Procedures, p. 1037. Prentice-Hall, Englewood Cliffs (1995)
30. Berge, P., Berryman, J., Blair, S., Pena, C.: Scaler properties of transversely isotropic tuff from images of orthogonal cross sections. In: Proc. 2nd Int. Conf. on Imaging Technologies: Techniques and Applications in Civil Engr., ASCE, pp. 185–196 (1997)
31. Bergfelt, A.: Loading tests on clay. Geotechnique 11, 15–31 (1956)
32. Berre, T., Iverssen, K.: Oedometer tests with different specimen heights on a clay exhibiting large secondary compression. Geotechnique 22(1), 53–70 (1972)
33. Bhatia, S., Soliman, A.: Frequency distribution of void ratio of granular materials determined by an image analyzer. Soil and Foundation 30(1), 1–16 (1990)
34. Bhatia, S., Huang, Q., Smith, J.: Application of digital image processing in morphological analysis of geotextiles. In: Proc. Digital Image Processing, Hawaii, pp. 95–108. ASCE (1993)
35. Bird, R., Stewart, W., Lightfoot, E.: Transport Phenomena. Wiley (1960)
36. Bishop, A.: Shear strength parameters for undisturbed and remolded soil specimens. In: Parry, R., Foulis, G.T. (eds.) Proc. Roscoe Memorial Symposium: Stress-Strain Behavior of Soils, Henley-on-Thames, England, pp. 3–58. Cambridge University (1971)
37. Bishop, A., Henkel, D.: The Measurement of soil properties in the triaxial test, 2nd edn. Edward Arnold Publishers, London (1962)
38. Bjerrum, L.: Geotechnical properties of Norwegian marine clays. Geotechnique 4(2), 49–69 (1954a)
39. Bjerrum, L.: Theoretical and experimental investigations on the shear strength of soils. Norwegian Geotechnical Institute Publication No. 5, 111 (1954b)
40. Bjerrum, L.: The effective shear strength parameters of sensitive clays. In: Proc. 5th Int. Conf. on Soil Mech. & Found. Engr., Paris, France, vol. 1, pp. 23–28 (1961)

41. Bjerrum, L., Simons, N.: Comparison of shear strength characteristics of normally consolidated clays. In: Proc. Res. Conf. Shear Strength of Cohesive Soils, pp. 711–726. ASCE (1960)
42. Bjerrum, L., Wu, T.: Fundamental shear strength properties of Lilla Edet clay. Geotechnique 10(3), 101–109 (1960)
43. Blair, S., Berge, P., Berryman, J.: Estimates of permeability of porous materials using image analysis of cross sections. In: Proc. Digital Image Processing, Hawaii, pp. 109–116. ASCE (1993)
44. Borden, R., Shao, L., Gupta, A.: Dynamic properties of piedmont residual soils. J. of Geotechnical Engr. 122(10), 813–821 (1996)
45. Bouchafa, S., Aubert, D., Bouzar, S.: Crowd motion estimation and motionless detection in subway corridors by image processing. In: IEEE Conference on Intelligent Transportation System, ITSC 1997, pp. 332–337 (1997)
46. Bouguet, J.: Camera calibration toolbox for Matlab. Computer Vision Research Group, Department of Electrical Engr. California Institute of Technology (2006), http://www.vision.caltech.edu/bouguetj/calib_doc/index.html
47. Bourdeau, P.: Radiographic visualization in experimental soil mechanics. In: Proc. of the Conference on Digital Image Processing, pp. 125–134. ASCE (1993)
48. Bourdeau, P.: Radiographic visualization in experimental soil mechanics. In: Proc. Digital Image Processing, Hawaii, pp. 125–134. ASCE (1993)
49. Bransby, P., Milligan, G.: Soil deformation near cantilever sheet pile. Geotechnique 25(2), 175–195 (1975)
50. Brinkgreve, R., Vermeer, P.: Plaxis finite element code for soil and rock analysis plaxis manual Version 7. Balkema, Rotterdam (1998)
51. Brock, D., Orr, F.: Flow visualization of viscous fingering in heterogeneous porous media. In: SPE Annual Technical Conference and Exhibition, pp. 211–222 (1991)
52. Broms, B.: Stabilization of very soft clay using geofabric. Geotextiles and Geomembranes 5(1), 17–28 (1987)
53. Buisman, A.: Results of long duration settlement tests. In: Proc. 1st Int. Conf. Soil Mech. & Found. Eng., vol. 1, pp. 103–105 (1936)
54. Burns, S., Yiacoumi, S., Frost, D., Tsouris, C.: Application of digital image analysis for distribution measurement of microbubbles. In: Proc. 2nd Int. Conf. on Imaging Technologies: Techniques and Applications in Civil Engr., pp. 100–107. ASCE (1997)
55. Butterfield, R., Andrawes, K.: The visualization of planer displacement field. In: Parry, R. (ed.) Proc. Roscoe Memorial Symposium: Stress Strain Behavior of Soils, pp. 467–475. Cambridge University (1971)
56. Casagrande, A.: Characteristics of cohesionless soils affecting the stability of slopes and earth fills. J. of Boston Soc. of Civil Engr., 1925–1940 (1936)
57. Casagrande, A.: Liquefaction and cyclic deformation of sands, a critical review. In: Procs. Fifth Pan-American Conference on Soil Mechanics and Foundation Engr., Bue-nos Aires, Argentina, January 1976. Harvard Soil Mechanics Series, vol. 88. Prentice Hall, Englewood Cliffs (1975)
58. Casagrande, A., Fadum, R.: Notes on soil testing for Engr. purposes. Harvard Univ. Grad. School of Engr., Publ. 268 (1940)
59. Chambon, P., Molaiard, L., Garnier, J.: An attempt to compare image processing and stereo photgrammetry in geotechnical centrifuge models. In: Proc. 2nd Int. Conf. on Imaging Technologies: Techniques and Applications in Civil Engr., pp. 30–39. ASCE (1997)

60. Charbeneau, R.: Groundwater Hydraulics and Pollutant Transport. Prentice-Hall, NJ (2000)
61. Charrier, J., Moliard, J., Garnier, J.: Numerical image processing in centrifuge testing. In: Proc. 2nd Int. Conf. on Imaging Technologies, pp. 20–29. ASCE (1997)
62. Chen, J., Wada, N.: Visualization of immiscible displacement in a three-dimensional transparent porous medium. Experiments in Fluids 4(6), 336–338 (1986)
63. Cheng, C., Atkinson, J., VanBenschoten, J., Bursik, M., DePinto, J.: Image-based system for particle counting and sizing. J. of Environmental Engr. V 126(3), 258–266 (2000)
64. Chermant, J.: Why automatic image analysis? An introduction to this issue. Cement and Concrete Composites 23(2-3), 127–131 (2001)
65. Chou, E., Sethi, V.: Machine vision based traffic-adjusted intersection signal control. In: Proc. of the Conference on Digital Image Processing, Hawaii, pp. 150–156. ASCE (1993)
66. Chu, T., Ranson, W., Sutton, M., Peters, W.: Application of digital image correlation techniques to experimental mechanics. Experimental Mechanics 25(3), 232–244 (1985)
67. Cloud, G.: Optical methods of engineering analysis. Cambridge University Press, Cambridge (1998)
68. Corapcioglu, M., Fedirchuk, P.: Glass bead micromodel study of solute transport. J. of Contaminant Hydrology 36(2), 209–230 (1999)
69. Corapcioglu, M., Chowdhury, S., Roosevelt, S.: Micromodel visualization and quantification of solute transport in porous media. Water Resource Research 33(11), 2547–2558 (1997)
70. Cousset, P.: Pore size NMR imaging. Magnetic Resonance Imaging 16(5/6), 621–623 (1998)
71. Dally, J., Riley, W.: Experimental stress analysis, 3rd edn. McGraw-Hill, New York (1991)
72. Dantec Measurement Technology, Flowmap®particle image velocimetry instrumentation-installation & user's guide, Tonsbakken 16-18, DK-2740 Skovlunde, Denmark (2000)
73. Das, B.: Principles of geotechnical engr, 4th edn. PWS Publishing Company (1998)
74. De Jong, D.: Consolidation models consisting of an assembly of viscous elements of a cavity channel network. Geotechnique 18(2), 195–228 (1968)
75. Desrsues, J., Mokni, M., Mazerolle, F.: Tomodensitometry and localization in sands. In: Proc. X ECSMFE: Deformation of Soils and Displacements of Structures, Florence, pp. 61–64 (1991) (in French)
76. Dhowian, A., Edil, T.: Consolidation behavior of peats. ASTM Geotechnical Testing J., GTJODJ 3(3), 105–114 (1980)
77. Diamond, S.: Interactive digital image feature analysis as applied to cement and concrete microstructural investigation. In: Proc. of the Conference on Digital Image Processing Techniques and Applications in Civil Engr., pp. 158–164. ASCE (1993)
78. Dodt, J., Genske, D., Kappernagel, T., Noll, P.: Digital evaluation of contaminated sites. In: Proc. of the Conference on Digital Image Processing, pp. 165–172. ASCE (1993)
79. Domenico, P., Schwartz, F.: Physical and chemical hydrogeology. Wiley, NY (1990)
80. Donohoe, G., Bacobella, M., Gill, J.: Segmentation and object tracking for the microstructure analysis of soil. IEEE 1-2(2), 1215–1219 (1991)
81. Drescher, A., De Jong, D.: Photoelastic verification of a mechanical model for the flow of granular material. J. of Mechanics and Physics of Solids 20, 337–351 (1972)

82. Dupre, L., Lagarde, A.: Photoelastic analysis of a three-dimensional specimen by optical slicing and digital image processing. Experimental Mechanics 37(4), 393–397 (1997)
83. Dwarakanath, V., Kostarelos, K., Pope, G., Shotts, D., Wade, W.: Anionic surfactant remediation of soil columns contaminated by non-aqueous phase liquids. J. of Contaminant Hydrology 38(4), 465–488 (1999)
84. Dzuy, N., Boger, D.: Direct yield measurement with the vane method. J. of Rheology 29(3), 335–347 (1985)
85. Edil, T., Haan, E.: Settlement of peats and organic soils. In: Proc. Settlement 1994, Vertical & Horizontal Deformation of Foundations and Embankments, pp. 1543–1572. ASCE (1994)
86. Eggstad, A.: Deformation measurements below a footing on the surface of dry sand. In: Proc. Eur. Conf. on Soil Mech. and Found. Eng., Wiesbaden, Germany, vol. 1, pp. 233–239 (1963)
87. Erf, R.: Speckle metrology. Academic Press, New York (1978)
88. Fakher, A., Jones, C., Clarke, G.: Yield stress of soft clays. J. of Geotechnical and Geoenvironmental Engr., 499–509 (June 1999)
89. Fetter, C.: Contaminant hydrogeology. Prentice-Hall, NJ (1999)
90. Finno, R., Chung, C.: Stress-strain-strength response of compressible Chicago glacial clays. J. Geotech. Eng. 118(10), 1607–1625 (1992)
91. Fleet, D.: Measurement of image velocimetry. Kluwer Academic Publ. (1992)
92. Fomin, N.A.: Speckle photography for fluid mechanics measurements. Springer, Heidelberg (1998)
93. Force, E.: Factors controlling pore-pressure generating during ko-consolidation of laboratory tests, MS Thesis. Dept. of Civil and Environmental Engr. MIT, Cambridge (1998)
94. Freeze, R., Cherry, J.: Groundwater. Prentice-Hall, Englewood Cliffs (1979)
95. Frost, J., Kuo, C.: Automated determination of the distribution of local void ratio from digital images. ASTM Geotechnical Testing J., GTJODJ 19(2), 107–117 (1996)
96. Gaganis, P., Skouras, E., Theodoropoulou, M., Tsakiroglou, C., Burganos, V.: On the evaluation of dispersion coefficients from visualization experiments in artificial porous media. J. of Hydrology 307(1-4), 79–91 (2005)
97. Ghalib, A., Hryciw, R.: Soil particle size distribution by mosaic imaging and watershed analysis. J. of Computing in Civil Engr. 13(2), 80–87 (1999)
98. Ghandehari, M., Krishnaswamy, S., Shah, S.: Technique for evaluating kinematics between rebar and concrete. J. of Engr. Mechanics 125(2), 234–241 (1999)
99. Giachetti, A.: Matching techniques to compute image motion. IEEE Image Vision Computing 18, 247–260 (2000)
100. Gill, D.: Experimental and theoretical investigations of pile and penetrometer installation in clay, Ph.D. Thesis, Trinity College, Dublin, UK (1999)
101. Gill, D., Lehane, B.: An optical technique for investigating soil displacement patterns. ASTM Geotechnical Testing Journal, GTJODJ 24(3), 324–329 (2001)
102. Gimenez, D., Allmaras, R., Huggins, D., Nater, E.: Prediction of the saturated hydraulic conductivity–porosity dependence using fractals. Soil Science Society of America J. 61(5), 1285–1292 (1997)
103. Gokceoglu, C., Aksoy, H.: Landslide susceptibility mapping of the slopes in the residual soils of the Mengen region (Turkey) by deterministic stability analyses and image processing techniques. Engr. Geology 44(4), 147–161 (1996)
104. Goodman, J.: Statistical properties of laser speckle patterns. In: Dainty (ed.) Laser Speckle and Related Phenomena, pp. 9–75. Springer, Berlin (1975)

105. Grantham, S., et al.: Study of internal deformation fields in granular materials using 3-D digital speckle X-ray flash photography. In: Proc. of SPIE - The International Society for Optical Engr., vol. 4101 (B), pp. 321–328 (2000)
106. Guler, M., Edil, T., Bosscher, P.: Measurement of particle movement in granular soils using image analysis. J. Computing in Civil Engr. 13(2), 116–122 (1999)
107. Gustafsson, L., Gustafsson, P.: Studying mixed granular flows by image analysis. In: Proc. 11th Conference ASCE Engr. Mechanics Division, pp. 614–617 (1996)
108. Haas, C., Kim, Y., Greer, R.: A model for imaging assisted automation of infrastructure maintenance. In: Proc. 2nd Int. Conf. on Imaging Technologies, pp. 108–117. ASCE (1997)
109. Hall, C., Colston, S., Jupe, A., Jacques, S., Livingston, R., Ramadan, A., Amde, A., Barnes, P.: Non-destructive tomographic energy-dispersive diffraction imaging of the interior of bulk concrete. Cement and Concrete Research 30(3), 491–495 (2000)
110. Hanrahan, E.: An investigation of some physical properties of Peat. Geotechnique 4, 108–123 (1954)
111. Hardin, B., Drnevich, P.: Shear modulus and damping in soils: measurement and parameter effects. J. Soil Mechanics and Foundations 98(6), 603–624 (1972)
112. Hart, D.: PIV error correction. Experiments in Fluids 29(1), 12–22 (2000)
113. Head, K.: Manual of soil laboratory testing. In: Permeability, Shear Strength And Compressibility Tests, 2nd edn. Pentech Press (1994)
114. Hecht, E.: Optics, 3rd edn. Addison-Wesley Longman, Inc., New York (1998)
115. Heikkila, J., Silven, O.: A four-step camera calibration procedure with implicit image correction. In: IEEE Computer Society Conference on Computer Vision and Pattern Recognition (CVPR 1997), San Juan, Puerto Rico, pp. 1106–1112 (1997)
116. Heiler, M., McNeil, S.: Pavement condition assessment using ground penetration radar. In: Proc. 2nd Int. Conf. on Imaging Technologies, pp. 128–137. ASCE (1997)
117. Heiler, M., Motazed, B., McNeil, S.: Ground penetration radar: images of subsurface characterization. In: Proc. Conference on Digital Image Processing, pp. 173–178. ASCE (1993)
118. Heinz, W.: Discussion of papers in session 1. In: Proc. Roscoe Memorial Symposium: Stress-Strain Behavior of Soils, pp. 120–122. Cambridge University, G.T.Foulis, Henley-on-Thames, Cambridge (1971)
119. Hendron, A.: The behavior of sand in one-dimensional compression, Ph.D. Dissertation, The University of Illinois at Urbana-Champaign (1963)
120. Henkel, D.: The effect of overconsolidation on the behavior of clays during shear. Geotechnique 6, 139–150 (1956)
121. Herrmann, K., Pohlmeier, A., Gembris, D., Vereecken, H.: Three-dimensional imaging of pore water diffusion and motion in porous media by nuclear magnetic imaging. J. of Hydrology 267, 244–257 (2002)
122. Himmelblau, D., Bischoff, K.: Process analysis and simulation. Deterministic systems. Wiley (1968)
123. Hobbs, N.: Mire morphology and the properties and behavior of some British and foreign peats. Quarterly J. of Engr. Geology, London 19(1), 7–80 (1986)
124. Holtz, R., Kovacs, W.: An introduction to geotechnical engr. Prentice-Hall, NJ (1981)
125. Horn, H., Deere, D.: Frictional characteristics of minerals. Geotechnique XII, 319–335 (1962)
126. Horri, H., Takamatsu, K., Inoue, J., Sasaki, N.: Measurement of displacement field by matching method and observation of strain localization in soft rock. In: Proc. 2nd Int. Conf. on Imaging Technologies, pp. 10–19. ASCE (1997)

127. Hossain, I., Liu, J., You, J.: Tropical cyclone pattern recognition for intensity and forecasting analysis from satellite imagery. In: IEEE International Conference on Systems, Man, and Cybernetics, IEEE SMC 1999, vol. 6, pp. 851–856 (1999)
128. Hryciw, R., Irsyam, M.: Behavior of sand particles around rigid ribbed inclusions during shear. Soils and Foundations 33(3), 1–13 (1993)
129. Hryciw, R., Ghalib, A., Rasckle, S.: Methods for soil characterization for images of grain assemblies. In: Proc. 2nd Int. Conf. on Imaging Technologies, pp. 88–99. ASCE (1997)
130. Huang, T., Tsai, R.: Image sequence analysis: motion estimation. In: Huang, T.S. (ed.) Image Sequence Analysis, pp. 1–18. Springer, New York (1981)
131. Huang, W., Smith, C., Lerner, D., Thornton, S., Oram, A.: Physical modeling of solute transport in porous media: evaluation of an imaging technique using UV excited fluorescent dye. Water Research 36(7), 1843–1953 (2002)
132. Hunt, R.: Geotechnical engr. investigation manual. McGraw-Hill, Inc., New York (1984)
133. Iler, R.: The colloid chemistry of Silica and Silicates. Cornell University Press, Ithica (1955)
134. Inoue, T.: Behavior and treatment of dredged clay, Ph.D. Thesis, National University of Singapore (1990)
135. Iskander, M.: A transparent material to model the geotechnical properties of soils. In: Proc.s 14th ICSMFE, vol. 1, pp. 315–319 (1997)
136. Iskander, M.: Transparent soils to image 3D flow & deformation. In: Proc. 2nd Int. Conf. on Imaging Technologies, pp. 255–264. ASCE (1998)
137. Iskander, M., Lai, J., Oswald, C., Mannheimer, R.: Development of a transparent material to model the geotechnical properties of soils. ASTM Geotechnical Testing J., GTJODJ 17(4), 425–433 (1994)
138. Iskander, M., Liu, J., Sadek, S.: Transparent amorphous silica to model clay. ASCE J. Geotech. & Geoenv. Eng. 128(3), 262–273 (2002)
139. Iskander, M., Sadek, S., Liu, J.: Optical measurement of deformation using transparent silica gel to model sand. Int. J. of Physical Modeling in Geotechnics 2(4), 13–26 (2003)
140. Jacquot, P., Facchini, M.: Interferometric imaging using holographic and speckle techniques, I) fundamentals and basic characteristics. In: Proc. 2nd Int. Conf. on Imaging Technologies: Techniques and Applications in Civil Engr., pp. 217–234. ASCE (1997a)
141. Jacquot, P., Facchini, M.: Interferometric imaging using holographic and speckle techniques, II) impact in civil Engr. In: Proc. 2nd Int. Conference on imaging technologies: techniques and applications in Civil Engr., pp. 235–254. ASCE (1997b)
142. James, R., Bransby, P.: Experimental and theoretical investigations of a passive earth pressure problem. Geotechnique 20(1), 17–37 (1970)
143. Jia, Z., Shah, S.: Two-dimensional electronic-speckle-pattern interferometry and concrete-fracture processes. Experimental Mechanics 34(3), 262–270 (1994)
144. Jia, C., Shing, K., Yortsos, Y.: Visualization and simulation of non-aqueous phase liquids solubilization in pore networks. J. of Contaminant Hydrology 35, 363–387 (1999)
145. Jin, M.: A study of nonaqueous phase liquid characterization and surfactant remediation, Ph.D. Dissertation, University of Texas at Austin, Austin, USA (1995)
146. Jones, R., Wykes, C.: Holographic and speckle interferometry, 2nd edn. Cambridge University Press, Cambridge (1989)
147. Kato, A.: Detection of fatigue damage in steel using laser speckle. Optics and Laser in Engr. 33, 323–333 (2000)

148. Keane, R., Adrian, R.: Optimization of particle image velocimeters. Part I: Double pulsed systems. Measurement Science and Technology 1, 1202–1220 (1990)
149. Keane, R., Adrian, R.: Theory of cross-correlation of PIV images. In: Nieuwstadt (ed.) Flow Visualization and Image Analysis, pp. 1–25. Kluwer Academic (1993)
150. Kenny, T.: Discussion. J. Soil Mechanics and Foundations Div. 85(SM3), 67–79 (1959)
151. Ketchman, R., Carlson, W.: Acquisition, optimization and interpretation of x-ray computed tomographic imagery: application to geosciences. Computer and Geoscience 27(4), 381–400 (2001)
152. Kihm, K.: Laser speckle photography technique applied for heat and mass transfer problems. Advances in Heat Transfer 30, 255–311 (1997)
153. Kirpatrick, W., Belshaw, D.: On interpretation of triaxial test. Geotechnique 18(3), 336–350 (1968)
154. Konagai, K., Tamura, C., Rangelow, P., Matsushima, T.: Laser-aided tomography: a tool for visualization of changes in the fabric of granular assemblage. Proc. JSCE No: 455 I-21, Structural Engr. / Earthquake Engr. 9(3), 193–201 (1992)
155. Konagai, K., Rangelow, P., Sato, T.: Real-time observation of dynamic changes in the fabric of granular material structures through laser-aided tomography. In: Proc. 10th European Conf. on Earthquake Engr., pp. 459–466 (1994)
156. Kostarelos, K.: A new concept: natural buoyancy for surfactant flooding, Ph.D. Dissertation, University of Texas at Austin, Austin, USA (1998)
157. Koustsopoulos, H., Mishalani, R.: Interpretation and aggregation of spatially detailed sensor data on pavement condition. In: Proc. 2nd Int. Conf. on Imaging Technologies, pp. 148–159. ASCE (1997)
158. Kuo, C., Frost, J., Chameau, J.: Image analysis determination of stereology based fabric tensors. Geotechnique 48(4) (1998)
159. Kuo, C., Frost, J.: Revised methodology to minimize bias in determining the porosity and void tensor of particulate media. In: Proc. Digital Image Processing, pp. 186–194. ASCE (1993)
160. Kuo, C., Frost, J., Lai, J., Wang, L.: Three-dimensional image analysis of aggregate particles from orthogonal projections. TRR No. 1526, National Research Council, Washington, DC, USA, pp. 98–103 (1996)
161. Kuraray America Inc., Water absorbent resin KI-GEL 201K-F2, Technical document, Kuraray America, Inc., 200 Park Avenue, New York, NY-10166 (1998)
162. Ladd, C.: Stress-strain behavior of saturated clay and basic strength principles. Research Report R64-17, Dept. Civil Eng., MIT (1964a)
163. Ladd, C.: Stress-strain modulus of clay from undrained triaxial tests. In: Proc. ASCE, vol. 90(SM5) (1964b)
164. Ladd, C.: Stress-strain behavior of anisotropically consolidated clays during undrained shear. In: Proc. 6th Int. Conf. of Soil Mech. and Found. Engr. (1964c), Available as Professional Paper No. P64-21, Soils Publication No. 160, Dept. Civil Engr., MIT
165. Ladd, C., Foott, R.: New design procedure for stability of soft clays. J. Geotech. Eng. Div. 100(GT7), 763–786 (1974)
166. Ladd, C., Lambe, T.: The strength of undisturbed clay determined from undrained tests. Laboratory Shear Testing of Soils, ASTM STP No. 361, pp. 342–371 (1963)
167. Ladd, C., Foott, R., Ishihara, K., Schlosser, F., Poulos, H.: Stress-deformation and strength characteristics, State of the Art Report. In: Proc. 9th Int. Conf. on Soil Mech. & Foundation Engr., Tokyo, Japan, vol. 2, pp. 421–494 (1977)
168. Lamb, T., Whitman, R.: Soil Mechanics. John Wiley (1969)

169. Landreth, C., Adrian, R.: Measurement and refinement of velocimetry data using high image density analysis in particle image velocimetry. In: Adrian (ed.) Applications of Laser Anemometry to Fluid Mechanics, pp. 484–497. Springer, Heidelberg (1989)
170. Lange, A., Abbell, A., Willis, K.: Powell: Characterization of cement pore structure using wood's metal and mercury porosimetry. In: Proc. 2nd Int. Conf. on Imaging Technologies: Techniques and Applications in Civil Engr., pp. 197–206. ASCE (1997)
171. Lee, H.: Survey: fundamental pavement crack imaging algorithms. In: Proc. Digital Image Processing: Techniques and Applications in Civil Engr., Hawaii, pp. 195–202. ASCE (1993)
172. Lee, K.: Triaxial compressive strength of saturated sand under seismic loading conditions, Ph.D. Dissertation, The University of California, Berkeley (1965)
173. Lee, K., Seed, H.: Drained strength characteristics of sands. J. of The Soil Mechanics and Foundation Division 93(SM6), 117–141 (1967)
174. Lefebvre, G., Langlois, P., Lupien, C.: Laboratory testing and in-situ behavior of peat as embankment foundation. Canadian Geotechnical J. 21, 322–337 (1984)
175. Lehane, B., Gil, D.: Displacement fields induced by penetrometer installation in an artificial soil. Int. J. Physical Modeling in Geotechnics 1, 25–36 (2004)
176. Leng, X., Hounslow, M., Bai, X., Luo, D., Costa, D., Xue, X., Tovey, N., Smart, P.: Image analysis of clay microstructure. In: Proc. Digital Image Processing: Techniques and Applications in Civil Engr., Hawaii, pp. 77–86. ASCE (1993)
177. Leuck, H., Nagel, H.: Automatic differentiation facilitates of-integration into steering-angle-based road vehicle tracking. In: Proc. IEEE Computer Society Conference on Computer Vision and Pattern Recognition (CVPR 1999), vol. 2, pp. 360–365 (1999)
178. Levine, E.: Physical chemistry, 4th edn. McGraw-Hill (1995)
179. Liberzon, A., Gurka, R., Shavit, U.: URAPIV™ Particle Image Velocimetry Research Group (2000), `http://www.openpiv.net/` (last accessed October 2009)
180. Liddell, P., Boger, D.: Yield stress measurements with the vane. J. of Non-Newtonian Fluid Mechanics 63(2-3), 235–261 (1996)
181. Liner, L.: Geophysics and NIH image. Computers & Geosciences 25(4), 403–414 (1999)
182. Little, J., Verri, A.: Analysis of differential and matching methods for optical flow. In: Proc. IEEE Visual Motion Workshop, pp. 173–180 (1989)
183. Liu, J., Iskander, M., Sadek, S.: Consolidation and permeability of transparent amorphous silica. ASTM Geotechnical Testing J. 26(4), 390–401 (2003)
184. Liu, J., Iskander, M.: Adaptive Cross Correlation for Imaging Displacements in Soils. ASCE J. Computing in Civil Engr. 19(2), 219–222
185. Loncaric, S., Majcenic, Z.: Optical flow algorithm for cardiac motion estimation. In: Proc. of the 22nd Annual EMBS Int. Conference, Chicago IL, pp. 415–417 (2000)
186. Lorenz, H., Heinz, W.: Change of density in sands due to loading. In: Proc. 7th Int. Conf. Soil Mech. & Found. Eng., vol. 1, pp. 267–273 (1969)
187. Luerkens, D.: Theory and application of morphological analysis: fine particles and surfaces. CRC Press (1991)
188. Lunati, I., Kinzelbach, W., Sorensen, I.: Effect of pore volume-transmissivity correlation on transport phenomena. Contaminant Hydrology 67(1-4), 195–217 (2003)
189. Macari-Pasqualino, E., Costes, N., Parker, J.: Digital image techniques for vol. change measurements in triaxial tests. In: Proc. Digital Image Processing: Techniques and Applications in Civil Engr., Hawaii, pp. 211–219. ASCE (1993)

190. Mandava, S., Watson, A., Edwards, C.: NMR imaging of saturation during immiscible displacements. AIChE Journal 36(11), 1680–1686 (1990)
191. Mannheimer, R.: Slurries you can see through. Technology Today, Southwest Research Institute, p. 2 (1990)
192. Mannheimer, R., Oswald, C.: Development of transparent porous media with permeabilities and porosities comparable to soils, aquifers, and petroleum reservoirs. Ground Water 31(5), 781–788 (1993)
193. Marangos, C.: Vertical and horizontal deformations for foundations on sand: an experimental study. In: Proc. of the Conf. on Vert.& Horiz. Deformations of Foundations and Embankments, Part 1, pp. 179–189 (1994)
194. Markandey, V., Reid, A., Wang, S.: Motion estimation for moving target detection. IEEE Transactions on Aerospace and Electronic Systems 32(3), 866–874 (1996)
195. Maser, K.: Condition assessment of transportation infrastructure using ground penetration radar. J. of Infrastructure Systems 2(2) (1996)
196. Matsubayashi, U., Devkota, L., Takagi, F.: Characteristics of the dispersion coefficient in miscible displacement through a glass beads medium. J. of Hydrology 192, 51–64 (1997)
197. McKenna, J., Sherlock, D., Evans, B.: Time-lapse 3-D seismic imaging of shallow subsurface contaminant flow. J. of Contaminant Hydrology 53(1-2), 133–150 (2001)
198. Mesri, G., Olson, R.: Mechanism controlling the permeability of clays. Clays and Clay Minerals 19(3), 151–158 (1971)
199. Mesri, G., Feng, T., Shahien, M.: Coefficient of consolidation by inflection point method. ASCE J. of Geotechnical and Geoenvironmental Engr. 125(8), 716–718 (1999)
200. Meyerhof, G.: Ultimate bearing capacity of footings on sand layer overlying clay. Canadian Geotechnical J. 11(2), 224–229 (1974)
201. Mitchell, J.: Fundamentals of soil behavior, 2nd edn. Wiley (1993)
202. Moore, C., Donaldson, C.: Quantifying soil microstructure using fractals. Geotechnique 45(1), 105–116 (1995)
203. Muhunthan, B., Chameau, J.: Mathematical characterization of fabric and its use in mechanics of geomaterials. In: Proc. of Engr. Mechanics, pp. 725–728. ASCE (1992)
204. Muhunthan, B., Masad, E.: Determination of void fabric tensor of soils without radial sampling bias. In: Proc. 2nd Int. Conf. on Imaging Technologies, pp. 40–55. ASCE (1997)
205. Multisorb Technologies Inc., 325 Harlem Road Buffalo, NY 14224-1893 USA (2001), `http://www.multisorb.com`
206. Nallamathu, S., Wang, K.: Automated license plate recognition with a neural net based accelerator. In: Proc. 2nd Int. Conf. on Imaging Technologies, pp. 166–174. ASCE (1997)
207. Nemat-Nasser, S., Okada, N.: Radiographic and microscopic observation of shear bands in granular materials. Geotechnique 51(9), 753–765 (2001)
208. Ni, Q., Hird, C., Guymer, I.: Physical Modeling of Pile Penetration in Clay using Trapnsparent Soil and Particle Image Velocimetry. Geotechnique 60(2), 121–132 (2010)
209. Ng, T., Kelley, M., Sampson, J.: MRI studies of direct shear tests on round particles. In: Proc. of Engr. Mechanics, vol. 1, pp. 572–575. ASCE (1996)
210. Oh, H., Garrick, N., Achenie, L.: Segmentaion algorithm using iterative clipping for processing noisy pavement images. In: Proc. 2nd Int. Conf. on Imaging Technologies: Techniques and Applications in Civil Engr., pp. 138–147. ASCE (1997)

211. Orsi, T., Anderson, A., Leonard, J., Bryant, W., Edwards, C.: Use of X-ray computed tomography in the study of marine sediments. In: Proc. CEO V, College Station, TX, pp. 968–982. ASCE (1992)
212. Paikowsky, S., Xi, F.: Particle motions tracking utilizing a high-resolution digital CCD camera. ASTM Geotechnical Testing J. 23(1), 123–134 (2000)
213. Paikowsky, S., Di, R., Xi, F.: Inter-particulate contact force analysis and measurement using photoelastic techniques. In: Proc. Int. Conf. Discrete Element Methods, pp. 449–462 (1993)
214. Penumadu, D., Hazan, I.: Geo material characterization using digital micrograph analysis. In: Proc. 2nd Int. Conf. on Imaging Technologies, pp. 66–75. ASCE (1997)
215. Perret, J., Prasher, S., Kontazas, A., Hamilton, K., Longford, C.: Preferential solute flow in intact soil columns measured by SPECT scanning. Soil Science Society of America J. 64(2), 469–477 (2000)
216. Pervizpour, M., Pamukcu, S., Moo-Young, H.: Magnetic resonance imaging of hydrocarbon-contaminated porous media. J. of Computing in Civil Engr. 13(2), 96–102
217. Posadas, D., Tannus, A., Panepucci, H., Crestana, S.: Magnetic resonance imaging as a non-invasive technique for investigating 3-D preferential flow occurring within stratified soil samples. Computers and Electronics in Agriculture 14(4), 255–267 (1996)
218. PPG Industries Product Specification Data, One PPG Place, Pittsburgh, PA 15272 (1996)
219. PPG Industries Inc, Silica Products, One PPG Place, Pittsburgh, PA 15272 (2000)
220. Prevost, J.: Undrained shear test on clays. J. Geot. Eng. Div. 105(GT1), 49–61 (1979)
221. Raschke, S., Hryciw, R.: Vision cone penetrometer (V-CPT) for direct subsurface soil observation. J. of Geotechnical and Geoenvironmental Engr. 123(11), 1074–1076 (1997)
222. Raschke, S., Hryciw, R., Donohoe, G.: Microdeformations in sands by digital image processing and analysis. Transportation Research Record 1548, 31–37 (1996)
223. Rechenmacher, A., Finno, R.: Digital image corelation to evaluate shear banding in dilative sands. ASTM Geotechnical Testing J. 27(1), 1–10 (2004)
224. Repin, D., Pavlis, G.: Interactive three-dimensional visualization of seismic array data. Computers & Geosciences 23(10), 1079–1092 (1997)
225. Robinsky, E., Morrison, C.: Sand displacement and compaction around model friction piles. Canadian Geotechnical J. 1(2), 81–91 (1964)
226. Rollins, K., Evans, M., Diehl, N., Daily, W.: Shear modulus and damping relationships for gravels. J. of Geotechnical and Geoenvironmental Engr. 124(5), 396–405 (1998)
227. Roscoe, K., Arthur, J., James, R.: The determination of strains in soils by an X-ray method. Civil Engr. and Public Works Review 58, 873–876, 1009-1012 (1963)
228. Saada, A., Liang, L., Figueroa, J., Cope, T.: Bifurcation shear band propagation in sands. Geotechnique 49(3), 367–385 (1999)
229. Sadek, S.: Soil structure interaction in transparent synthetic soils using digital image correlation, Ph.D. Dissertation, Polytechnic University, Brooklyn, NY (2002)
230. Sadek, S., Iskander, M., Liu, J.: Geotechnical properties of transparent silica. Canadian Geotechnical J. 39(1), 111–124 (2002)
231. Sadek, S., Iskander, M., Liu, J.: Accuracy of digital image correlation for measuring deformations in transparent media. ASCE J. Computing in Civil Engr. 17(2), 88–96 (2003)

232. Santamarina, J., Fam, M.: Changes in dielectric permittivity and shear wave velocity during concentration diffusion. Canadian Geotechnical J. 32(4), 647–659 (1995)
233. Santamarina, J., Klein, K., Wang, Y., Prencke, E.: Specific surface: determination and relevance. Canadian Geotechnical J. 39(1), 233–241 (2002)
234. Santogata, M., Bobet, A., Johnston, C., Hwang, J.: One-dimensional compression behavior of a soil with high organic matter content. J. Geotech. and Geoenvir. Engr. 134(1), 1–13 (2008)
235. Scarano, F., Riethmuller, M.: Iterative multi-grid approach in PIV image processing with discrete window offset. In: Experiments in Fluids, vol. 26, pp. 513–523. Springer (1999)
236. Schnell, U., Piot, J.: Dändliker: Detection of movement with laser speckle patterns: statistical properties. J. of the Optical Society of America A 15(1), 207–216 (1998)
237. Scott, R.: Silica gel and boded phases. Wiley (1993)
238. Seed, H., Idriss, I.: Soil moduli and damping factors for dynamic response analysis. Report No. EERC 70-10, Earthquake Engr. Research Centre, U. California, Berkeley (1970)
239. Shchepinov, V., Pisarev, V., Novkov, N., Balalov, V., Odintsev, I., Bondarenko, M.: Strain and stress analysis by holographic and speckle interferometry. Wiley (1996)
240. Shi, B., Murakami, Y., Wu, Z.: Orientation of aggregates of fine-grained soil: quantification and application. Engr. Geology 50(1-2), 59–70 (1998)
241. Shi, B., Murakami, Y., Wu, Z., Chen, J., Inyang, H.: Monitoring of internal failure evolution in soils using computerization x-ray tomography. Engr. Geology 54(3-4), 321–328 (1999)
242. Shikada, M., Kusaka, T., Kawata, Y., Miyakita, K.: Extraction of characteristics properties in landslide areas using surface temperature. In: Int. Geoscience and Remote Sensing Symposium (IGARSS), vol. 1, pp. 103–105. IEEE (1993)
243. Simons, N.: Comprehensive investigation of the shear strength of an undisturbed Drammen clay. In: Proc. Conf. Shear Strength of Cohesive Soils, pp. 727–746. ASCE (1960a)
244. Simons, N.: The effect of overconsolidation on the shear strength characteristics of an undisturbed Oslo clay. In: Proc. Conf. Shear Strength of Cohesive Soils, pp. 747–763. ASCE (1960b)
245. Sjödahl, M., Benckert, L.: Electronic speckle photography: analysis of an algorithm giving the displacement with sub-pixel accuracy. Applied Optics 32, 2278–2284 (1993)
246. Sjödahl, M.: Some recent advances in electronic speckle photography. Optics and Lasers in Engr. 29, 125–144 (1998)
247. Skempton, A.: The pore-pressure coefficients A and B. Geotechnique 4, 143–147 (1954)
248. Skempton, A.: Long-term stability of clay slopes. 4th Rankine Lecture. Geotechnique 14(2), 77–102 (1964)
249. Slavik, J.: Fluorescent probes in cellular and molecular biology. CRC Press, Boca Raton (1994)
250. Smith, G., Lin, C.: Fuzzy logic distress classifier for pavement imaging. In: Proc. 2nd Int. Conf. on Imaging Technologies, pp. 118–127. ASCE, DAVOS (1997)
251. Smith, S.: The scientist & engineer's guide to digital signal processing, Technical Pub., CA (1998)
252. Song, Z.: Pullout Behaviour of Suction Embedded Plate Anchors in Clay. Ph.D. Thesis, Curtin University of Technology, Australia

253. Song, Z., Hu, Y.: Plate anchor keying under inclined pullout in clay: observation and estimation. In: Proc. XVII Int. Conf. Soil Mechanics & Geotechnical Engr., pp. 708–711 (2009)
254. Song, Z., Hu, Y., O'Loughlin, C., Randolph, M.: Loss of Anchor Embedment during Plate Anchor Keying in Clay. J. Geotechnical and Geoenvironmental Engr. 135(10), 1475–1485 (2009)
255. Sphinx Adsorbent Inc., Granular silica gel data sheet, 53 Progress Ave. Springfield, MA 01104 (1993)
256. Spiesberger, W., Tasto, M.: Processing of medical image sequences. In: Huang, T.S. (ed.) Image Sequence Analysis, pp. 381–428. Springer, Heidelberg (1981)
257. Stainer, S.: Geotechnical modeling using a transparent synthetic soil, M.Sc. Report, University of Sheffield (Supervisor Dr. C. Hird) (2006)
258. Sutton, M., Wolters, W., Peters, W., Ranson, W., Mcneill, S.: Determining of displacements using an improved digital correlation method. Image Vision Computing 1(3), 133–139 (1983)
259. Sveen, J.: An introduction to MatPIV, v.1.6.1. Eprint No.2, Dept. of Math. University of Oslo (2004) ISSN 0809-4403
260. Tabe, K.: Aquabeads to model the geotechnical behavior of natural soils, Ph.D. Dissertation, Polytechnic Institute of NYU, New York, USA (2009)
261. Taylor, R., Grant, R., Robson, S., Kuwano, J.: An image analysis system for determining plane and 3D displacements in soil models. In: Proc. Centrifuge 1998, Tokyo, pp. 73–78 (1998)
262. Terzaghi, K.: Theoretical soil mechanics. Wiley (1943)
263. Terzaghi, K., Peck, R., Mesri, G.: Soil mechanics in engr. practice, 3rd edn. Wiley (1996)
264. Theodoropoulou, M., Karoutsos, V., Kaspiris, C., Tsakiroglou, C.: A new visualization technique for the study of solute dispersion in model porous media. J. of Hydrology 274(1-4), 176–197 (2003)
265. Toiya, M., Jacco, H., Losert, W.: 3D imaging of particle motion during penetrometer testing, from microscopic to macroscopic soil mechanics. In: Granular Matter, pp. 323–329. Springer (2007)
266. Tokumaru, P., Dimotakis, P.: Image correlation velocimetry. Experiments in Fluids 19, 1–15 (1995)
267. Tomlinson, M.: Foundation design and construction, 4th edn. Pitman, London (1986)
268. Tsuruya, H., Nakano, S., Tanaka, J.: Investigation in rheological properties of soft muds with a rotary viscometer. Rep. No. 566, Port and Harbor Research Institute, Japan (1986)
269. Udupa, J., Herman, G.: Medical image reconstruction, processing, visualization, and analysis: the MIPG perspective. IEEE Transactions on Medical Imaging 21(2), 281–295 (2002)
270. Wang, L., Lai, J., Frost, D.: Fourier morphological descriptors of aggregate profiles. In: Proc. 2nd Int. Conf. on Imaging Technologies, Davos, pp. 76–87. ASCE (1997)
271. Weast, R.: Handbook of chemistry & physics, 66th edn. CRC Press, Boca Raton (1986)
272. Weber, L., Schumate, E., Mawer, N.: Experiments on flow at a 90° open-channel junction. J. of Hydraulic Engr. 127(5), 340–350 (2001)
273. Weil, G., Graf, R., Forister, L.: Non-destructive remote sensing of hazardous waste sites. In: Proc. of the Conference on Digital Image Processing, Hawaii, pp. 279–284. ASCE (1993)

274. Welker, A.: In situ remediation of contaminated soils using prefabricated vertical drains. Dissertation, The University of Texas at Austin, USA (1998)
275. Welker, A., Bowders, J., Gilbert, R.: Applied research using transparent material with hydraulic properties similar to soil. Geotechnical Testing J. 22(3), 266–270 (1999)
276. Welker, A., Bowders, J., Gilbert, R.: Using a reduced equivalent diameter for a prefabricated vertical drain to account for smear. Geosynthetics Int. 7(1), 47–57 (2000)
277. Westerweel, J.: Efficient detection of spurious vectors in particle image Velocimetry data. Experiments in Fluids 16, 236–247 (1994)
278. Westerweel, J., Dabiri, D., Gharib, M.: The effect of a discrete window offset on the accuracy of cross-correlation analysis of digital PIV recording. Experiments in Fluids 23(1), 20–28 (1997)
279. White, D., Randolph, M., Thompson, B.: An image-based deformation measurement system for the geotechnical centrifuge. Int. J. of Physical Modeling in Geotechnics 5(3), 1–12 (2005)
280. White, D., Take, W., Bolton, M.: Soil deformation measurement using particle image velocimetry (PIV) and photogrammetry. Geotechnique 53(7), 619–631 (2003)
281. Whorlow, R.: Rheological Techniques, vol. 460. Ellis Horwood, Chichester (1992)
282. Willersinn, D., Enkelmann, W.: Robust obstacle detection and tracking by motion analysis. In: IEEE Conf. on Intelligent Transportation System, ITSC 1997, pp. 717–722 (1997)
283. Wong, R.: Mobilized shear strength components of Athabasca sand in triaxial compression. Canadian Geotechnical J. 23(4), 413–422 (1999)
284. Wong, R., Wibowo, R.: Tomographic evaluation of air and water flow patterns in soil column. Geotechnical Testing J., GTJODJ 23(4), 413–422 (2000)
285. Woods, R., Barnett, N., Sagesser, R.: Holography – a new tool for soil dynamics. J. of Geotechnical Engr. Division 100(GT11), 1231–1247 (1974)
286. Wykham: WF23500 laboratory vane equipment handbook. Wykeham Farrance Int. (1999)
287. Xing, Z., Wells, L., Yaping, J., Shearer, S.: Using image analysis to measure two-dimensional soil deformation. J. of Terramechanics 34(2), 73–82 (1997)
288. Yaling, P.: Optical tomographic imaging using finite element method, Ph.D. Dissertation, Polytechnic University, Brooklyn, New York (1999)
289. Yamaguchi, H., Ohira, Y., Kogure, K.: Volume change characteristics of undisturbed fibrous Peat. Soils and Foundations 25(2), 119–134 (1985)
290. Yamamoto, K., Kusuda, K.: Failure mechanism and bearing capacities of reinforced foundations. Geotextiles and Geomembranes 19(3), 127–162 (2001)
291. Yang, R., Buenfeld, N.: Binary segmentation of aggregate in SEM image analysis of concrete. Cement and Concrete Research 31(3), 437–441 (2001)
292. Yano, K., Watari, Y., Yamanouchi, T.: Earthwork on soft clay grounds using ropenetted fabrics. In: Proc. Recent Developments in Ground Improvement Tech., pp. 225–237 (1985)
293. Yudhbir, R.A.: Image analysis applied to geotechnical engr. In: Proc. of the Conference on Digital Image Processing, Hawaii, pp. 291–298. ASCE (1993)
294. Zenina, A., Dupre, J., Lagrade, A.: Separation of isochromatic and isoclinic patterns of a slice optically isolated in a 3D photoelastic medium. European J. of Mechanics A - Solids 18(4), 633–640 (1999)
295. Zhao, H.: Development and assessment of transparent soil and particle image velocimetry in dynamic soil-structure interaction. Dissertation, University of Missouri-Rolla (2007)

Index

W

X

Y

GPSR Compliance

The European Union's (EU) General Product Safety Regulation (GPSR) is a set of rules that requires consumer products to be safe and our obligations to ensure this.

If you have any concerns about our products, you can contact us on ProductSafety@springernature.com

In case Publisher is established outside the EU, the EU authorized representative is:

Springer Nature Customer Service Center GmbH
Europaplatz 3
69115 Heidelberg, Germany

Batch number: 10614940

Printed by Printforce, the Netherlands